Literacies

This book is part of the Peter Lang Education list.
Every volume is peer reviewed and meets
the highest quality standards for content and production.

PETER LANG
New York • Washington, D.C./Baltimore • Bern
Frankfurt • Berlin • Brussels • Vienna • Oxford

Colin Lankshear & Michele Knobel

Literacies

Social, Cultural and Historical Perspectives

PETER LANG
New York • Washington, D.C./Baltimore • Bern
Frankfurt • Berlin • Brussels • Vienna • Oxford

Library of Congress Cataloging-in-Publication Data

Lankshear, Colin.
Literacies: social, cultural, and historical perspectives /
Colin Lankshear, Michele Knobel.
p. cm.
Includes bibliographical references and index.
1. Literacy—Social aspects. 2. Literacy—Research.
3. Language arts—Social aspects. 4. History—Philosophy.
5. Multicultural education. I. Knobel, Michele. II. Title.
LB1576.L337 302.2'244—dc22 2011000027
ISBN 978-1-4331-1024-5 (hardcover)
ISBN 978-1-4331-1023-8 (paperback)

Bibliographic information published by **Die Deutsche Nationalbibliothek.**
Die Deutsche Nationalbibliothek lists this publication in the "Deutsche Nationalbibliografie"; detailed bibliographic data is available on the Internet at http://dnb.d-nb.de/.

The paper in this book meets the guidelines for permanence and durability of the Committee on Production Guidelines for Book Longevity of the Council of Library Resources.

29 Broadway, 18th floor, New York, NY 10006
www.peterlang.com

Printed in the United States of America

To dedicate this book to each other would be tacky and cheesy.
But we're going to do it anyway.
To CJL from MJK
To MJK from CJL

Contents

Acknowledgments

We want to thank Chris Myers and colleagues at Peter Lang, New York, for the privilege of being able to publish a collection of essays that encapsulate the main currents of our work since we first began researching and writing in the area of literacy. The question we asked ourselves at the beginning of this endeavor was "which essays do we think best capture what we have tried to do and who we have been as individuals and as a team working in literacy studies over the years?" As we began addressing this question the list of essays grew beyond the scope we originally pitched for the book. Chris, characteristically, said "include the ones you most want and we' ll look at word numbers later." That generosity is typical of what we have experienced over the years working with Lang—as authors and series editors alike. We appreciate it greatly, along with the careful and cheerful manner the Lang team, to a person, bring with them to their work. Special thanks also go to Bernadette Shade for her exemplary work in the production phase and at so many other points of contact in our work with Lang. In this context we want also to recognize and honor the tremendous work done by Shirley Steinberg and the late Joe Kincheloe in helping to build the success and standing Peter Lang enjoys among U.S. publishers of academic work. Were it not for Shirley and Joe's unstinting efforts, our own series and several of our books would not exist. We esteem their contribution to academic publishing, and thank them for their role in enabling us to be part of the life of Peter Lang Publishing.

Many other colleagues have encouraged us along the way. Michael Apple gave a mighty push to the paper, "Ideas of functional literacy," encouraging its development from a conference presentation to a platform for subsequent work. Libby Limbrick invited "Dawn of the people" as a keynote address for a teacher professional development conference, thereby getting this early work into the sights of a very substantial professional readership. Over several pints of beer, Ivor Goodson worked with an early career academic to conceive a project that became the book, *Literacy, Schooling and Revolution*. He persuaded the late Malcolm Clarkson, founder of Falmer Press, to take a chance on a book by an unknown New Zealand academic. The generosity and faith of Ivor and Malcolm is deeply appreciated. And once the book was published, Michael Apple got behind it in the best tradition of academic mentoring, doing his very best to promote it in North America. Peter McLaren and Henry Giroux, in the U.S., Jack Shallcrass in New Zealand, and Kevin Harris in Australia were also generous and encouraging in ways that went far beyond any possible call of academic duty. Allan Levett was pivotal in bringing a perspective on literacy and the changing world of work to our thinking, and Geoff Bull and Michele Anstey were instrumental in getting us together and working as a team and in providing wonderful mentoring for Michele' s early publishing and postgraduate work. Wendy Morgan supervised Michele's doctoral study with bounteous care and attention and nurtured her independence as a researcher. Jim Gee has been there, constantly, since the early 1990s, as the best inspiration and role model that anyone working in literacy studies could possibly have. His contribution to literacy studies is second to none, and he does collegial friendship with incomparable style and grace. He is all through the pages of this book. We cannot thank him enough for his contribution to our work and sense of purpose. We also want to acknowledge the valued contributions to our work on new literacies themes made over many years by Chris Bigum, Leonie Rowan, Michael Doneman and Donna Alvermann. When our will to go on has weakened we have always been inspired by their ideas and commitments.

A work of collected essays cannot exist without the goodwill of those who have published work in the first place and who allow it to be republished elsewhere. We are grateful to the following journals and publishers for permitting us to republish our previous work, as follows:

Lankshear, C. (1985). Ideas of functional literacy, *New Zealand Journal of Educational Studies*, 20(1): 5–19. Reprinted with permission of the editor, New Zealand Journal of Educational Studies.

Lankshear, C. (1986). The dawn of the people: The right to literacy in Nicaragua. *Reading Forum* NZ (2 and 3): 15–20 & 12–19. Reprinted with the permission of the New Zealand Reading Association.

Lankshear, C. (1987). *Literacy, Schooling and Revolution*. London: Falmer Press, pages 80–81 and 93–111. Reprinted with the permission of Taylor & Francis Books (UK).

Knobel, M. (1993). Simon says see what I say: Reader response and the teacher as meaning maker. *Australian Journal of Language and Literacy*. 16(4): 295–306, © ALEA. Reprinted with permission from the Australian Literacy Educators' Association.

Lankshear, C. and Knobel, M. (1997). Critical literacy and active citizenship, in P. Freebody, S. Muspratt, and A. Luke (eds), *Constructing Critical Literacies*. Norwood, NJ: Hampton Press. Reprinted with the permission of Hampton Press.

Lankshear, C. (1997). Literacy and empowerment. Chapter 3 of *Changing Literacies*, Buckingham, UK: Open University Press, pages 63–79. Reprinted with permission of Open University Press.

Lankshear, C. (1997). Language and the new capitalism. *International Journal of Inclusive Education* 1(4), reprinted by permission of the publisher (Taylor & Francis Ltd, http://www.tandf.co.uk/journals).

Knobel, M. (2001). "I'm not a pencil man": How one student challenges our notions of literacy "failure" in school. *Journal of Adolescent and Adult Literacy*. 44(5): 404–419. Reprinted with permission of the International Reading Association.

Lankshear, C. and Knobel, M. (2007). Researching new literacies: Web 2.0 practices and insider perspectives. *E-Learning and Digital Media*, 4(3): 224–240 (http://dx.doi.org/10.2304/elea.2007.4.3.224). Reprinted with permission of the publisher.

Finally, we want to thank those students who have kindly given us permission to use their work, and the many cohorts of students who, since 2004, have "come along for the ride" as we have conjointly explored ways of learning that have been new and different for all of us so far as formal educational contexts are concerned. In particular we want to thank Shannon Donovan, Susan Whitty, Jenny Hawley, Karey Lee Donovan, Marie Barry, Rhonda Currie, Patricia Donovan, Janice Ciavaglia, Holly Stone, Ann Landry, Sonja Beck, Carolyn Coley, Krista Conway, Deborah Hoven, Paula Maynard, Tanya Hunt, Sherry Healey-Jennings, Joy Seaward, Michelle Patey, Anne Payne, Michele Dawson, Dara Best-Pinsent, Robyn Hillier, Lori Deeley, Heather Wood, Donna Powers-Toms and Rosena Dunphy for graciously allowing us to make explicit reference to their work.

Preface

In his *History of Western Philosophy* (1961: 463), Bertrand Russell affirms the Socratic value of "following wherever the argument may lead," Of course, what the argument is and, hence, where it may lead, has a lot to do with contingency. Arguments are situated. They are impacted by frames and by evidence, by what is considered important and relevant in different times and places. As a person's circumstances change, so it is likely that the arguments they follow and the places they follow them to will change. Underlying core principles may remain intact—such as trying to keep an open mind, trying to support what seems right and fair, and so on—but maintaining such principles in conjunction with following wherever the argument may lead is consistent with one's views and positions changing over time and place. To invoke John Dewey (1944: Ch. 4), along with Russell, one may hope—and it may be one's best hope—that the changes that come from following arguments wherever they may lead will reflect growth (in Dewey's sense). That has been the enduring hope—indeed, aim—we have individually and jointly sought to maintain throughout our academic lives.

This book contains sixteen essays, all about literacies, ranging over a period of twenty-five years and over diverse circumstances, places and influences. We have both written individually as well as together, although the great majority of our work, since first meeting in 1992, has been joint. "Wherever the argument may lead" has predominantly been a matter of negotiation. Indeed, for each of us, meeting the other has significantly influenced what we see as worthy of argument in the first place, as well as the considerations that shape an argument's "leading."

In many ways, where this book ends up could scarcely be more different from where it began. Both of us experience not a little embarrassment over some of the positions taken and the tracks our arguments have taken. But those will surely be the facts of life for any thinking person who sustains an interest in a particular topic or concern over a long period of time. The chapters in this book collectively comprise an ongoing argument about the natures, roles and significance of literacies understood as social practices—as social phenomena. They trace a path that follows the leads of the individual constitutive arguments, which in turn follow leads shaped by times and places and circumstances. When these essays began there was still a Cold War, an East and a West. Poststructuralist thinking was yet to significantly impact thinking in the social sciences and humanities in New Zealand. In Australia and New Zealand policy was gearing up to respond to local "discoveries" of a "literacy crisis" that just happened to emerge around the very time politicians and economists perceived a need to begin responding to evident deep changes in the conditions and practices of economic production; changes that would require—in places like New Zealand and Australia—getting used to the fact that the days of full employment might be over and that for many people the quality of work they aspired to and the kind of work that was available might increasingly diverge.

Even at that time "literacies" did not have to reckon with the digital electronic revolution—at least, for most people and most schools. But that was just around the corner. Colin, in New Zealand, did not "drive" a computer until 1988. By contrast, Michele, in an Australian private school that was already alert to what would become rapidly changing conditions of reading and writing, first "drove" a computer—at school—in 1980–81, and began programming in Logo within her teacher education program in 1984. But we both had to wait until 1993 to "get on the internet." Not surprisingly, however, our respective experiences of "getting on the internet" were very different. It was Michele who grasped what the arrival of a graphic internet browser interface would likely mean—and who snail mailed a money order to the U.S. to buy Mosaic, only to find that by the time Mosaic arrived it had already been scooped by Netscape Navigator, and we could download that (but s.l.o.w.l.y).

Such were some of the ingredients that shaped where arguments began leading by the time we got to what now comprises the middle sections of the book. Issues of access to "powerful literacies" that might help make "learning outcomes" "more equitable," were the order of the day in Australian literacy education and literacy studies, within and across print and digital literacies, respectively. How well prepared were teachers to integrate new technologies into classroom learning?; how well prepared were teachers to help students learn how to master "genres of power"? What needed to be done to enhance literacy education in these respects? New arguments; new places for arguments to lead.

Six years of that proved to be more than enough for our temperaments and dispositions, and we left Australia for the Americas, south of the Rio Grande. Mexico was wrestling with the "new capitalism" and, as part of this, pulling out all the stops to keep abreast of the "revolution" in computing and communications technologies. The internet assumed a new significance in our lives—indeed, we experienced it increasingly as a necessity; an everyday mediator between lives we had known and future lives we were having to forge on a moment-by-moment basis. Reflecting on this, and on the relationships between our own appropriations of the internet and those of other people—including the scores and hundreds of young Mexicans who lined sometimes three and four deep awaiting turns at the machines in their preferred internet cafes—tuned us into the theme of "new literacies," in all their variety. The argument took a new turn. This was not so much within the pedagogies of the educational institutions we interacted with in Mexico as within the everyday lives we encountered.

By 2003 we were both back into regular contact with Australian and North American universities as well as having regular contact with our Mexican lives. Once again the argument became open to new shapers, most notably, the opportunities and constraints associated with integrating internet technologies into learning within formal educational settings. Like many other academics interested in new literacies we found ourselves surfing the interfaces between nonformal kinds of learning mediated by the internet—especially within popular cultural affinity spaces—and our daily workbound experiences of formal learning mediated by the internet. The interactions and intersections between these varying experiences have dominated our thinking and researching and writing over the past several years. This is where we have got to today.

Assembling this book has been a personally interesting and intellectually challenging experience for us both, and we hope that readers can in turn share some of this interest and challenge as they follow the argument we trace through the sixteen chapters below.

— Colin Lankshear and Michele Knobel
Mexico City

Bibliography

Dewey, J. (1944). *Democracy and Education: An Introduction to Philosophy of Education.* New York: The Free Press.

Russell, B. (1961). *A History of Western Philosophy.* London: Allen and Unwin.

Part One

ONE

Ideas of Functional Literacy: Critique and Redefinition of an Educational Goal (1985)

Colin Lankshear

Biography of the Text

This was my first publication in the area of literacy, being published in the *New Zealand Journal of Educational Studies* in 1985. Some years earlier, whilst working on my doctoral thesis on freedom and education, I had begun reading the work of Paulo Freire. His position on freedom as liberation/emancipation seriously challenged the line of argument I had been developing over several years. Rather than abandon those years of work I decided to return to Freire after completing my thesis. The opportunity to do this came during 1980, following a motorcycle accident that laid me up for several months. Nineteen-eighty was the year in which Nicaragua instigated a mass literacy campaign, based broadly on concepts, values, and principles developed by Paulo Freire, and Freire himself was an adviser for the campaign. From that point Freire's intellectual work became tied to my material interest in what was happening in Nicaragua, and the nexus between the two drove my reading, teaching, and civic engagement throughout the 1980s. In 1984 a colleague at Auckland University, knowing my interest in and ideas on literacy, passed on to me a copy of the report of the Adult Performance Level study undertaken some years previously (in 1975) by the Office of Continuing Education, University of Texas at Austin, on behalf of the U.S. Office of Education. This coincided with the onset of a period during which governments throughout the industrialized Western world became intensely interested in adult literacy levels, as the "long post-war boom" ended and modern economies entered a period of massive restructuring accompanied by escalating levels of structural unemployment and underemployment. The gulf between my evolving personal "philosophy" of literacy and the official line on "functionality" being taken by governments in countries like my own motivated me to develop a position in this area. An annual conference of the Philosophy of Education Society of Australasia in 1984 presented the immediate pretext. The text that follows was my formal initiation into literacy studies in education.

Adult Illiteracy in New Zealand—An Emerging Disorder?

New Zealanders have long been aware of the problem of adult illiteracy (not to mention that of youth) within the Third World. Some of us with a special interest in literacy have also become familiar with a variety of post-war responses to the Third World challenge—including UNESCO programs, government initiative projects (as, for example, in India), the literacy campaigns following popular revolutions in Cuba, Nicaragua and elsewhere, and the work of Paulo Freire among peasant illiterates in Brazil and Chile, to mention but a few (Ahmed, 1958; Cardenal and Miller, 1981; Freire, 1972, 1974, 1981; Morales, 1981; UNESCO, 1949, 1976). However adult illiteracy has seemingly been a non-existent problem in New Zealand—at least until recently. Certainly there was no acknowledged problem in 1956 when, as Gray reports, data relating to the extent of illiteracy—based on the most recent reports available to UNESCO at the time—revealed that "in some of the northern European countries, Australia, New Zealand and Japan . . . little or no illiteracy exists" (Gray, 1956: 29).

Whether or not New Zealand was indeed virtually exempt from adult illiteracy in 1956 is a moot point. Certainly adult illiteracy at that time was by no means exclusively a Third World disorder. In America, for instance, the drafts for each world war revealed considerable illiteracy. President Roosevelt reported in 1942 that 433,000 men graded as eligible and fit for immediate service had been deferred on the grounds that they could not meet the army's literacy requirement (Cook, 1977: 48). At the time the American army defined illiterates as "persons who were incapable of understanding the kinds of written instructions that are needed for carrying out basic military functions or tasks" (Harman, 1970: 227). But whatever the New Zealand reality was with respect to illiteracy in 1956, it is a fact that during the last ten years activity directed at overcoming illiteracy among adult New Zealanders has increased progressively.

To date [i.e., 1985] there has been no widespread testing of literacy levels in adult New Zealanders, and there are no figures claiming accurately to portray the extent of adult illiteracy in this country.

Because our population composition and education system are similar to Britain's, it has been assumed that the standards of literacy in the adult population are the same. Any tests done at Secondary School level in New Zealand have tended to reinforce this assumption. Consequently, we base our estimation of the extent of adult illiteracy almost entirely on British statistics (Adult Reading and Learning Assistance Federation, 1983).

But that there is significant adult illiteracy in New Zealand is widely acknowledged today within educational circles, including the Department of Education. Furthermore, an organized practical response to the reality of adult illiteracy is evident, and the dimensions of this organized response suggest at least something about the extent of adult illiteracy here.

> Partly as a result of the awareness of illiteracy generated in Britain, Adult New Reading Schemes were started up in various centers throughout New Zealand during the 1970s and early 1980s. . . . There are now 120 schemes operating throughout New Zealand (some of which are small and may have as few as 5 student/tutor pairs). . . . It happens that there are 133 student/tutor pairs working together within the Hawkes Bay Scheme, but there is little meaning to this figure on its own. (Adult Reading and Learning Assistance Federation, 1983, personal communication)

While precise figures concerning the actual extent of adult illiteracy within New Zealand are not available, they are not essential for the purpose of this chapter. What is important is that in recent years adult illiteracy has increasingly been recognized locally as an educational problem in need of corrective attention and one that will continue in (at least) the near future.

Confronting the Problem—Toward Functional Literacy?

The increasing awareness of significant adult illiteracy in this country poses the question (among others) of what comprises an appropriate standard of achievement for corrective programs to aim at. It is most important that local educationists give serious consideration to this question. There is a danger that we in New Zealand might simply follow American and British responses to adult illiteracy, and these responses in my view contain some undesirable features.

This judgment will be supported here by investigating the concern of official policies in America and Britain with promoting *functional literacy* among illiterates adults. That there is reason to believe local policies are likely to follow American and British lines, unless alternative models are produced here to counter this likelihood, is easily argued. As has already been noted, recognition of and response to adult illiteracy in New Zealand leading to the Adult New Reading Schemes has closely followed British initiative. And as will shortly be shown, the evidence is that recent British trends have in turn closely followed the American pattern. Furthermore, adult illiteracy has achieved official recognition locally, within the Department of Education and the NZCER, as an educational concern. The fact that official policy in both America and Britain is to promote functional literacy, and that educational trends in New Zealand are often influenced markedly by Anglo-American theory and practice, further support this view.

Before turning to an extended critique of functional literacy it is necessary first to demonstrate official American and British commitment to functional literacy as the appropriate standard of attainment in corrective programs. Levine reports that functionality was adopted by the *Right to Read* campaigns initiated in America (in 1969) and Britain (in 1973) (Levine, 1982: 256). Indeed, The Economic Opportunity Act of 1964 [U.S.] enshrined a conception of functionality into legislation when it referred to adult basic education for those whose

> inability to read and write the English language constitutes a substantial impairment of their ability to get or retain employment commensurate with their real ability, so as to help eliminate such inability and raise the level of education of such individuals with a view to making them less likely to become dependent on others, to improving their ability to benefit from occupational training and otherwise increasing their opportunities for more productive and profitable employment, and making them better able to meet their adult responsibilities. (Levine, 1982: 256)

In like vein the U.S. Office of Education defined the literate person as "one who has acquired the essential knowledge and skills in reading, writing and computation required for effective functioning in society, and whose attainment in such skills makes it possible for him to develop new attitudes and to participate actively in the life of his times" (Levine, 1982: 256). Furthermore, Levine claims that virtually the same view of functional literacy was assumed in Britain: The early manifesto of the adult literacy movement, *A Right to Read,* quoted with approval the U.S. National Reading Center's (USNRC) definition: "A person is functionally literate when he has command of reading skills that permit him to go about his daily activities successfully on the job, or to move about society normally with comprehension of the usual printed expressions and messages he encounters" (Levine, 1982: 256). Thus construed, functional literacy comprises an ethically reprehensible ideal for adult attainment. In order to justify this judgment let us turn now to an extended analysis and critique of "functional literacy."

Functional Literacy: An analysis

Although purely formal definitions of functional literacy have—and are designed to have—applications across different cultures and societies, just what counts *in empirical terms* as being functionally literate is culturally and socially variant. According to Gray, a person may be regarded as functionally literate "when he has acquired the knowledge and skills in reading and writing which enable him to engage effectively in all those activities in which literacy is normally assumed in his culture or group" (Gray, 1956: 24). However such activities are by no means universal. Since they may vary considerably across groups, a criterion will be needed "as a guide in determining the training essential to produce functional literacy within a particular group" (Gray, 1956: 24). Gray reports actual examples of literacy training in the field which indicate that "widely different levels of reading and writing may be needed in achieving the goals of different groups" (Gray, 1956: 22).

This raises the question: what does functional literacy entail for societies like America, Britain or New Zealand? Two dimensions are evident here. (a) The essential—i.e., what is the *essence* of functionality? (b) The empirical—i.e., what are the *actual skills or competencies* that constitute being functionally literate?

On the question of essence I will be brief, since elaboration and confirmation of my claims here can be derived from discussion of the empirical aspects of functional literacy. The essence of being functionally literate is that it comprises a minimal, survival-oriented (and hence, negative) and passive state. The person who is functionally literate can survive; can cope with the world; can manage to fill in job interview forms having read an advertisement for the job; can get the job (perhaps), and having got it survive in it; can locate medical and health services within the community; read a bus timetable, etc. To be functional is to be not unable to cope—it is essentially a negative state, in the sense that it represents nothing that can be regarded as optimal, as a *positive* expression of one's human life. Furthermore, it comprises *a passive* state, since being able to survive, being able to cope effectively in one's social or cultural setting, is a matter of understanding or following, not of commanding and leading. (Compare: "being able to move about society normally with comprehension of the usual printed expressions and messages one encounters"; Levine, 1982: 256).[1] In this context the omitted reference to writing, the stress on the right to *read* in the manifesto of the British literacy campaign, is most interesting. Writing has an active dimension that is not so obviously apparent in reading—a point well appreciated by many people who, in 19th-century England, pondered the extent to which working-class children should be educated. Whereas teaching reading to working-class children "attracted almost universal approval, especially when associated with explicit moral and religious instruction and due stress on subordination," writing was "more controversial, and unqualified approval of teaching all 3 Rs was a mark of a liberal or even radical outlook" (Wardle, 1974: 90; see also Braithwaite, 1982).

With regard to the empirical dimension, what are the sorts of skills or competencies taken as constitutive of being functionally literate within societies like our own? The most elaborate attempts made thus far to establish "a selection of print-mediated activities on behalf of a 'prototypical' citizen in an 'average' structural location with a 'standardized' lifestyle" (Levine, 1982: 261) have been undertaken in North America.[2] Perhaps the best known of these is the Adult Performance Level Study (APL)[3] upon which we draw in order to amplify functional literacy in empirical terms.

In *Adult Functional Competency: A Summary* (Adult Performance Level Project, 1975), APL indicates the sorts of skills or competencies regarded by experts in the field as constitutive of functional literacy. They actually speak of functional *competency* (of what it means for an American adult to be functionally competent) rather more than of functional *literacy*. The former is largely, but not totally, exhausted by the latter. The match, however, is so close and the connection sufficiently obvious for us to proceed without further clarification. APL constructed a model of functional literacy on two dimensions:

(a) the *content* of functional literacy (the kind of information one needs to be able to get at or the knowledge one must be able to generate in order to function competently); and
(b) the *skills* of functional literacy.

On the content side they produced a taxonomy of human needs, presented as five general knowledge areas. Compare:

These general areas, which may be considered as the content of adult literacy, are . . . known as (1) consumer economics, (2) occupational (or occupationally-related) knowledge, (3) community resources, (4) health, and (5) government and law. Turning to the skills dimension, APL claims that

> . . . four primary skills seemed to account for the vast majority of requirements placed on adults. These skills were named (1) communication skills (reading, writing, speaking, and listening), (2) computation skills, (3) problem solving skills, and (4) interpersonal relations skills. (Adult Performance Level Project, 1975: 2)[4]

These two dimensions of functional literacy yield a closer specification of the functionally literate adult as follows. First, for each general knowledge area a broad *goal* is specified. To the extent that an adult achieves these goals s/he is functionally competent. The goals are:

(a) consumer economics—to manage a family economy and to demonstrate awareness of sound purchasing principles;
(b) occupational knowledge—to develop a level of occupational knowledge that will enable adults to secure employment in accordance with their individual needs and interests;
(c) health—to insure good medical and physical health for the individual and his family;

(d) government and law—to promote an understanding of society through government and law and to be aware of government functions, agencies and regulations that define individual rights and obligations;
(e) community resources—to understand that community resources, including transportation, are utilized by individuals in society in order to obtain a satisfactory mode of living. (Adult Performance Level Project, 1975: Appendix)

Second, each goal statement is in turn more closely defined by a series of *objectives.* There are 20 objectives for consumer economics, 10 for occupational knowledge, 13 for health, 6 for government and law, and 16 for community resources (including 9 relating to transportation). Examples include:

Consumer economics

(i) to build an oral and written consumer economics vocabulary,
(ii) to be able to count and convert coins and currency and to convert weights and measures using mathematical tables and mathematical operations,
(iii) to understand the concepts of sales tax and income tax,
(iv) to develop an understanding of credit systems;

Occupational knowledge

(i) to identify sources of information (e.g., radio broadcasts, newspapers, etc.) that may lead to employment,
(ii) to be aware of vocational testing and counseling methods that help prospective employees recognize job interests and qualifications, (iii) to prepare for job applications and interviews, (iv) to know attributes and skills that may lead to job promotion, (v) to know standards of behavior for various types of employments;

Government and law

(i) to develop an understanding of the structure and functioning of the federal government,
(ii) to investigate the relationship between the individual citizen and the government,
(iii) to understand the relationship between the individual and the legal system,
(iv) to explore the relationship between government services and the U.S. tax system. (Adult Performance Level Project, 1975: Appendix)

Third, at the most specific level of definition each objective is in turn described by a series of situation-specific requirements called *tasks.* The APL team did not publish examples of tasks since these were undergoing "thorough revision" at the time of publication in 1975. They do, however, note that any tasks that may eventually be specified within developed curricula should be considered as "paradigms or general guidelines" and not be "interpreted as being engraved in stone". This is because "it is the *objective* that is the most important element in the requirements for functional literacy" (Adult Performance Level Project, 1975: Appendix).

So much then for an official view of what functional literacy comprises in empirical terms within a complex, developed society (such as New Zealand is). To what extent, however, can functional literacy, thus conceived, be regarded as a desirable and appropriate objective for adult literacy schemes? To make good the earlier judgment, a critique of functional literacy follows.

A Critique of Functional Literacy

This critique takes the form of elucidating briefly the following claims: (a) that adult literacy schemes built around such ideas and values of functionality as outlined above are politically naive (or else willfully perverse);[5] (b) that the attempt to make (illiterate) adults functionality literate is an exercise in domestication; (c) that the values underlying programs aiming at functional literacy are dehumanizing.

Let us examine first the charge of political naivety. Thus far I have said little about the *values* underlying quests for functional literacy in "developed" societies like our own. Typical statements from proponents of functional literacy stress the functionality of literacy for the individual person concerned (and her/his dependents). The APL team, for instance, speak of the importance of literacy in meeting the requirements of adult living, in fostering the competencies "which are functional to economic success [although 'survival' would be a better word] in today's society" (Adult Performance Level Project, 1975: 1; square brackets mine). Both material and intangible advantages are supposedly opened up to the individual.

On the material side are aspects bound up with an enhanced standard of living—a better paid job (or a job at all), improved health care, a planned family, sound housing, etc. Intangible benefits include such things as enhanced job satisfaction, heightened self-esteem, psychological well-being (reduced anxiety, feelings of optimism, competence and security), and the like.

Put like this the political naivety involved becomes transparent. The logic here is similar to that which underlies current recommendations that young people who will soon be seeking work, or who are having difficulty finding work, should make themselves *employable:* whereas the reality of the situation is that there are rather few jobs available, and that thousands of eminently employable New Zealanders cannot find work (just as many thousands of employable Americans, Australians, and Britons cannot). Unemployment, subsistence wages, inadequate housing for a proportion of the population, unequal access to health care, etc., are material conditions that are part of the very structure of capitalist economies. There is simply no chance that making all people functionally literate can put them in the way of a job (let alone a well-paid job), or of housing adapted to their needs (let alone their wants), or quality health care. Access to these goods (or have they become *privileges*?) is systematically denied to many people by the very structure of a capitalist economy. This applies *a fortiori*—as we can see at present—in periods of crisis within capitalist economies. And, of course, with increasingly greater use being made of technology within the labor process, the forecast for even near-full employment in the future is gloomy (see, for example, Braverman, 1974; Gorz, 1982; Jones, 1982). To see functional literacy for present illiterates as a rescuing savior is the height of naivety.

The same holds for the alleged intangible benefits of being functionally literate. The progressive deskilling of labor increasingly erodes the possibilities of job satisfaction in the future for all but a privileged minority. The ever-widening gulf between conception and execution in the labor process, and the location of conception within increasingly smaller groups of "executives" bodes ill for future job satisfaction among the great majority who (will) have no effective control over their labor (see Braverman, 1974). Put this together with the reality of subsistence level wages, the possibility of youth rates, and increased competition for even the "worst" jobs, and the alleged intangible benefits of functional literacy begin to look like a bad joke.

However this still leaves the important possibility of heightened self-esteem as a consequence of attaining a functional level of literacy. Unfortunately the chances of any *real* gains here are also suspect. The argument here is simple. Self-esteem is to an important extent a relative affair. In reality we tend to value our performances relative to what other people can do. Bringing more and more people up to a functional level of literacy is rather like those with an inferior view at a football match standing up to see the game—it is not an effective move once everyone starts doing it. People overall end up at a higher elevation, but the original superiorities and inferiorities are maintained. Levine makes this point in the following passage:

> An individual needs a minimum level of mastery in order to "pass" as literate in public and keep intact his or her self-respect; as schools and literacy programmes become more effective in equipping their students with these skills, the effective threshold of acceptability will be raised accordingly. There is, quite simply, no finite level of attainment, even within a specific society, which is capable of eliminating the disadvantages of illiteracy or semi-literacy by permitting the less literate to compete on equal terms of employment and enjoy parity of status with the more literate . . . As long as there are people credited with special or

> superior literacy skills, the least competent will remain vulnerable to discrimination. (Levine, 1982: 259–260)

Those people who *do* see themselves initially as more adequate, more complete, and hence more worthy of their own esteem in virtue of their functional literacy will nevertheless—as argued above—be unable en masse to realize the *material* gains they aspire to. The most likely consequence of this is that the focus of their self-evaluation will then shift to these other (material) considerations, where they *are* objectively "inferior." It is possible, we suppose, that they might—on the basis of their new found self-esteem—begin actively to seek these material gains. Finding themselves unable to achieve them they might conceivably ask themselves why this is so and thereby move toward a more critical, political literacy. But while this may be possible in principle it is most unlikely in practice. For it presupposes a collective experience and subsequent questioning rather than an individual one, and this is unlikely to eventuate. Even if scattered individuals did feel dissatisfied and disenchanted they are unlikely, without a direct stimulus to perceive this as a shared experience, to view the situation as anything other than a consequence of personal inadequacy. Such is the effect on consciousness of ideologies stressing equal opportunity and individual merit. After all, our society contains many literate individuals who seek more in material terms than they are able to get. There is no evidence whatsoever that significant numbers of them—who presumably enjoy the confidence and self-esteem that literacy affords—are seeking a critical political awareness with a view to improving their lot. And there is no reason to believe that individual recipients of functional literacy programs will respond any differently. The fact is that in general people have to be *invited* to seek a different (heightened?) consciousness. They have to be invited, stimulated, to identify themselves as a *group* and *then* to question the adequacy of their consciousness, as a prelude to achieving shifts in consciousness and engaging in social action. Functional literacy programs are not characterized by any such invitation. Such considerations as these underscore our doubts that functional literacy will even promote abiding self-esteem. I conclude, then, that those who espouse functional literacy as a route (even if not the *only* route) to personal and economic success are, insofar as they are well meaning, open to the charge of being politically naive.[6]

To claim, next, that functional literacy programs are exercises in *domestication* is to note that they may be functional not for the clients but more for those whose interests are best served by maintaining the status quo (Postman, 1970). Of the many important themes that can emerge here the following discussion touches briefly on just three.

(a) Ensuring that people can read is indispensable to controlling their behavior. This objective fuels the urgency currently attached to second-language (English) learning in New Zealand, and particularly in Auckland, where there is a large Polynesian immigrant population. Postman notes that in a complex society a person cannot be *governed* unless s/he can read rules, signs, regulations, procedures, forms and directives. "Thus," claims Postman, "some minimal reading skill is necessary if you are to be a 'good citizen,' but 'good citizen' here means one who can follow the instructions of those who govern him" (1970: 246).

This, however, goes beyond the mere mechanics of controlling *behavior*. For the process of facilitating behavioral control through literacy has important hegemonic implications. Those who are enabled to follow directives, obey requirements, and measure up to standards by acquiring the ability to read, are simultaneously introduced to a set of values: namely, those underlying these very directives and requirements. Consequently, their perception of the world, their consciousness, is shaped in the same process.

(b) This latter point is graphically illustrated by the objectives actually specified by the APL team. To meet the objectives they lay down is precisely to be inducted into the values of a competitive capitalist-consumer society. Compare here: "to be aware of the basic principles of money management, including the basics of consumer decision-making"; "to know the attributes and skills which may lead to promotion"; "to know basic procedures for the care and upkeep of personal possessions (home, furniture, car, clothing, etc.) and to be able to use resources relating to such care" (Adult Performance Level Project, 1975: Appendix). No further comment seems necessary.

(c) We should note that accounts of functional literacy attainment typically place minimal emphasis on *writing* skills—at least beyond their most rudimentary forms (cf. Sticht 1972).[7] The significance for domesticating people of the difference between being able to read and being able to write is crucial. Levine argues that on the whole it is

> writing competencies that are capable of initiating change. Writing conveys and records innovation, dissent, and criticism; above all it can give access to political mechanisms and the political process generally, where many of the possibilities for personal and social transformation lie. (Levine,1982: 262)

Let us recall that it was only when the working class in Britain became capable of writing as well as reading that a real threat was presented to ruling-class interests through literacy. This point has always been well understood by representatives of ruling interests. Certainly it did not escape Hannah More who, having established Sunday Schools in the Mendips mining area during the 1790s, came under attack for extending education to the working class. Defending her practice against attack, she made it perfectly clear that her schools were intended to keep the working class in their subordinate state—and for this reason writing, although not reading, was prohibited:

> My plan of instruction is extremely simple and limited . . . They learn, on weekdays, such coarse works as may fit them for servants. *I allow of no writing for the poor.* My object is not to make them fanatics, but to train up the lower classes in habits of industry and piety. (Cited in Simon, 1960: 133; emphasis added)

The argument, finally, for the claim that the values underlying programs aiming at functional literacy are *dehumanizing* is largely implicit in what has already been said. By "dehumanizing" is meant "negating our essence as human beings," drawing on Freire's view of what is involved in being (truly) human (see, for example, Freire, 1974: 3–20). According to Freire, as humans we have the ontological and historical vocation of becoming more fully human (Freire, 1972: Ch. 1). Becoming more fully human, in Freire's view, involves becoming increasingly aware of one's world and increasingly in creative control of it. The more that one engages in conscious action to understand *and transform* the world—one's reality—the more fully human one is. Our ontological and historical vocation as human beings is to each become an active *Subject* who enters into the process of creating reality—creating history and culture—rather than merely existing as a passive *object* accepting the world/reality as ready-made by other forces (and where this ready-made world infringes against one's best human interests).

Any humanly contrived phenomenon that tends to negate or obviate our pursuit of our ontological and historical vocation must, on this view, be dehumanizing.

Correspondingly, values comprising an ideology, or part of an ideology, which maintain people at the level of passive objecthood, immersed in false consciousness, must equally be regarded as dehumanizing. The values underlying programs aiming at functional literacy very clearly comprise such an ideology. The APL program is little more than an induction into the practices and values of consumerism and capitalism—engaging the disadvantaged in a mode of life where most of them will remain disadvantaged (and their children after them), while a few may achieve advantage but only at the expense of others who become downwardly mobile. The procession to functional literacy is an initiation into the same (distorted) consciousness of capitalist reality as informs and motivates such programs as APL in the first place (for an extended argument that would support this charge of distortion see, for example, Harris, 1979).

Clearly the argument presented here invites more detailed development than space permits. Also I have thus far avoided explicitly adopting a particular political-ideological theory or stance. Yet, ultimately a justified theoretical stance is necessary if we are fully to understand the prevailing view of functional literacy (and move beyond it) and to grasp the importance for New Zealand education of critiquing functional literacy as an educational goal. We will shortly attempt (albeit sketchily) to correct this omission. However even as it stands my critique is sufficient to suggest the following paradox: whereas functional literacy is presented as being functional for illiterate disadvantaged persons it may well be *dysfunctional* for them, *and functional instead* for those persons/classes whose interests are best served—at the expense of the disadvantaged—by maintaining the economic, political, and cultural status quo. Exploring this paradox in detail and developing a strongly defended sociopolitical analysis to facilitate this exploration would be a major undertaking. But by way of a concluding section I will try briefly to suggest a direction I think is promising. Readers familiar with the work of Marx, as well as of such contemporary writers as Michael Apple, Sam Bowles and Herb Gintis, Martin Carnoy, Paulo Freire, Henry Giroux and Kevin Harris, to name but a few of those better known, will appreciate the broad direction in which I would go (see, for example, Apple, 1979, 1982; Bowles and Gintis, 1976; Carnoy, 1974, 1975, 1977; Giroux, 1981; Harris, 1979, 1982).

Capital, Class and Functional Literacy

To claim that something is functional is elliptical. It invites the question: functional for what? or, for whom? or, in what ways? The intimation in the talk of functional literacy we have encountered so far is that primarily it will be functional for the hitherto illiterate person. It will, first and foremost, meet *their* purposes, enhance *their* quality of life, help *them* to actualize their human potential, etc. Of course functional literacy will be expected also to promote purposes or ends beyond those of the illiterate person alone. But the primary intimation is as I have described it. I want now to challenge this intimation by endorsing a particular theoretical frame, which will accommodate, sharpen, and give direction to the arguments already advanced. This will be a broadly Marxist frame.[8]

The clue to this concluding section is provided by Freire's claim that it is the *ontological and historical* vocation of human beings to become more fully human (Freire, 1972: 13). I want to develop two ideas that emerge from this claim. First, it is significant that Freire speaks of the human vocation as being *both* ontological *and* historical. Second, I want to suggest that talk of a human vocation is essentially the same as the old Greek notion of humans per se having a function (Lankshear, 1982: 142–144).

Our *ontological* vocation is that which we have by dint of our human essence; that we have as humans in the abstract. That is, our ontological vocation is a destiny or pursuit seen as somehow stamped upon our nature as being appropriate for humans per se: for humans in all times, in all places and under any conceivable circumstances. By contrast, the idea that our proper vocation is also *historical* in nature draws attention to the fact that humans do not exist or live in the abstract. They live—function, if you will—within structured settings: within some set of economic, social, political, cultural, institutional, etc., structures. Their (ontological) vocation must necessarily be pursued within some structured context or other. The ontological and historical dimensions of our human vocation merge, as is obvious if we recognize two things:

(i) that becoming more fully human is a call to engage actively and as far as possible in the creation of one's (structured) reality—a call to *create history,* in the sense that historical epochs are characterized by (among other things) distinctive material structures—institutions, practices, buildings, etc.—as well as distinctive structures of a non-material kind—systems of beliefs, values, goals or aspirations, etc.—and that these are human *creations;*[9] and

(ii) that some structures (individually or collectively) impede or negate for many human beings the pursuit of their ontological and historical vocation—effectively locating the power to shape reality in the hands of a few—while other structures will enable human beings generally to engage on an equal basis in shaping their shared world and controlling their related destinies.

This being so we are forced to recognize that some structured settings tend to thwart the attainment of the human vocation by all on an equal basis, *and that this is itself a historical, humanly produced phenomenon,* while others tend to promote general attainment of the human vocation. A marxist view of capitalist structures (that is, the structures that collectively comprise capitalism as a total context of human existence; a totality of social relations) entails, of course, the claim that they necessarily thwart attainment of the human vocation by all people on an equal basis. Capitalism produces distinct social classes standing in antagonistic relation to each other, and where one class in particular is debilitated in its power to create history *and* objectively disadvantaged materially by the history that *is* created and sustained.[10]

This provides a particular framework within which to (re)describe and evaluate functional literacy. Given a critical analysis of capitalist social relations it can be argued that functional literacy is in fact functional for those who are materially advantaged by the status quo since it has the effect of supporting and maintaining existing capitalist structures. On the one hand functional literacy appears to offer the promise of (at least) survival within a capitalist economy to those who, without basic literacy, might well fail to survive for want of employment, or efficient management of their (meager) resources, or ability to communicate with government departments,[11] etc. It offers this chance of survival (even, it may seem, the chance of "making it") without in any way requiring significant change within the capitalist formation. On the other hand, functional literacy is functional for the existing advantaged elites in that it systematically withholds exposure to ideas, beliefs, practices, etc., which might encourage illiterates to challenge capitalist structures. At the same time the availability of official functional literacy programs discourages illiterates from seeking literacy training within organizations, which might promote such challenging ideas, beliefs and practices.

Conversely, functional literacy, as currently conceived, is *dysfunctional* for those who are disadvantaged, exploited and oppressed within the existing capitalist order. While offering them the possibility of (mere) survival it impedes their potential pursuit of a more fully human life (as defined by the notion of our ontological vocation) and an enhanced material existence. These ends can, it seems, be achieved only through struggle: by confronting oppressive and exploitative structures within economic, social, political and cultural life.

This leads, finally, to advancing a different conception of functional literacy: one which is ethically desirable rather than reprehensible. In order to do this, however, it will be necessary first to link talk of a human *vocation* with the notion of humans per se having a *function*. In Greek thought the notion of goodness was commonly related to that of function. Something is good to the extent that it performs its function well. The Greeks extended this view to the ideal of living the Good Life. To live as one ought, was, in their view, to fulfill one's function as a human person. For the Greeks the function of human beings qua human beings was bound up with the distinctiveness or uniqueness of human beings. The question, "what is the function of a human being simply considered as a human being?", assumed in Greek thought the form, "what is distinctive of human beings simply as humans?" (Lankshear, 1982: 143; see also Taylor, 1970: Ch. 1). Of course for the Greeks human function was conceived as living one's life under the demands of Reason: living rationally is how humans ought to live, how they are supposed to function, or what they are "called on" by their essence to do

This idea of a human *function* is, then, essentially the same as Freire's idea that humans have an ontological *vocation*. But whereas the Greeks construed our vocation or function as living a life under the demands of Reason—rationality being for them the distinctive or unique quality of human beings—Freire construes it as engaging actively in creating the world: creating history and culture. Humans, and humans alone, are both *in* the world and *with* the world (Freire, 1974: 3). Humans alone can consciously understand their relationship with the world, the distinction between culture and nature, the way in which their action upon reality transforms the natural into the cultural and the historical: the way in which they can transform reality, "creating the world of men and women, which is the world of culture and history" (Freire speaking to Davis in Davis, 1980: 63). To create this world is the function, the calling, the vocation of human beings: of *all* human beings equally.

This notion of function, coupled with the previous account of social life under capitalism, leads directly to a radical alternative conception of functional literacy. On this view, for literacy to be functional is for it to enhance the uniquely human potential of every person to create the world of men and women, which is the world of culture and history. Furthermore, *all* literacy must, morally speaking, be functional in this sense—otherwise it is dehumanizing. It follows also from the above view of capitalist structures and social relations generally, that pursuit of functional literacy is necessarily a *revolutionary* process. It is a process whereby people acquire—in dialogue facilitated by reading and writing—intellectual access to their world and their place within that world, a conception of their unique human status and vocation, and the commitment to pursue their vocation. As such it will bring them to consciousness of oppressive, exploitative, and otherwise dehumanizing structures, and the need (in social justice) to challenge and transform these. Necessarily complementing this consciousness, and in dialectical relationship with it, is a commitment to *action:* to action aimed at replacing dehumanizing structures with humanizing structures within which people can, on an equal basis and with each other, pursue their function as human beings. Mackie (1980: 2) makes this point well when he speaks of Freire's

> recognition of the way language forms our perceptions of the world and our intentions towards it. In doing so he highlights the connections between language, politics and consciousness. Conceiving the task of literacy to be humanisation, Freire is led inevitably to an examination of the ways social and political structures impede this goal. As a consequence, his discussion of literacy and education has as one of its principal concerns the promotion of revolutionary social change. Freire's pedagogy focuses on human liberation from oppression, not only in Brazil, but everywhere oppression exists. So while his theory has situated origins, its applications are much wider.

Also, and finally, functional literacy, properly conceived, is an ongoing process. Just as the creation of history and culture is never complete, so equally are our understanding of the world and our transforming action upon it—in accordance with our understanding of the world—never complete. And integral to this on-going process of historical and cultural creation is the process of becoming ever *more functionally* literate.[12]

Endnotes

1. It might be counter-argued at this point that it is not really a criticism of functional literacy to claim that it is a minimal ideal. For surely its minimal status is acknowledged in the very idea of *functionality.* Furthermore, proponents of functional literacy are themselves likely to admit that it is a minimal goal and simply assert that it is nevertheless better than illiteracy. It is precisely the intuitive force of this challenge that suggests the importance of attempting a *political economy* of functional literacy—which is the purpose of this paper. For although it seems morally undesirable to present functional literacy and outright illiteracy as the only available options, these are in fact the only options that large numbers of people in societies like our own effectively have open to them—despite the possibility *in principle* of their attaining so much more. The fact is that, almost without exception, where official functional literacy programs are provided for disadvantaged adults it is with the view that this is all they need (and certainly all they are going to have provided). No provision is made for attainment of a more advanced or more critical literacy. Now if the goal of literacy provision is simply to enable people to survive in society *as it is,* then no doubt to be functionally literate is better than to be completely il-

literate. But if one seeks social change in the direction of greater economic justice and more active participation on the part of people generally in the shaping of their social reality, it might well be argued that functional literacy constitutes no significant advance over illiteracy. (Indeed it may actually constitute a hindrance in some settings. Certainly the Cuban and Nicaraguan revolutionary vanguards found it a positive advantage to have the chance to share their literacy skills with the illiterate peasants who were to fight alongside them.) My ultimate concern in this paper is with an ideal of functional literacy that would of its very nature be part of a process of social change. From this standpoint it clearly *is* a criticism of functional literacy, as currently conceived, to claim that it is a minimal ideal. For this, in Freire's language, is to claim that functional literacy domesticates people and thus dehumanizes them.

2. Of course *every* particular literacy scheme aiming at making people functionally literate will have its own (explicit or implicit) definition in empirical terms. Space does not permit investigation of multiple programs here. Indeed, often it is very difficult to obtain information about these schemes, largely for reasons relating to maintaining the privacy of illiterate "clients."
3. This study was undertaken by the Office of Continuing Education, University of Texas at Austin, on behalf of the U.S. Office of Education.
4. NB. Interpersonal relations skills seem less obviously an aspect of functional *literacy* than do, say, communication skills. This indicates how "functional literacy" does not completely exhaust "functional competence."
5. It could conceivably be argued that official support for functional literacy might be a ruse to mask the reality of social relations under capitalism and to keep the disadvantaged falsely aware—thereby gaining 'breathing space' for a capitalist economy currently in crisis. Such an interpretation would involve some form of conspiracy theory. While I believe there may be something to this view there is certainly not sufficient space to argue it here. Nevertheless this is the kind of idea I have in mind in suggesting that if the functional literacy drive is not politically naive it is willfully perverse. See also note 6 below.
6. See note 5 above. A further possibility, although I have no positive evidence for it, is that some reformers may still advocate a program like APL in the hope that it will produce disaffected people who will cause social unrest. In my view the chances of such an outcome appear remote. Compare for example, the reality of university and college graduates who prefer en masse to take their chances "on making it within the system" rather than to challenge that system. This seems especially to be true in times of economic recession.
7. NB. "REALISTIC" in Sticht's "Project REALISTIC" is derived from *re*ading, *lis*tening and arithmet*ic*.
8. In what follows I will lean heavily on Freire's marxism, but the importance and relevance of the accumulated corpus of marxist theory, and its application recently to education in capitalist societies, cannot be overemphasized.
9. Of course the material and non-material dimensions of structure cannot be separated completely. For example, particular lived material practices incorporate particular beliefs, goals, values, etc. See here Kevin Harris (1979).
10. I recognize that this is a superficial gloss of a marxist position and can do no better than to direct readers to the writings of Marx himself and marxist theorists. All I am trying to do here is to locate talk of functional literacy within a context of antagonistic class relations, and capitalist structures generally which are unequal in their effect on the capacity of human beings to pursue their ontological and historical vocation. 1 am satisfied that the description given, such as it is, is sufficient to this end.
11. The dimension of functional literacy for those learning English as a second language is important in this context, but cannot, unfortunately, be developed as a sub-theme here.
12. As Mackie (1980: 1) puts it: "literacy is a process which continues throughout life. To be literate is not to have arrived at some pre-determined destination, but to utilize reading, writing and speaking skills so that our understanding of the world is progressively enhanced."

Bibliography

Adult Performance Level Project. (1975). *Adult Functional Competency: A Summary*. (ED No. 114 609.) Austin: University of Texas, Division of Extension.

Adult Reading and Learning Assistance Federation (NZ) Inc, (ARLAF) (1983). Personal correspondence, July.

Ahmed, M. (1958). *Materials for New Literates.* New Delhi: Research Training and Production Centre.

Apple, M. W. (1979). *Ideology and Curriculum.* London: Routledge and Kegan Paul.

Apple, M. W. (1982). *Curriculum and Power.* Boston: Routledge and Kegan Paul.

Bowles, S. and Gintis, H. (1976). *Schooling in Capitalist America: Educational Reform and the Contradictions of Economic Life,* London: Routledge and Kegan Paul.

Braithwaite, E. (1982). Education and equality. *Access,* 1(1): 33–50.

Braverman, H. (1974). *Labour and Monopoly Capital: The Degradation of Work in the Twentieth Century.* New York: Monthly Review Press.

Cardenal, F. and Miller, V. (1981). Nicaragua 1980: The battle of the ABCs. *Harvard Educational Review,* 51: 1–26.

Carnoy, M. (1974). *Education as Cultural Imperialism.* New York: McKay.

Carnoy, M. (1975). *Schooling in a Corporate Society: The Political Economy of Education in America.* New York: McKay.

Carnoy, M. (1977). *Education and Employment: A Critical Analysis.* Paris: UNESCO.

Cook, W. (1977). *Adult Literacy in the United States.* Newark, DL: International Reading Association.

Davis, R. (1980). Education for awareness: A talk with Paulo Freire. In R. Mackie (ed.), *Literacy and Revolution: The Pedagogy of Paulo Freire.* London: Pluto Press. 57–69.

Freire, P. (1972). *Pedagogy of the Oppressed,* Harmondsworth: Penguin.

Freire, P. (1974). *Education: The Practice of Freedom,* London: Writers and Readers.

Freire, P.(1981). The people speak their word: Learning to read and write in Sao Tome and Principe. *Harvard Education Review,* 51: 27–30.

Giroux, H. A. (1981). *Ideology, Culture and the Process of Schooling.* Philadelphia: Temple University Press.

Gorz, A. (1982). *Farewell to the Working Class: An Essay on Post Industrial Socialism.* London: Pluto Press.

Gray, W. S. (1956). *The Teaching of Reading and Writing.* Paris: UNESCO.

Harman, D. (1970). Illiteracy: An overview. *Harvard Educational Review,* 40: 226–243.

Harris, K. (1979). *Education and Knowledge: The Structured Misrepresentation of Reality.* London: Routledge and Kegan Paul.

Harris, K. (1982). *Teachers and Classes.* London: Routledge and Kegan Paul.

Jones, B. (1982). *Sleepers Wake: Technology and the Future of Work.* Melbourne: Open University Press.

Lankshear, C. (1982). *Freedom and Education.* Auckland: Milton Brookes.

Levine, K. (1982). Functional literacy: Fond illusions and false economies. *Harvard Educational Review,* 52: 249–266.

Mackie, R. (ed.) (1980). *Literacy and Revolution: The Pedagogy of Paulo Freire.* London: Pluto Press.

Morales, A. (1981). The literacy campaign in Cuba. *Harvard Educational Review,* 51: 31–39.

Postman, N. (1970). The politics of reading. *Harvard Educational Review,* 40: 244–252.

Simon, B. (1960). *Studies in the History of Education 1780–1870.* London: Lawrence and Wishart.

Sticht, T. G. et al. (1972). Project REALISTIC: Determination of adult functional literacy levels. *Reading Research Quarterly,* 7(3): 424–465.

Taylor, R. (1970). *Good and Evil.* New York: Macmillan.

UNESCO (1976). *The Experimental World Literacy Programme: A Critical Assessment.* Paris: UNESCO.

UNESCO (1949). *World Illiteracy at Mid-Century.* Paris: UNESCO.

Wardle, D. (1974). *The Rise of the Schooled Society.* London: Routledge and Kegan Paul.

Two

The Dawn of the People: The Right to Literacy in Nicaragua (1986)

Colin Lankshear

Biography of the Text

My continuing interest in the politics of literacy and, in particular, Freire's ideas about literacy in the interests of emancipation, raised many empirical questions. To what extent was there empirical evidence for claims that literacy education could be part of emancipatory politics, particularly under conditions where popular movements for social justice have succeeded in gaining the power of government? During the mid 1980s, Nicaragua provided an interesting "test case" for such questions, and between 1985 and 1992 I made regular visits to urban and rural settings and spent a lot of time collecting documents and doing field work. As might be expected, the data told complex and often contradictory stories. At the same time, I was politically "onside" with grassroots efforts to maintain the work of the 1980 Nicaragua Literacy Crusade and became involved in attempts to raise money and to support literacy, health and popular education initiatives in Nicaragua, as well as to contribute to the academic and professional literature about the ongoing grassroots literacy work there.

In New Zealand a Labour Government had been elected in 1984, and some key members of that government had very strong social democratic convictions that made them sympathetic to social initiatives in Nicaragua. In her foreign affairs ministry role, Helen Clark—who, two decades later, became New Zealand's longest serving Prime Minister to date—was especially supportive. She and colleagues worked to ensure generous (3 : 1) subsidies for fund-raising initiatives for small-scale development projects in Nicaragua. My academic work in the area of literacy became fused with civic activist work in fund-raising and publicity. In 1985 a colleague at Auckland University invited me to present a plenary address at a professional development event for teachers. The theme was on the politics of literacy, with an emphasis on how Freire's ideas had been interpreted within Nicaragua's Literacy Crusade and subsequent popular literacy initiatives. This chapter is the version of the address published by the New Zealand Reading Association.

Introduction

In September 1980, the following message was conveyed from the UNESCO offices in Paris:

> *The panel of judges designated by the Director General of UNESCO to grant the 1980 prizes for distinguished and effective contributions on behalf of literacy . . . has unanimously chosen for first prize—the National Literacy Crusade of Nicaragua.*

I do not recall any mention in our popular media of this award being made, or, indeed, any news whatsoever of the Crusade itself. To this day proportionately very few New Zealanders have any significant awareness of the character of the Nicaraguan Revolution, far less of the importance placed by Nicaraguans on literacy as a crucial factor in the process of human liberation.

My hope is that what I say here will help to foster at least some awareness of these matters.

The Literacy Crusade itself represents but one phase—albeit the most dramatic—in what the overall process of *alfabetización* in Nicaragua.

"Alfabetización" is the name (in Latin American Spanish) for the process of helping people to become literate. Three distinct but related phases are evident in the drive for popular literacy *as of right* in Nicaragua.

(a) The FSLN (Sandanista Liberation Front)'s practice of teaching illiterate recruits to read and write during the insurrection period of 1961–1979.
(b) The National Literacy Crusade of 23rd March to 23rd August, 1980.
(c) The program of Popular Adult Education that has operated since the completion of the Literacy Crusade.

The thread that connects these three phases is the ideal of human liberation—the vision and values articulated by the FSLN and which, following the elections in November 1984, can be said clearly to have the active support of the overwhelming majority of Nicaraguans today. Throughout its existence, the FSLN has emphasized the necessity of universal education for genuinely democratic participation in Nicaraguan life, and recognized universal literacy as *the absolute baseline* for anything approaching democratic process. Their guiding view of liberation is of people, all people equally, having a direct and active voice in shaping their own destinies. If we compare it broadly with Paulo Freire's ideal of human liberation we will be close to the mark.

What 1 want to do is to trace these three phases in turn, seeking in the process to say something about Nicaraguan reality as well as addressing the more specific theme of promoting literacy. I will begin with some brief comments about *"Alfabetización."*

Alfabetización: Literacy "Plus"

In recent years it has been increasingly accepted in societies like our own that literacy teaching is not and cannot be a "pure" process, by which people acquire mastery over symbols and nothing else is transmitted in the learning process.

Rather, it is acknowledged that the symbols—words, texts, primers—employed in teaching and learning communicate ideological values to the learner in the very process of their becoming literate. (Values, that is, besides the worthwhileness of becoming literate). Thus it is, for example, that people have scanned school readers in our own country to purge them of unwanted—*and previously unrecognized*—ideological content: sexism, racism, elitism, stereotypings, etc. (We may compare here also the work of such organizations abroad as the Council on Interracial Books for Children).

By literacy "plus" I mean, then, that the process of making people literate is also, and inevitably, a process of communicating values, assumptions, habits, traditions, practices, prejudices, etc., *whether this is recognized or not.* In addition to learning to read and write, the learner is also inducted into or further confirmed in some kind of world-view, an ideological framework, a consciousness.

As yet the politics of literacy is a radically underdeveloped area of inquiry in New Zealand. We have not progressed far—if anywhere—beyond purging some children's books of the more gross manifestations of sexism and racism.

Certainly we have not made a concerted effort to discover just to what extent our language—oral and written—embodies and reflects a definite system of values: political, ethical, social, cultural, spiritual, economic. Consequently, we have rather little appreciation of the power of literacy as an active inductor into a system of political and social values.

By contrast, educators within the ranks of the FSLN have been acutely aware of the power of literacy teaching as a *shaper of consciousness.* In each of the phases of alfabetización, the ideological—and particularly the political—character of language has been recognized and consciously deployed.

The process of transmitting literacy has at the same time been an invitation to a given view of the world and of human values. It is an invitation to scrutinize the world in a certain way and to consider how that world might be changed: changed in the direction of breaking down privilege, gross exploitation and oppression, and replacing these realities with the pursuit of social justice in a society where the voice of each person would be heard, and the equal right of each person to live with dignity and to fulfill their human potential honored.

Both prior to the overthrow of Somoza in 1979 and subsequently, this deliberate fusion of literacy teaching with an explicit ideological position has had a dual aspect. In some cases—i.e., with illiterate people who had actively joined the revolutionary forces—the literacy process combined teaching the skills of reading and writing with helping the "pupils" to analyze and understand more clearly something they had already intuitively grasped: namely, that oppression, poverty, disease, and ignorance were not aspects of a "natural" or "given" reality but, rather, consequences of how Nicaraguan society operated under Somoza, and that a new social order could be built once the old had been overthrown. In other cases—indeed, the majority—the literacy process was/is more an exercise in trying to get people committed to a set of revolutionary democratic values; to engender commitment to a new set of values and to building a new society on the basis of these values.

We are now in a position to consider in turn the three phases of alfabetización in Nicaragua identified here.

1. The Insurrection Period

The ideal of popular literacy was quite irrelevant to Nicaragua under Somoza in the same way that it remains irrelevant to much of the predominantly rural Third World. As one writer puts it:

> It is no accident that so many of the Third World's rural poor are illiterate. When lands are in the hands of a few, and profits from their plantations depend on large numbers of docile and dependent field hands, there is neither incentive nor encouragement for education; certainty, few peasants would be given technical or managerial responsibilities. Furthermore, the paternalistic, authoritarian political structures of such countries have little use for the participation of the rural or urban work force. Its demands tend to seriously conflict with those few landowners and businessmen in whose interests such systems have developed. (Hirshon, 1983: 6–7)

Indeed, popular literacy was not merely irrelevant. It would also have been completely undesirable as far as the ruling interests in Nicaragua were concerned. The kind of society that Nicaragua had become under two generations of Somoza dictatorships positively *demanded* ignorance and docility from the great majority of the population. The economy was operated in the interests of a very small minority of Nicaraguans—the Somoza family and its close connections, the business and professional elite (industrialists, importers/exporters, lawyers, doctors, bankers, etc.), and foreign (largely U.S.) multi-national corporations.

For the protection of the interests of his own family, Somoza had at his disposal a large, well-armed, and notoriously brutal National Guard. The pattern of life which had developed by the 1970s was as bad as the contemporary world has to offer—given that Nicaragua had a small population, ample arable land and fertile soils for its food needs, a climate sympathetic to agriculture, and good fishery potential.

According to an official government study of the decade 1966–1975, 83 percent of Nicaraguan children suffered some degree of malnutrition—i.e., just 17 percent were of average height and weight for their age (as cited in Black and Bevan, 1980: 17). Infant mortality averaged 120/1000, rising to 333/1000 in poorer neighborhoods. Most infant deaths resulted from preventable diseases such as diarrhea, measles and intestinal parasites. Living conditions militated against health and educational development alike. According to the 1971 census, 47 percent of Nicaraguan homes had no sanitation whatsoever, 90 percent of houses in Managua (then a city of over 500,000) had no running water, 61 percent of houses had dirt floors and 55 percent of houses had no electricity (cited ibid.). Disadvantages were typically disproportionate between urban and rural settlement, in favor of the urban.

As has been implied, many of these problems, and certainly the intensity of them, resulted from radically unequal distribution/ownership of resources. At the time that Somoza fled, his family's wealth was estimated at $400–500 million. The family had holdings in more than 500 corporations, owned two T.V. stations, a radio station, a newspaper, the national airline and steamship company, and controlled half the sugar mills, two-thirds of the commercial fishing, 40 percent of rice production, the largest milk-processing plant, and cement/concrete production (cf. Weissberg, 1981: 8–9*)*. A very small minority of Nicaraguans shared in this lopsided ownership of resources.

> Fewer than 200 families—1.8 percent of those owing land—owned almost half of all farmland, while the poorest 50 percent of the farmers had only 3.4 percent of the land. Few small agricultural producers were able to survive by farming alone, but there was little steady work except for the three or four-month harvest period. In 1972 the average annual income of the poorer half of the rural population was thirty-five dollars. (ibid.: 10)

Somoza's cynicism toward his countryfolk plumbed the depths following the devastating earthquake in Managua, the capital and most developed city, in 1972. In this disaster, half the buildings in Managua were destroyed, 10,000 people were killed, 50,000 injured, and 200,000 left homeless. Despite the proportions of the disaster and the world attention and sympathy it commanded, Somoza diverted the international aid which poured into the country to the wealth and business interests of the Somoza clan.

Given this background it is scarcely surprising to find that at the time of Somoza's demise just 5 percent of the population had completed elementary school, and 53 percent of the population were illiterate—with figures reaching 90 percent in some rural areas. Half the secondary schools were private, and even the public schools charged fees that were beyond the reach of the average Nicaraguan (ibid.). Indeed, the most basic literacy, let alone a sound formal education, would have been beyond the wildest dreams of many in a population where 50 percent of the sick received no medical treatment at all, where 20,000 in a population of less than 3 million suffered from advanced tuberculosis, and where 4,000 cases of malaria were reported among the urban population alone between January and April 1974. Average unemployment was 30 percent, rising to 50 percent seasonally.

This, coupled with supply and demand, kept wages very low—that is, where people could find work.

It was these circumstances of gross social injustice, oppression, and exploitation that prompted the emergence of the FSLN in 1961 and, eventually, fomented the civil upheaval of 1978–1979, which resulted in Somoza's overthrow. The concern of the FSLN with promoting literacy in the context of armed struggle during the insurrection period is illustrated by two characteristic instances: one relating to illiterates who had already joined the fight, and the other focusing on potential recruits to the revolutionary cause.

The first concerns Carlos Fonseca, one of the founders of the FSLN and its main theoretician. Fonseca is credited with inspiring the central role that literacy was to assume in the political program of the FSLN. Tomás Borge recalls a crucial conversation between Fonseca, Borge, and Germán Pomares—the latter a peasant who had himself learned to read and write during the struggle. Borge and Pomares were training a group of peasant fighters. In Borge's words: "We were training them to dismantle and reassemble the Garand, the M-1 carbine and the .45 caliber pistol. Carlos arrived and instructed us, *'And also teach them to read'* " (cited in Hirshon, 1983: 4).

In this Fonseca was keeping alive a tradition begun by Sandino himself in the 1930s, during his forces' seven-year battle against U.S. marines in Nicaragua. Sandino strongly encouraged his troops to learn to read and write and was proud of the fact that "among all his officers the number of those illiterate could *"be counted on fewer than the fingers of one hand"* (cited in Cardenal and Miller, 1981: 3).

The second example involves an impressive case of subversive infiltration of a *bone fide* educational structure, as recalled by a senior official in the Nicaraguan Ministry of Education. In 1976 a group of priests with considerable funds available established a program of literacy teaching in rural areas, using university students as teachers. The FSLN successfully infiltrated the scheme—of course, many university students were active FSLN members or sympathizers—to the point where it was effectively controlled by the FSLN. It apparently took until 1978 for the priests to discover what was happening—although given the sympathies of many clergy at the time it is possible that they may have been turning a blind eye—by which time considerable important ideological work had been done. In early 1979, when the National Guard began raiding the central offices of the program, the priests closed down. The program by then, of course, had its own momentum, had done its work, and the overthrow of Somoza was but a few months away. This, we might say, was a program of functional literacy. But unlike the functional literacy with which we are familiar—i.e., functional relative to the way the existing society operates—this was functional to the overthrow of existing structures and to the building of a new society. This feature of being functional to the building of a new society is characteristic of all literacy teaching that has occurred in Nicaragua since the overthrow.

One final point must be made concerning the insurrection period. In 1969—ten years before the final triumph—the FSLN produced a political program to be adopted upon victory. The first item on the program's culture and education agenda was to be "a massive campaign to immediately wipe out illiteracy" (cited in Hirshon, 1983: 40).

2. The National Literacy Crusade of 1980

Victory over Somoza came on July 19, 1979. Before leaving, Somoza ordered his air force to bomb Managua, and he escaped with all the reserves of the Central Bank less just $3.5 million. He left behind him an international debt of $1.6 billion—the largest per capita debt in Latin America. The country's infrastructure was in ruins following the civil upheaval and Somoza's last-minute bombings. Buildings were wrecked, crops neglected and destroyed, agricultural land laid waste. The country was bankrupt and there was, so to speak, no cash in the till. The human cost of the war was astronomical—especially so, given the small total population. Between 40,000 and 50,000 had been killed, 100,000 injured (in a country with just 5,052 hospital beds), and 40,000 orphaned. Most of the victims by far were young people.

Needless to say, no one among the victors had any experience of governing in the conventional sense of the word. Shortly before victory, and when victory seemed imminent, a Junta of National Reconstruction was formed in exile in Costa Rica.

> It created several working groups to prepare a governmental plan, as well as the plans and programs of the various ministries. Among these was the educational working group which drafted a program for the literacy crusade. Within days

> after the victory, Father Fernando Cardinal . . . was charged with responsibility for the crusade. (Hirshon, 1983: 5–6)

Despite the unbelievable circumstances Nicaragua was in, the official announcement that the Literacy Crusade would go ahead was made just a month after victory. The crusade itself began eight months after Somoza fled. The United Nations subsequently described the crusade as the most important social movement of this generation. I cannot hope here to describe it in anything like the style and detail it deserves. The following sketchy overview must suffice as description of this wonderful human story. We may begin by considering some of the problems facing the project of making Nicaragua a literate population. To begin with there was no money—the country was bankrupt. There was no expertise—at least in the terms we are familiar with. According to Hirshon (ibid.),

> a team of five young people with little pedagogical training but with experience in the liberation struggle set about studying different methodologies of basic adult education. They also analyzed the experiences of literacy campaigns throughout the world.

The number of illiterate persons was not known. The first task was to locate those who were illiterate. Who were they, and where did they live? A census was needed. This would seemingly require time, money and person power that could not be afforded. UNESCO offered to help with technicians and money, but time was short. There were only 15 computers in the entire country and all were needed for mainstream economic activity. In the events the census was carried out in one month at a total cost of US $10,000. By the time a UNESCO specialist arrived to help with planning the census it was completed. Her evaluation "indicated that the technical limitations had had minimal effect on the accuracy of the data" (Hirshon, 1983: 10), The results revealed around 722,000 illiterate persons over 10 years of age—just over half those questioned, 21 percent were between 10 and 14 years. How was the census carried out? The answer gives a clue to the nature of the Literacy Crusade itself. Thousands of school pupils literally tramped the countryside, "recording names, ages, education levels, occupations, interest in learning convenient times for teaching, and, simultaneously, if among the people there were those who wanted to teach, what day, what time and where" (ibid.) The results were processed by hand on the floor of an auditorium. Throughout, the work of the school pupils was complemented by the efforts of the Rural Workers' Association in the countryside and the Sandinista Trade Union Federation in the cities.

The total number of illiterate persons confirmed what had been suspected: that a vast army of literacy teachers would be required and that the great majority of them would be needed in rural areas—many of them remote and inhabited by armed remnants of the National Guard.

Moreover, promoting literacy means more than just ensuring that people can read and write on a particular day—the day of an exam or assessment. If newly won skills are not to be lost quickly, it is necessary to establish a basis for ongoing practice, daily use. and further development of these skills. This requires developing a context in which literacy skills are relevant and *perceived as being relevant* by those newly literate, as well as building educational structures affording ongoing learning.

According to Fernando Cardenal and Valerie Miller (1982), eliminating illiteracy was the first goal of the campaign. This would involve reducing illiteracy to around 10–15 percent initially, establishing a nationwide system of adult education, and expanding primary schooling throughout the country. Basic skills in reading, writing, maths, and analytical thinking were sought, along with elementary knowledge of history and civics. These goals, however, implied others as part of a total process of development and independence. Literacy under Somoza was irrelevant for most people precisely because Nicaragua was an *underdeveloped and dependent* country. The very meaningfulness of a country like Nicaragua striving for universal literacy is grounded in the pursuit of *development and independence:* of the nation as a whole and of the people as individuals.

What, otherwise, would have been the point of all the bloodshed, death and ruin? At least, this is how the campaign organizers conceived of literacy in the new Nicaragua. The specific pursuit of literacy entailed other important goals: namely,

> to encourage an integration and understanding among Nicaraguans of different classes and backgrounds; to increase political awareness and critical analysis of underdevelopment, to nurture attitudes and skills related to creativity, production, co-operation, discipline, and analytical thinking; to forge a sense of national consensus and of social responsibility; to strengthen channels of economic and political participation; to acquaint people with national development programs; to record oral histories and recover popular forms of culture; and to conduct research in health and agriculture for future development planning. (Cardenal and Miller, 1981: 6)

These wider goals, of course, explain why elements of analytical thinking, history and civics were conceived as part of literacy alongside the 3 Rs. For it is only if people have a sense of (national) history, the ability to analyze situations and problems, and some commitment to civic responsibility, that the nature and problems of the past can be understood and commitment to building the future in a spirit of co-operation, integration, and equality established. The new social goals, in other words, presupposed attitudes, habits, and ways of thinking that many people either did not have at all or else had only in an embryonic way.

To put it bluntly, this is a very wide and ambitious notion of literacy in any context, let alone in a bankrupt country lacking teachers and technical expertise, and where 5 months only had been set aside for the literacy campaign (remembering that full production was to be sustained). Some international expertise and technical support was sought and obtained. Twenty-four personnel from Argentina, Canada, Chile, Colombia, Costa Rica, Cuba, El Salvador, Honduras, Mexico, Peru, and Spain joined the Crusade's National Office.

Organizations like UNESCO, the World Council of Churches, the Organization of American States, and the Cuban Ministry of Education provided technical assistance. Advisers and volunteer teachers from a number of other countries also participated in the field. (Paulo Freire was an important influence and visited Nicaragua at an early stage in the planning process.) Most of the financial costs were met from within Nicaragua itself—somehow—but donations from organizations and governments of other countries were also significant. Most generous in this respect were West Germany, Switzerland, Sweden, Holland and England. The success of the campaign depended largely on two obvious factors: sufficient effective teachers and appropriate methodology and materials. A teaching force would have to be recruited and trained. Assuming an ideal ratio of one teacher to five pupils, this would require 140,000 teachers.

Officials were reluctantly prepared to make do with less. It is estimated that approximately 100,000 were trained in the four months leading up to the Crusade. Some 60,000 of these operated in the rural areas. Who were they and how was it all done?

The group that could most easily spend 5 months in rural areas without upsetting the economic and social fabric beyond a minimum were, of course, literate adolescent students. They, however, would require supervision, support, and time off school. The most obvious source of effective supervision and support were the nation's teachers. The logical step was taken. The 10,000 Nicaraguan teachers were diverted—some would say drafted—to the Literacy Crusade. Schools were closed for the duration of the campaign, and students were invited to train and serve as literacy teachers, preferably in the countryside. Students were given an incentive to participate. They were offered promotion to the next school grade if they proved successful literacy teachers. Any pupil/student of 12 years and over who gained parental consent would spend 5 months in a rural area living with a peasant (campesino) family. This is precisely what 50,000 Nicaraguan youth eventually did. Others who could not leave home taught in their local area. Teachers also joined in enthusiastically.

Training the popular literacy teachers proceeded on a multiplier effect within a workshop setting. In the first phase the seven national trainers taught 80 selected personnel: 40 teachers and 40 university students. Of these, 40 were selected in the next phase to train approximately 600 teachers and students. They, in turn, prepared 12,000 more, mainly teachers, in late February. Then in March the schools were closed, and these 12,000 conducted an 8–day intensive workshop course for the many thousands of volunteers. Pupil volunteers trained 11 hours a day during this course. Volunteers in the workforce— factory workers, housewives, government employees, professionals, etc.—were trained outside of work hours: 3 hours each evening, 6 hours on Saturday and 8 on Sunday.

By March 23, the starting date the teaching corps was ready—at least, as ready as possible—and the campaign began.

The Practicalities of the Crusading Campaign: Materials and Methodology?

The materials were few and very simple: the objective being to promote maximum learning at minimum expense. Each pupil received a primer-workbook—*Dawn of the People*—and arithmetic workbook—*Math and Economy Reconstruction: One Single Operation*—and a pencil. Teachers had a manual that accompanied the primer—*Teacher's Guide for Literacy Volunteers*—as well as chalk and a portable blackboard.

The Primer: The Text for the Lesson

The clue to the pedagogical process is the primer. It was designed to involve the learners as much as possible as active participants in the learning process through the dialogue which introduced each lesson that and in the task of actively constructing words by combining different syllables as these were learned.

Besides the actual symbols of written language, the primer contained material intended to stimulate the analytical thinking, creative insight, historical perspective, co-operative attitude and social commitment mentioned earlier. The primer contained 23 lessons organized into 3 broad parts:

(1) the history and development of the Revolution;
(2) the socio-economic programs of the Government of National Reconstruction;
(3) civil defense (see Miller, 1982: 252).

For each lesson there was a photograph and topic sentence, jointly establishing a clear theme for that lesson. The inherently political nature of each theme is self-evident: for example, "Sandino, guide of the Revolution;" "The popular masses made the insurrection;" "The Sandinista defense committees defend the revolution;" "Spend little, save resources and produce a lot—that is Revolution;" "Our democracy is the power of the organized people," etc.

The entire pedagogy, in fact, was grounded in the reality of the Nicaraguan Revolution. The actual life experience of the learners was related as closely and as far as possible to the reality, goals and values of the Revolution as a dynamic on-going process.

The Primer: Means of Recording the Lesson

Prior to the first lesson each pupil had been given a crash course in printing—in using a pencil. They had 'learned' to print the alphabet, their own name, and the name "Carlos." In the primer, dotted guidelines for printing letters were provided until, eventually, printing unaided was possible.

The first lesson focused on vowels alone, but thereafter syllables became the basic unit of learning. The highly regular nature of Spanish—the language of most Nicaraguans—allowed for a phonetic approach built around syllables extracted from key words, which were in turn extracted from the topic sentences.

> Because Spanish is such a highly regular language phonetically—one letter, one sound—a method based on *syllable* recognition will eventually permit the student to read virtually every word in the language. (Hirshon, 1983: 50)

Absolutely crucial to the active dimension of the learning process was the learner grasping the mechanism of combining syllables to make words. When the learner had grasped that this is the basis of written (and, indeed, spoken) language, and had memorized the syllables, they could read and write virtually any word. They could *construct* words, and thereby create their own texts, messages, ideas-in-print. That is, they could approach written language as active creators of words, phrases, and sentences—ideas expressed in written language—rather than as passive recipients of a vocabulary and ideas-in-print donated by others (cf. Freire, 1972). The secret, then, was to get the principle of word construction out of syllables grasped as quickly as possible. Thereafter all that would be needed was memory.

The Lesson: Means of Conducting the Lesson

A typical lesson followed this broad procedure. The photograph was presented and a group dialogue was initiated by the teacher. This discussion went on for a quarter to a half-hour, during which time the learners would ideally communicate their own experience, understanding, interpretation, and aspirations relating to the lesson theme. Part of the point of the dialogue was to secure an emotional commitment on the part of the learners to the lesson content. In addition, it also offered the skillful teacher a chance to promote critical analysis by the learners of their own circumstances and, where appropriate, of their intuitive understanding of their circumstances. At the completion of the discussion, attention shifted to the topic sentence, which the teacher would ideally have drawn out as a conclusion from the discussion. This being so, the learners would sense a direct relationship between the words of the topic sentence—the words from which they would learn to read and write—and their own reality and circumstances.

Word-attack?

The teacher would write the topic sentence and the learners then repeated it several times. Next the teacher selected a key word from the sentence, wrote it on the board and read it several times with the learners. The teacher then read the word very slowly emphasizing one syllable: *po* co ("poco"—little).

Pupils were then asked to identify the same syllable in other words containing it, and to separate the syllable from the rest of the word. The syllable families (i.e., the consonant in the syllable with each of the 5 vowels) were then produced: *pa—pe—pi— po—pu;* and the pupils would copy these in print.

They were then invited to join these syllables with others from earlier lessons to *create* words. Following this, the different variations on the syllables were produced—*inverted:* pa—ap; pe—ep; pi—ip; po—op; pu—up; and *a consonant added:* e.g., pa—pab; pe—peb; pl—pib; po—pob, pu—pub.

In the next step, a range of words and sentences containing these syllables and others already known were read. A short dictation followed. Finally, a phrase or motto containing the syllables was copied in printing.

Each lesson contained at least 2 sections comprising the same steps, but featuring different key words and syllables (see appendices). Working on literacy for 2 or 3 hours a day after a full quota of normal labor—the teachers worked with the peasants by day—each group could take a week or more on a single lesson. As might be expected, progress initially was slow but became rapid toward the end of the campaign.

The Results

With respect to results, some 406,000 people passed the final examination. In 5 months illiteracy had been reduced from around 53 percent to around 12 percent. The first hurdle on the way to a universally educated population had been passed. Thus far, we have considered the origins of the insurrection and crusade. Now we can move to the third phase of literacy development in Nicaragua.

√ 3. Popular Adult Education

The Literacy Crusade provided essential resources, and learning structures subsequently became the program of Collective Popular Education for adults.

The group of learners associated with each literacy teacher (or "brigadista") would become a collective of popular education. From each collective, the most advanced or the most willing person (and hopefully they would be the same) would become the "coordinator," and assume the leadership or teaching role. From every 10 or so coordinators one would be chosen as a "promoter." Promoters would do liaison work with the education commission, organize learning materials, conduct workshops, and also encourage adults in the community to continue with their learning or enroll as new learners.

Collective Popular Education currently [i.e., 1986] offers a literacy course, a post-literacy course, and an adult primary education course up to school-leaving level in maths, science, geography, history and Spanish/English. In other words, it goes beyond merely maintaining the literacy skills acquired in the 1980 campaign. It promotes these same skills among those who missed out on them originally. But in addition it takes basic literacy as a foundation on which to build progressively higher levels of achievement.

Of particular significance is the way adult education involves learners in an *active* process, just as the conception of literacy in the crusade involved a vital active dimension. While the Education Ministry provides basic materials and ensures channels of communication from the grass roots "classrooms" to the Ministry itself, ultimately it is the people themselves who must take responsibility for keeping learning alive. To a large extent, "the people themselves" are those who first became literate in 1980. By way of suggesting that the Literacy Crusade was successful in fostering independent, active, creative human beings committed to personal and national development, let me finish with a brief account of the structure of adult education in Nicaragua.

The Structure of the Program

Adult education *is* structured in a way that permits a maximum of dialogue or communication between the Ministry of Education and the grass roots collectives.

In Freire's terms, the emphasis is on dialogue and not monologue; on communication and not communiques (Freire, 1973: 45–46). The channels facilitate two-way communication so that the needs, difficulties, and challenges of the collectives are as familiar to the Ministry as the needs, difficulties, and challenges of the Ministry are to the collectives. The success of the system is utterly dependent on the willingness of people at the grass roots level to commit considerable time and energy to organizing, liaising, coordinating, teaching and learning. This activity is wholly voluntary.

There are five levels in the network of adult education. These are the national, regional and zonal levels, the promoters, and the popular education collectives themselves. Presented diagrammatically it looks like this:

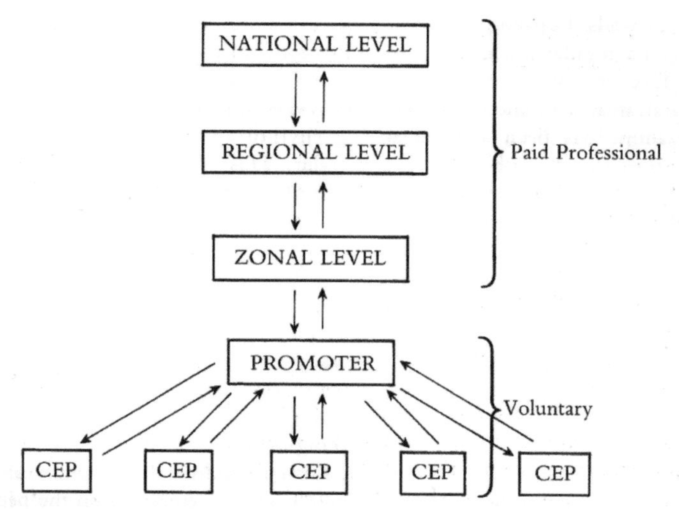

Figure 2.1: The Network of Adult Education

Personnel at the national, regional, and zonal levels only are paid professional staff. Nicaragua is divided into nine regions and each region is further divided into zones. The zones are further broken down into sectors, although a sector does not comprise a separate administrative level. Rather, the number of sectors within a zone determines the number of professional staff allocated to that zone. If there are six sectors in a zone, there are six professional staff for that zone. There may be up to four (unpaid) promoters per sector. Each promoter is responsible for a given number of collectives. The aim is for each farm and factory to have a promoter, although this is not always possible.

How the Networks Function

The communication (or dialogue) network operates very simply. Each month there is a workshop meeting between the national and regional levels and a further meeting between the regional and zonal levels. The zone representatives (who spend most of their time in the community rather than in the zone office) and promoters meet twice monthly. In addition, the promoter meets with the coordinators (or popular teachers) twice monthly. Each meeting at each level permits a two-way exchange of information and ideas, with the result that each level is integrated with the others and none becomes isolated. Most importantly, those doing the work in the collectives know that they have a voice that will be heard regularly at regional and national levels. The importance of this for sustaining a high level of voluntary educational activity cannot be overestimated.

A Transformation of Human Beings

The main point to grasp is the *existential meaning* of this adult education system: its significance in *human* terms. What we find here are people who were illiterate five years ago, but who have now won the confidence to teach and learn with their peers, to articulate their educational needs and aspirations, and to communicate with them.

As Freire might put it:

In the new Nicaragua the people are *authentically* literate; they are *naming* the world; speaking *true* words; writing the text that is their *life*.

Bibliography

Baez, E. (1985). Personal communication at the Nicaraguan Ministry of Education, Managua, 8th February.

Black, G. and Bevan, J. (1980). *The Loss of Fear: Education in Nicaragua,* London: World University Service.

Cardenal, F. and Miller, V. (1981). Nicaragua 1980: The battle of the ABCs. *Harvard Educational Review*, 51: 1–26.

Cardenal, F. and Miller, V. (1982). Nicaragua: Literacy and revolution. *Prospects*, 12(2): 201–212.

Council on Interracial Books for Children (1981) The Literacy Crusade in Nicaragua. *Interracial Books for Children Bulletin*. 12(2): 12–14.

Freire, P. (1972). *Pedagogy of the Oppressed.* Harmondsworth, England: Penguin.

Freire, P. (1973). *Education: The Practice of Freedom.* London: Writers and Readers.

Hirshon, S. (1983). *And Also Teach Them to Read.* Westport, Connecticut: Lawrence Hill & Co.

Mackie, R. (ed.) (1980). *Literacy and Revolution: The Pedagogy of Paulo Freire.* London: Pluto Press.

Miller, V. (1982). The Nicaraguan Literacy Crusade. In T.W. Walker (ed.) *Nicaragua in Revolution.* New York: Praeger Press. pp. 241–258.

Nicaragua Ministry of Education (1979). *The Sunrise of the People.* Managua: Government Printer.

Weissberg, A. (1981). *Nicaragua: An Introduction to the Sandinista Revolution.* New York: Pathfinder Press.

APPENDIX 1

The Method by Stages

1. An evocative photo from the primer designed to stimulate dialogue within the class is presented to participants. This leads to a conclusion expressed by a short following sentence;
2. Learners focus on the key word from the sentence;
3. This word is separated into syllables, and a specific syllable is selected to be the lesson's objective;
4. The consonant sound of this syllable is presented along with the five possible vowel combinations (for example, *la le li lo lu*);
5. These syllables are copied and later written using small and capital letters;
6. New words are formed by combining the new syllables with others learned in previous lessons;

7. All possible variations of these syllables are then presented—for example, the syllables are inverted (*al el il, ol, ul*) or have an ending consonant sound added to them (*las les lis los lus*);
8. Words and sentences that contain the known syllables are then read;
9. There is a short dictation to test participants' mastery;
10. The lesson concludes with a *muestra,* or demonstration, in the form of a phrase or motto to be copied in the student's best handwriting.

Spanish is a highly regular language phonetically—one letter one sound. As Freire recognized, a method based on syllable recognition will eventually enable learners to read almost any word in the language. It was hoped that this formula, together with an open dialogue, would develop competent readers who would actively seek personal growth and community development. (See: Hirshon, 1983, p. 50.)

APPENDIX 2—A typical exercise from the primer workbook

Exercise C

1. We read the slogan:
   1980, ANO DE LA GUERRA CONTRA EL ANALFABETISMO
   (1980, year of the war against illiteracy)
2. We read the word:
   ANALFABETISMO (illiteracy)
3. We separate the word

fa fo fi fu fe

Fa Fo Fi Fu Fe

6. We build and write words by combining known syllables

7. We read the syllables:
   fa fi fo fu fe
   fes fis fal fer

8. We read and write:
   faro fusil café fanega
   fósforo Fernando falda finca
   Nicaragua tiene muchas fincas de café
   El cafe de Matagalpa es famoso
9. We have dictation
10. We write clearly

Venceremos al analfabetismo

(Nicaraguan Ministry of Education, 1979)

THREE

Literacy and the Struggle for the Working Class Press in England, 1816–1836 (1987)

Colin Lankshear

Biography of the Text

When I began reading in the area of the politics of literacy it wasn't long before I ran into Wayne O'Neil's 1970 *Harvard Educational Review* paper, "Properly Literate." O'Neil distinguished between "being able to read" and "being literate." He argued that being able to read means "that you can follow words across a page, getting generally what's superficially there." By contrast, "being literate means you can bring your knowledge and your experience to bear on what passes before you." O'Neil suggested that we think of the latter as "proper literacy" and the former as "improper literacy." This resonated with Freire's idea of literacy involving reading the word *and* the world—of keeping words and world together in ways that enhance our capacity to *name* the world. It also resonated with C. Wright Mills' (1959) concept of "sociological imagination"—a way of thinking and using information that helps people understand relationships between what is going on in the world and what is happening in their own lives and beings and/or the lives and beings of others they may know, care about, or otherwise be interested in. I was interested in the possibility of tweaking O'Neil's distinction in accordance with ideals of political, economic and social justice. I was also interested in claims that history offers many examples of marginalized groups developing non-formal literacy and popular education initiatives that approximate to the kind of proper literacy I believed was central to an educational ideal. Being contracted to write a book presented an excellent opportunity and incentive to pursue this interest.

In this chapter I present the second of three case studies comprising my account of literacy and working class politics in England during the period from 1790 to 1850 in Chapter 3 of *Literacy, Schooling and Revolution*.

Introduction

Throughout this period the English working people bore a triple yoke of oppression. They were oppressed politically (in the *formal* sense, within the sphere of formal politics), economically, and culturally. Organic working class struggle against oppression emerged on all three fronts between 1790 and the 1830s—the decades which, according to E.P. Thompson, witnessed "the making of the English working class" (Thompson, 1963: 213).

There were two major dimensions to this "making": the growth of working-class consciousness, whereby the diverse groups of working people began in numbers to perceive an identity of interest among themselves, as distinct from and opposed to the interests of other classes; and the emergence of distinct forms of political and industrial organization corresponding to and reflecting this perceived identity of worker interests.

By 1832 there were strongly based and self-conscious working class institutions—trade unions, friendly societies, educational and religious movements, political organizations, periodicals—working class intellectual traditions, working-class community patterns, and a working class structure of feeling (Thompson, 1963: 213).

Thompson stresses the importance of recognizing the active role of working class people in making themselves as a class. The working class was not *forged,* in the manner of a casting, in the crucible of the Industrial Revolution—with external forces operating on inert human raw material, pressing it into shape, and turning it out "ready made" at the other end. Rather, the making process was one in which working people acted and created, as well as being acted upon and responding to externally imposed economic, political, and cultural forces. "The working class made itself as much as it was made" (Thompson, 1963: 213). Moreover, the pursuit, attainment, and practice of proper forms of literacy was a vital galvanizing element within the active role played by working folk in making themselves as a class.

Literacy and Struggle for the Working Class Press

While precise figures are not available, the working class was increasingly a reading public from the late eighteenth century (Webb, 1955: 167, note 34). Furthermore, during the 1790s the Corresponding Societies had demonstrated that, through association, individuals did not personally require reading skills in order to gain access to ideas and critiques through which to achieve enhanced understanding of their circumstances and pursue greater control over their lives (Webb 1955). The end of the war with France in 1815 ushered in a period during which the battle for access, via print, to the minds of working people assumed major proportions and significance, in the context of struggle for a popular press.

Wickwar (1928) identities 1816 as a landmark in the struggle for freedom of the press. The peace with France had brought continued economic distress rather than improvement to the lower orders and fanned disillusionment within the middle class. Politically aware critics identified a government that had palpably failed to promote the welfare of its subjects. They saw electoral corruption in "borough mongering, pensions, sinecures and patronage" (1928: 19). They saw general political and economic corruption, privilege, and vested interest in

> a corn law passed to keep the price of wheat up to eighty shillings a quarter, on the assumption that rents would thus be kept high and that the high rents were necessary to keep up the landed interest on which the government of church and state was assumed to depend. (Wickwar, 1928: 49)

Political malcontents again turned their attention to reform. Change to the constitution was the precondition of improved social and economic conditions, they argued. Unless the people shared the power of government, they could scarcely hope to share its benefits (Wickwar, 1928: 49).

Under conditions of intense and prolonged economic hardship, discontent had become widespread among the working class, and 1811 brought violence with the outbreak of the Luddite Revolts. Initially much of the violence was directed against machinery and industrial property, in the belief that machines and profiteering were the source of economic distress. There were, in addition, sporadic food riots and outbreaks of mob violence triggered by unemployment, high prices, and wage reductions. Commentators generally describe the widespread and regular outbreaks of violence and disorder between 1811 and 1816 as spontaneous and unorganized—in the sense that they were not individually parts of a unified, informed, orchestrated program of political agitation. Moves were taken to alter this in 1816 when, under two main influences, political reform re-emerged as a unifying theme and educational activity aimed at promoting widespread commitment to the reform cause was once more in evidence.

The first influence was the revival of reform societies in the wake of a reform tour by Major Cartwright. In the tradition of corresponding societies the Hampden, Union, and Spencean Clubs promoted discussion and political activity within regular class meetings (cf. Wearmouth, 1948: Ch. 1 and 2. See also Simon, 1960: 186–189, and Thompson, 1963). This initiative was soon accompanied by the development of radical Sunday schools—secular and political in nature and focus. The second influence was the emergence of a politically critical popular press directed specifically at working people with the intention of educating them as to the real cause of their distress: namely, political and constitutional evils. It is with the emergence of the popular press that I am mainly concerned here.

Three points can be made by way of background to trends which evolved in popular publishing from 1816.

(i) Since 1712 publications had been subject to tax. In that year parliament taxed printed papers, pamphlets and advertisements and required a stamp to be placed on anything it deemed to be a newspaper. During George II's reign the stamp on newspapers was set at a penny per sheet, and penalties for transgressions of the Stamp Tax were extended from publishers to include vendors as well. In 1815 the Stamp Tax was set at fourpence per newspaper. Clearly, the cost of the tax had to be included in the sale price of newspapers, unless publishers and vendors chose to risk publishing and selling unstamped papers. The Stamp Tax thereby placed legitimate newspapers beyond the pockets of working class individuals.

(ii) Strict legislation to control the content of publication was in place. This began in 1637, when all books and papers were required by law to be licensed and registered before publication. During the eighteenth and early nineteenth centuries tough laws governing sedition and libel were added. As Wickwar summarizes:

> The publication of anything with a malicious intention of causing a breach of the peace was a misdemeanour at Common Law. Anything that it was thus illegal to circulate was called a criminal libel, and the same term was commonly applied to the act of circulating it. Criminal libels were distinguished as defamatory, obscene, blasphemous or seditious libels, according to as they treated of personal, sexual, religious, or political matters. (Wickwar, 1928: 19. See also pp. 18–20 for legal descriptions of malicious intent, breach of the peace, etc.)

Of particular relevance here is seditious libel. This law legislated against public expressions of discontent with the established government. A publication was a seditious libel if it (a) tended to bring into contempt or hatred either the monarchy (including heirs and successors), the government and constitution, parliament, or the administration of justice; or (b) tended to incite subjects of the realm to attempt to change any matter of church or state by other than lawful means. Given that there was no provision for popular participation in government, there *were* no lawful (or, by definition, peaceful) means by which the people could change either government or constitution. Hence the law covering seditious libel presented a powerful and wide-ranging control on the political content of the press. Since Britain's rulers in the early nineteenth century "were generally satisfied with the working of the constitution and the Christianity of the day," and "saw no reason why the whole nation should not be united in . . . respect for Christianity and in contentment with the constitution they had inherited" (Wickwar, 1928: 19), there was both an incentive and a tendency for the laws covering libel and sedition to be employed—particularly when public unrest reached crisis point, as it did frequently between 1790 and 1816 (and after) (Wickwar, 1928: 19).

(iii) By 1815 an impressive range of types of publication existed for middle and upper class readers. Besides books, these included newspapers—containing national, local, and foreign news, and comprehensive reports of legal and parliamentary proceedings—monthly and quarterly reviews and magazines, and an increasingly popular (typically) weekly form known as political registers. Together with certain books and pamphlets, these registers mainly fell outside "the respectable part of the press" (Wickwar, 1928: 51). While some had a literary and dramatic focus, most were political in content and anti-establishment in bias. They were usually published by a single individual and reflected that person's viewpoint. Whereas newspapers sought to describe or record events, registers aimed explicitly to shape them. And whereas monthly reviews and magazines reviewed general policies, the registers reviewed and evaluated current events. They evolved as potent instruments of political and, often, religious critique. And they were subject to the Stamp Tax.

Perhaps the most celebrated of the register genre was *Cobbett's Weekly Political Register,* founded by William Cobbett in 1802. Cobbett reflects *par excellence* the spirit and intent of the register writers as described by Wickwar. These individuals expressed in their publications what they "took to be the interests of the otherwise unrepresented people"—and particularly the working class. For Cobbett, the ills of the working class flowed from political corruption; their interests called for parliamentary reform. But if Cobbett argued and agitated on behalf of the interests of others, he required in turn their support. "He had to try to make his opinion their opinion, so that they might together accomplish what he could never do alone" (Wickwar, 1928: 52). The way in which Cobbett sought to make his opinion working class opinion led to a further chapter in the chronicle of literacy as a force and an outcome of struggle between competing interest groups.

Because of the heavy stamp duty Cobbett's *Register* sold at 1s 0½d—with only "a very small portion . . . left to the author" (Cobbett, cited in Wickwar, 1928: 53–54). Despite the high price it was read by workmen who grouped together to buy copies and read them in public-houses. When Cobbett heard of publicans objecting to "meetings for reading the *Register* being held at their houses for fear they should lose their licences," he decided to make available a cheaper edition (ibid.). He was (legally) able to do this by means of a loophole in the Stamp Tax law. Printing on open sheets (i.e., sheets printed without the intention of folding them) required no stamp. And so "the whole of one of my *Registers* might be printed in rather close print upon the two sides of one sheet of foolscap paper" (ibid.).

Cobbett's unstamped version of the *Register* sold for twopence—the original "Twopenny Trash." It sold 44,000 copies inside a month, and more than 200,000 in all. A score of political periodicals followed Cobbett's lead: notably, Wooller's *Black Dwarf,* Sherwin's *Political Register* (later transformed by Richard Carlile into the *Republican),* and the penny *Gorgon,* edited by John Wade. Political corruption and the pressing need for reform was the central theme of the twopenny trash. This theme was expressed in articles analyzing and commenting upon current circumstances and events, exposing the motives and interests of opponents of reform, advocating and documenting the advantages of lawful association, and generally expounding elements of radical political theory. The politicizing influence of this literature upon the working class was enormous. It performed a role akin to that of the literacy engendered earlier by Corresponding Societies, providing a focus for class meetings within the Hampden Clubs and other reform societies. The role of Cobbett's work in particular was widely acclaimed. According to Samuel Bamford, Cobbett's writings were

> read on every cottage hearth in the manufacturing districts of South Lancashire . . . Leicester, Derby, and Nottingham . . . Their influence was speedily visible; he directed his readers to the true cause of their sufferings—misgovernment; and to its proper corrective—parliamentary reform. Riots soon became scarce, and from that time they have never obtained their ancient vogue with the labourers of this country . . . Instead of riots and destruction of property, Hampden Clubs were now established in many of our large towns . . . The labourers read [Cobbett's works] and thenceforth became deliberate and systematic in their proceedings. (Bamford, 1984: 13–14)

Thompson cites a reformer who attributed the emergence of political knowledge and fixed political principles among Manchester's poor to "Mr. Cobbett's masterly essays, upon the financial situation of the country, and the effects of taxation, in reducing the comforts of the labourer" (Thompson, 1963: 679).

Cobbett's original twopenny *Weekly Political Pamphlet,* "Address to the journeymen and labourers," is an exemplar of the genre. Cobbett argues that despite being smaller in population and poorer in soil and climate than many other countries, England was (in 1816) the most wealthy and powerful nation in the world. This wealth and power, he says, spring from the laboring classes. Moreover, the same laboring classes as produce the nation's wealth also secure its safety. While military and naval commanders receive the titles and the financial rewards, it is the people who actually win the victories. What do working people receive in return for producing wealth and ensuring security? They are denigrated by their "betters"—referred to as the Mob, the Rabble, the Swinish Multitude—and reduced to abject misery.

Cobbett asks after the cause of this misery and how it might be remedied. The main cause, he says, is excessive taxation. But do "the friends of corruption" recommend reduced taxes for the poor? Not a bit. Instead they complain about being levied for the Poor Rate. They would seek even to deny poor relief to the laboring classes—despite the fact that poor relief is the only tangible return workers might see for the taxes they pay. Even less do these friends of corruption propose political reform that would admit the real creators of wealth and security to the body politic. The same political corruption that reduces the poor to misery ensures that sinecure placemen and pensioners receive from twenty to forty thousand pounds a year—in return for producing and securing nothing! Having attacked Malthus' "remedy" for the situation, Cobbett proceeds to his own. The only remedy is to give every person who pays direct taxes the right to vote for MPs at annual elections. A reformed parliament would redress economic injustices and ensure the most democratic electoral procedures. He ends by exhorting working people to pursue political reform with zeal and resolution—by peaceful and lawful means. And

> if the *Skulkers* will not join you, if the "decent fireside" gentry still keep aloof, proceed by yourselves. Any man can draw up a petition, and any man can *carry* it up to London, with instructions to deliver it into trusty hands, whenever the House shall meet. (Cobbett, cited in Cole and Cole, 1944: 216)

The revival of reform societies, initiated by Major Cartwright, had its heyday between 1816 and 1823. The twopenny trash was an important part of the literature read and discussed by working-class folk in the various Hampden Clubs, political Protestant Unions, secular Sunday schools, and other reform associations. The literacy practised within the class meetings of these societies reflects pedagogical approaches and a range of educational concerns, which are interesting and important in their own right (see, for example, Simon, 1960: 186–93; Thompson, 1963: 712–36; and Wearmouth, 1948: 31–49).[1] Unfortunately, they are beyond our scope here. For my concern in this section is not with the overall context and practice of a particular literacy. Rather, I wish to focus more narrowly on the emergence of an important *medium and expression* of working class literacy: namely, the working class press. In this I will emphasize the dynamic between efforts and initiatives taken to establish a distinctively working class press and the many obstacles presented to these efforts.

The twopenny trash was the first step toward a genuinely working class press. By "a working class press," I mean (i) a press which offered working people access to information and comment on their daily reality at a price (more or less) within their economic grasp; and (ii) a press which reflected working class interests and was committed to promoting those interests. This, of necessity, was a press increasingly under the control of working people themselves.[2]

We may think of (i) and (ii), crudely, as cost and content dimensions. Pursuit of a working class press involved *struggle* against oppositional forces on both of these dimensions. The Stamp Tax militated against a working class press on the cost dimension. Laws covering sedition and libel imposed powerful obstacles on the content dimension. So too did initiatives undertaken by representatives of middle- and upper-class interests to make cheap literature available to working-class readers with the intention of diverting them from authentically working-class critiques of social, economic, and political conditions, and accommodating them to the status quo—thereby promoting the interests of the privileged at the expense of working class interests. To carry this part of the argument forward it it is necessary to outline some key aspects of the struggle that ensued following Cobbett's intervention in 1816. Once again, my description here is intended to be illustrative rather than exhaustive, and will be limited to selected aspects of struggle between 1816 and 1836.

The struggle for a working class press echoes the earlier struggle for *proper* worker literacy within the Corresponding Societies, in that it too was confronted by both coercive and ideological/hegemonic forms of opposition. Examples of coercion include the 1819 legislation covering Stamp Duty and Sedition, and measures employed against Hetherington and others in the 1830s. Hegemonic opposition is represented in attempts by various publishers, organizations, and even the government itself, to "write Cobbett down" and establish a cheap anti-reform literature. It is also to be found in the activities of such organizations as the Society for Promoting Christian Knowledge and the Society for the Diffusion of Useful Knowledge.

Coercion

In the midst of heightened and critically informed activity for political reform, the government passed the notorious Six Acts of 1819. Two of these were explicitly directed against the low cost reform press. The "Act for the more effectual prevention and punishment of Blasphemous and Seditious Libels" reminded the public of what constituted criminal libel (as outlined above), and established mechanisms for administering the law more effectively than before and for frightening would-be offenders. Greater powers of search and arrest were given to magistrates and constables, and penalties for a second offence under the Act included banishment from the Empire or, alternatively, transportation for up to fourteen years. The "Act to subject certain Publications to the Duties upon Newspapers, and to make other Regulations for restraining the Abuses arising from the Publication of Blasphemous and Seditious Libels" (or Publication Act, for short!), closed Cobbett's loophole by bringing the twopenny trash within the definition of a newspaper, thereby subjecting all such publications to the Stamp Tax of fourpence per copy. The complex definition of a newspaper written into this Act—that is, the lengths the Act went to in order to prevent the proliferation and accessibility of the reform press to working people—is recorded by Wickwar (1928: 137).

The two Acts, then, undermined a free press in two ways: by economic constraint and by controlling content. The battle for the free press during the 1820s was mainly a battle against the restriction on content. It was not, to this extent, a battle for a working class press *per se*. Rather, people like Richard Carlile, his shopmen, and his army of vendors, fought for the right to express political beliefs and criticism freely. While the battle was fought on this front the twopenny trash collapsed. Wooller, Cobbett, and others conformed to the Stamp Tax requirement, and their circulation fell away under the resulting price increase—with Wooller folding in 1824. A system of reading rooms, coffee shops, and other networks, at which people could peruse papers they could not afford to purchase, continued throughout the decade. In general, however, "the working-class press struggled under the crushing weight of the stamp duties" until 1830 (Thompson, 1963: 799). In the meantime the most authentic expressions of working class interests available in print had been effectively moved beyond the economic means of individual working class readers.

For a decade the Publication Act of 1819 checked development of a cheap popular press. The legislation of that year brought in its wake prosecutions for seditious publishing rather than for defiance of the Stamp Tax. But in 1830 the struggle moved to the other front, with the battle of the "great unstamped." By 1830 the battle for free expression had been largely won. The courage and defiance of Carlile's army of persecuted and punished had defeated—morally and practically—those who would suppress political critique in the name of preventing sedition. The barrier that remained against a politically informed and critical working-class press was the Stamp Tax—which remained at the prohibitive level of fourpence set in 1815.

After several years of relative quiet, clamoring for political reform broke out anew in 1829, when "the widespread depression afflicting various sections of the community found voice and passion" (Wearmouth, 1948: 50). A host of political unions quickly emerged, based upon the principle of middle and working class collaboration in the pursuit of reform. Some working class leaders, however, anticipating the subsequent betrayal of working class efforts for reform in the 1832 Reform Act, formed associations explicitly concerned with promoting working-class interests. The most important of these was the National Union of Working Classes and others—formed in early 1831 when London workers broke away from the middle class dominated Metropolitan Political Union. Its leaders included a printer named Henry Hetherington, and it was he who became the central figure in the struggle against the Stamp Tax on behalf of a working class press.

Through October and November 1830 Hetherington published a series (twenty-five in all) of penny daily papers, entitled *Penny Papers for the People.* These were written in letter form, addressed to their intended audience, in an attempt to evade the Publication Act (and, thus, the Stamp Tax) whilst at the same time providing "cheap political information for the people" (Barker, 1938: 5). The first *Penny Paper* was addressed to the people of England, and subsequent issues were addressed to such as the Duke of Wellington, the King, and the Archbishop of Canterbury. In December 1830 Hetherington shifted to a weekly format with *A Penny Paper for the People* by the *Poor Man's Guardian,* containing "a comprehensive digest of all the political occurrences of the week" (Barker, 1938: 5–6; see also Lovett, 1920: 60). This new format brought Hetherington to court, and to conviction, for defiance of the Stamp Tax. He was sentenced to six months imprisonment, appealed, but had the appeal disallowed. Hetherington's response to his conviction and sentence was to produce (on 9 July 1831) the first issue of *The Poor Man's Guardian.* Instead of the official stamp it bore the emblem of a hand-press. Its motto was "Knowledge is Power," and it was headed "Published contrary to 'law' to try the power of 'Might' against 'Right.'"

Hetherington was uncompromising; his aim in defiance of the law was absolutely explicit. His opening address stated the intention to protect and uphold the freedom of the press, "the press, too, of the ignorant and the poor" (Barker, 1982: 8). He served notice that *The Poor Man's Guardian* will contain "news, intelligence, and occurrences," and "remarks and observations thereon" and "upon matters of church and state tending to excite hatred and contempt of the government and constitution of the tyranny of this country, as by law constituted" (Barker, 1983: 9). One by one he cited the clauses of law his paper was to defy (see also Collett, 1933: Ch. 2). Gone was any attempt to evade the law by loopholes—as in the earlier format of a letter addressed to an "intended" audience. Hetherington was confronting the Publication Act head on in the cause of a working class press.

Other unstamped newspapers appeared, including Carlile's *Gauntlet,* Hobson's *Voice of the West Riding,* Doherty's *Poor Man's Advocate,* O'Brien's *Destructive,* and a paper called the *Working Man's Friend* which, together with Hetherington's *Poor Man's Guardian* became the voice of the National Union of Working Classes. The working class press was born: a press by working class people, for working class people, expressing and promoting working class interests, and at a price working people could reasonably afford.

Some appreciation of the quality of ideas and thought accessible to working class readers via their own press can be gleaned from a typical example taken from the *Poor Man's Guardian,* 17 November, 1832. The background to this particular article concerned the formation of a separate Union of the Working Classes in the

Midlands. Faced with this development a council member of the Birmingham Political Union claimed that no sufficient reasons had been given that would justify the formation of a distinctively working class union. The *Poor Man's Guardian* published a reply to this charge (the original Address is reproduced in Hampton, 1984: 458–459). In it five grounds were advanced for establishing the new organization.

(i) Leaders of other political unions simply could not represent working class interests because their own interests conflict with those of workers. For example, men of property who live off rents would have an interest in preserving the Corn Laws. Yet abolition of the Corn Laws was absolutely basic to working class interests.

(ii) The most active members of existing political unions were interested in securing representation of *property* rather than of human beings. As with ruling classes from time immemorial, they seek power to make laws which will promote their own ends. It is precisely the creation of such interest-serving laws which has yielded "extreme wealth on the one hand, and the destitution and starvation of the artisans of our own town on the other" (Hampton, 1984: 458).

(iii) Working class distress has resulted from displacement of manual labor by machines and other inventions, which have forced workers to compete with each other for employment. This has resulted in low wages. Since masters and capitalists have an obvious interest in further mechanization if it brings still cheaper labor, they could hardly be expected to exercise power—inside or outside of parliament—with due consideration of working class interests.

(iv) Those "above" working class station seek to avoid involvement in productive labor. Consequently, they have an interest in securing privileged positions in the army, navy, church, or excise, for themselves, their families and connections. This makes them part of the very problem producing the heavy taxes, which cripple working people. To this extent they cannot represent working class interests, which directly conflict with their own.

(v) The working classes are sufficiently intelligent to discuss issues concerning their best interests, their rights and liberties, and to acquire enhanced knowledge of these matters, *among themselves*—without being dictated to or controlled by persons with opposing interests.

In purely economic terms, then, the working class had access through the work of Hetherington, Hobson, Doherty, and others to properly literate publications at a price they could afford. At a different level, however, the price of such publications was extremely high. As is self-evident from the article I have just described, it was very much in the interests of the ruling classes to have strong coercion brought against the development of a cheap press that politicized working people. Were such ideas to become prevalent among the masses, the social, economic and political order would surely be overthrown. And so the law moved against the working class press. Hetherington served multiple six month terms of imprisonment, and between these spent considerable time (publishing) on the run from the law. Watson served six months. A veritable army of vendors responded to Hetherington's advertisement calling for "some hundreds of poor men out of employ who have nothing to risk, some of those unfortunate wretches to whom distress has made prison a desirable home," to sell the *Poor Man's Guardian* in the face of the law. They sold; they were prosecuted in large numbers—up to 750 prosecutions according to one reliable estimate; they were jailed. Shortly after the *Poor Man's Guardian* finally ceased publication, at the end of 1835 with its 238th number, the Stamp Tax was reduced to a penny, "and the way had been opened for the Chartist press" (Thompson, 1963: 800).

The role of legal coercion against the working class in their struggle for a press that authentically expressed and aimed to promote their class interests is accentuated by the fact that publications reflecting ruling class interests, *and which ought to have been stamped but were not,* remained free from prosecution. This patently class-interested operation of the law, which opposed worker initiatives in search of a proper literacy and sided *with* church and state sponsored activities to perpetuate improper literacy among working people, is best articulated by Simon.

> Sellers of the *Poor Man's Guardian* were unmercifully persecuted up and down the country; James Watson was jailed with Hetherington, Cleave and his wife seized, Heyward of Manchester, Guest of Birmingham, Hobson and Mrs Mann of Leeds, and about 500 others suffered imprisonment, as sellers of the unstamped press. Yet at the same time prominent members of the government unctuously promoted the activities of the [Society for the Diffusion of Useful Knowledge] whose *Penny Magazine,* which had been launched in 1832 as part of the policy of providing innocent amusement for the workers, but which should equally have been stamped, circulated unchallenged. (Simon, 1960: 227)

This biased application of the law did not escape working class notice. And in the best tradition of informed struggle the National Union of Working Classes and the worker press together treated it as an issue through which to further politicize working people. As an example of this we may consider a letter from the Leicester Branch of the National Union of Working Classes to the *Poor Man's Guardian.* The branch formally expressed its "detestation and abhorrence" at the "base spite and vindictive malice" by which Hetherington had been singled out for persecution

> whilst Brougham, and a whole host of *lying* editors, proprietors, and publishers of the *Penny Magazines, Omnibus,* and others too numerous to mention, all equally offending against the *damnable and detestable taxes on knowledge,* are suffered to go on with impunity, and even rewarded with honour, expressly because they either basely abuse and deceive the people, or attempt to divert their

> attention from their true state, and the cause of their distress, instead of showing these. (cited in Simon, 1960: 228)

The letter ends by expressing the Branch's resolve to continue their efforts until tyranny is overthrown and Equal Rights and Equal Law established.

Besides denouncing selective coercion against working class publishers, this correspondence draws our attention to the role of popular publication as an ideological tool for preserving ruling class interests by fostering improper literacy among worker readers. To follow this theme further I turn now to the hegemonic dimension of the struggle surrounding the emergence of a working class press.

Hegemony

The state was actually involved in activity against the twopenny crash and those associated with it prior to the Acts of 1819. In part this was coercive activity. A Shropshire magistrate, for example, "caused two men to be apprehended under the Vagrant Act for distributing *Cobbett's Political Register,* and had them well flogged at the whipping post" (Aspinall, 1949: 47). Elsewhere hawkers were detained, prosecuted and, in some cases, fined with the option of imprisonment for non-payment. In addition, however, the state was implicated from 1816 in an extensive ideological campaign against Cobbett and others who published reform literature at working class prices. The primary object of this campaign was hegemonic: to maintain support—especially among the working class—for the existing political order by creating a cheap anti-reform press centering around a concerted anti-Cobbett campaign. The focus on Cobbett stemmed from the fact that he was universally acknowledged as the most effective and, therefore, most dangerous communicator of radical ideas to working class readers. Following a four-column assault on Cobbett published in *The Times,* and subsequently republished for sale at "a penny singly, or 6s. per 100" (ibid.), a correspondent of the *Morning Post* recommended that bastions of the status quo adopt the same approach to influencing political consciousness as that taken by their opponents. After all,

> if Cobbett's poisons are circulated in short pamphlets, at the expense of Jacobins, why not make their antidotes be circulated, in the same manner, at the expense of loyal men who can afford to give them away? (ibid.)

This correspondent recalled that during the 1790s "many excellent pamphlets were circulated by government and by individuals, *which gave a just tow to the public mind,"* and was at a loss as to why the same measures were not being adopted in 1816 (Aspinall, 1949: 155, italics mine).

In fact they were. Aspinall claims that the government was involved to the limits of its financial resources in assisting the publication and distribution of pamphlets "calculated to counteract the mischief done by 'incendiary' publications." Indeed, government thought so well of an anti-Cobbett pamphlet published in 1819 (called *The Beauties of Cobbett),* that it printed thousands of copies and assisted in its circulation. Lord Sidmouth, the Home Secretary, received much correspondence seeking subsidies for such anti-reform publications and personally issued the challenge (in 1818) that Cobbett "must be written down."

Anti-reform publication generally, and anti-Cobbett initiatives in particular, extended far beyond government activity. Webb notes that, in addition to the government, numerous publishers and organizations were involved. Cobbett's work and character were attacked in a host of low cost pamphlets, including *Anti-Cobbett, The Political Death of Mr William Cobbett, Politics for the People by William Cobbett,* and the *Letter to William Cobbett* published by the Birmingham Association for the Refutation and Suppression of Blasphemy and Sedition. Wooller's *Black Dwarf* was countered with Merle's *White Dwarf.* And W.H. Shadgett published a *Weekly Review of Cobbett, Wooller, Sherwin, and other Democratical and Infidel Writers,* "designed as an antidote to their dangerous and subversive doctrines" and to disseminate "just and sound principles, on all popular subjects." The wider body of anti-reform literature included a refurbished *Village Politics* and more than a dozen new tracts from Hannah More. Activity was feverish in the towns as well as the provinces.

> London publishers, like Hatchard and Seely, turned out numbers of cheap anti-reform pamphlets. George Cayley, a physician, published two addresses to pitmen and keelmen at Durham . . . Edward Walker in Newcastle published *A Word from the Other Side, The Friendly Fairy* . . . and reprinted Paley's *Reasons for Contentment* . . . The *Leeds Intelligencer* in 1819 published a penny *Reformers' Guide* and also issued a loyal paper called *The Domestic Miscellany,* and *Poor Man's Friend* . . . [In Manchester] a periodical called *The Patriot* appeared after Peterloo . . .The Pitt Club in 1817 distributed two [dialogues] by Canon C.D. Wray . . . and in the same year Francis Philips wrote *A Dialogue between Thomas, the Weaver, and His Old Master.* (Webb, 1955: 52)

This literature aimed to counter directly the reformist flavor of material, which had become increasingly accessible to working people after 1816. It confronted the ideology of radicalism and reform with an ideology grounded in the beliefs and values of the established order: that is, in the worldview of those whose interests were best served by existing political, social, and economic arrangements. It is true that the Publication Act of 1819 had been largely successful in restricting working-class access to radical ideas in print. As we have seen, however, seditious libels continued apace, and despite their reduced circulation among working-class readers it was clear that they would continue to exert upon working-class consciousness an influence hostile to ruling interests. After all, ideas in currency could be communicated orally, from those with direct access to published opinion to those without—providing a basis for discussion and further development of these ideas among those penalized by taxes on newspapers. Since the cause of reform embraced (to 1832) both middle class and working class activists—and radical societies with mixed class membership continued into the 1820s, and were revived again after 1829—dissemination of radical critiques among the working class would (and did) continue. Hence the considerable activity on the part of supporters of the status quo to develop and communicate as widely as possible a direct counter to reformist ideas: one reflecting their own interests and ideological position.

The policy of publishing a direct counter to reformist ideas was an *overtly* political strategy—an exercise in consolidating active support for maintaining the status quo by shaping and controlling political consciousness. This, however, was only one line of ideological attack available to the ruling classes. A second approach involved a more *covert* strategy, but one which would equally preserve the status quo. This was to make "safe" literature available at low cost to working class readers. "Safe" literature was of two main types: religious tracts, and material intended to inform, interest, and amuse, but which was powerless to stimulate political critique. Religious tracts would secure loyalty to Christian doctrine and, to that extent, help maintain the hegemony of church and state. Informative literature would operate (in political terms) on the logic of diversion—it would deflect the reading habits of workers out of the political field altogether; whether the politics of reform or anti-reform. It would effectively depoliticize working-class readers by channeling their reading energies into politically impotent content, with the effect of maintaining the status quo by failing to stimulate opposition to it.

Among the leading groups to employ the strategy of providing cheap but safe literature for worker readers were the Society for Promoting Christian Knowledge (SPCK), and the Society for the Diffusion of Useful Knowledge (SDUK).

The SPCK was formally established in March 1699. In part it was a response to its founders' perception of the deplorable moral and religious situation in England. More broadly, however, it was concerned to promote Christian knowledge at home as well as in His Majesty's Dominions. A subscription society, its major activities at home during the eighteenth century included promoting charity schools with a catechetical flavor, formulating policy (communicated by the bishops) for charity schools under the trusteeship of Anglican churchmen, publishing religious literature for sale at subsidized prices and for use within charity schools, establishing libraries for poor clergy and religious services for prisons, and producing bibles, prayer books, liturgies, etc., in Irish, Welsh and Gaelic. Records from 1815 establish that at this time the Society recognized three main tasks:

missions abroad; distribution of the Scriptures, the prayer book, and religious tracts; and the education of the poor "in the principles of our faith." Its publishing activities were conducted under eighteen separate headings, with major priorities including printing and distributing bibles, prayer books, commentaries and explanations (pitched at different levels for different readerships), sermons and tracts on catechetical themes, books for public and private devotion, guides to confession and absolution, and works concerning duties, vices, and the evils of popery (See Clarke, 1959: 148–152, and Allen and McClure, 1898: Ch. 5, especially pp. 188–189).

In the midst of the political turmoil of 1819, however, a new dimension was added to the work of the SPCK, and its activity took on a special urgency. Viewing with much concern and dismay "the efforts which the enemies of Christianity were making in disseminating the poisons of infidelity," and believing it proper to employ all its available means to counteract the evils being done by radical publications, the Society appointed a special committee charged with countering the infidel influence. This committee had instructions "to publish in a more popular form, and at a diminished price, suitable tracts then on the Society's Catalogue," and also to publish "such other works as might be deemed necessary" (Allen and McClure, 1989: 189). Large print runs were made of several existing works, and more than thirty new tracts were produced. According to SPCK records more than a million copies of books and tracts "against infidelity and blasphemy" were printed and distributed in less than a year—with expenses being met from the £7000 raised by appeal to supporters.

Webb suggests that this attempt (during 1819–1820) to counter the influence of the emerging radical press on working class readers brought some disappointments to the SPCK. Reports from Manchester, Bolton, and London's East End expressed great difficulties in getting the poor to take the tracts, even where the original policy of selling them cheaply was waived in favor of distributing them gratis in order to reach an audience. While sales were good in better-off areas, such as London's West End, the Society's real concern was to have an effect in the poorer neighborhoods. Whatever its true degree of success may have been, the SPCK expressed satisfaction "that the measures . . . pursued were productive of much good." So much so that the work of the anti-infidel committee was reactivated in 1830–1831 when, once more, "the infidel press teemed with the bitterest invectives against religion and the ministers of Christ," and publications "of the most pernicious kind, full of blasphemy . . . were circulated with unceasing activity" (Allen and McClure, 1989: 190).

This was the era of Hetherington and defiance of the Stamp. The SPCK again raised funds for publication. Many of the earlier tracts were reprinted and distributed, and no less than twenty-nine new titles were produced. Together these comprised *A Library of Christian Knowledge*. In 1832 the Society's rejuvenated publishing program took a further step with the formation of a Committee of General Literature and Education. This was a response not only to the "evil opinions being inculcated" in some parts of the popular "penny literature," but also to the fact that in other parts (where opinions were not in themselves "evil") the knowledge being diffused among the masses was "studiously separated from religion" (Webb, 1955: 73). Under the auspices of the new committee the SPCK entered popular publishing on several new fronts, including historical and biographical series, a scientific series with a "decided bias" towards revelation, and a penny weekly called the *Saturday Magazine.* Together with the work of the Religious Tract Society, the activities of the SPCK represent the most impressive attempts to foster safe literature on the model of religious content. On a political level their work was complemented by the efforts of publishers specializing in essentially secular knowledge. Foremost among these was the Society for the Diffusion of Useful Knowledge.

√The major figure behind the formation of the SDUK (in 1826) was Henry Brougham, a leading Whig and, subsequently, Lord Chancellor. Brougham's *Practical Observations upon the Education of the People* had been published in 1825. In this he noted two main impediments to a sound working class education. First, working people could not afford the books and instructors available to more affluent citizens. Second, even had they been able to afford the expense, workers lacked the necessary leisure time to plow through the kind of learning material as did exist within such areas of knowledge as science, literature and the arts. To overcome these impediments facing the provision of a sound education for the working class, Brougham advocated making available cheap publications adapted to the special learning circumstances of workers—this material to be available in the fields deemed useful knowledge. Even earlier, in 1821, Charles Knight had expressed his hope that "ignorant disseminators of sedition and discontent"—i.e., people such as Cobbett—would be "beaten out of the [publishing] field" by opponents with "better principles," who would thereafter "direct the secret of popular writing to a useful and righteous purpose" (cited in Simon, 1960: 159). The sentiments of Brougham and Knight (who became the main publisher for the SDUK) were reflected in the official aim of the Society: namely, "the imparting *of useful* information to all classes of the community, particularly to such as are unable to avail themselves of experienced teachers, or may prefer learning by themselves" (cited Webb, 1955: 67).

The ideological purpose of such activity was expressed very clearly by Knight himself, some years after the SDUK was founded. He insisted that

> the object of the general diffusion of knowledge is not to make men dissatisfied with their lot—to make the peasant yearn to be an artisan, or the artisan dream of the honours and riches of a profession—but to give the means of content

> to those who, for the most part, must necessarily remain in that station which requires great self-denial and great endurance; but which is capable of becoming not only a condition of comfort, but of enjoyment, through the exercise of those very virtues, in connection with a desire for that improvement of the understanding which to a large extent is independent of rank and riches. (Cited in Hollis, 1973: 334)

The early publications of the SDUK kept clear of explicit political themes, including political economy. The *Library of Useful Knowledge* specialized in biography and natural science. It was supplemented by the *Library of Entertaining Knowledge* which, as its title suggests, offered amusement on less esoteric matters. The two *Libraries* appeared in monthly issues and "were filled with miscellaneous scientific and cultural information, ranging from lepidophera to 'Autumnal Customs in Kardofan' " (Simon, 1960: 160). In 1832 the *Libraries* were joined by the *Penny Magazine,* edited by Charles Knight. Knight aimed to produce "a safe Miscellany, in which all classes might find much information and some amusement." Webb suggests that the proportions were rather more the reverse. The *Penny Magazine* was largely a compilation "of quaint facts and descriptions of various animals, buildings, and natural phenomena" with much of its initial popularity doubtless due to its woodcut illustrations. Even so, it was by no means entirely bereft of political implication. Consider, for example, the ideological message conveyed in "The Weaver's Song," published in an early number of the *Magazine.*

> Weave, brothers, weave!—Swiftly throw
> The shuttle athwart the loom,
> And show us how brightly your flowers grow,
> That have beauty but no perfume!
> Come, show us the rose, with a hundred dyes,
> The lily, that hath no spot,
> The violet, deep as your true love's eyes,
> And the little forget-me-not!
> Sing,—sing brothers! weave and sing!
> 'Tis good both to sing and to weave:
> 'Tis better to work than live idle:
> 'Tis better to sing than grieve.
> Weave, brothers, weave!—Toil is ours;
> But toil is the lot of men:
> One gathers the fruit, one gathers the flowers,
> One soweth the seed again!
> There is not a creature, from England's King,
> To the peasant that delves the soil,

That knows half the pleasures the seasons bring,
 If he hath not his share of toil!
So,—sing, brothers! etc. (Cornwall, cited in Hollis, 1973: 53)[3]

Despite such excursions into thinly veiled political comment on the virtues of accepting one's station with grace and serenity, comforted by the "insight" that toil is the lot of (all) men, the content of the *Libraries,* the *Penny Magazine,* and the *Penny Cyclopaedia* was diversionary rather than explicitly political in nature; an exercise in covert rather than overt strategy. This material was attacked from all sides: by Tories, middle class radicals, and the working class press itself; and from 1830 the SDUK published in addition to its program of safe literature a number of works steeped in the political economy of the bourgeoisie (See, for example, Simon, 1960: 160–161, and Hollis, 1973: 334–335, for reference to Tory and middle class attacks).

While there is not the space to develop the theme in depth here, it is worth noting that the attempt by the SDUK to diffuse political economy as useful knowledge among working class readers involved a shift to an overtly hegemonic strategy. Such publications sought to shape the consciousness of working class folk in accordance with an ideology grounded in middle class interests and that directly contradicted the interests of workers themselves. Quite simply, there is no other way in which to understand Charles Knight's *Results of Machinery,* Brougham's arguments on wages, consumption levels and employment, presented in the *Companion to the Newspaper,* or the SDUK's *Short Address to Workmen on Combinations to Raise Wages*—all of which insisted that attempts by workers to force higher wages through combined activity were futile. According to these arguments, the inexorable operation of supply and demand meant that wages must inevitably be set by market forces. The economy simply could not sustain wages above the level fixed by the labor market. The real choices facing laborers were, according to SDUK theory, strictly limited. They must either accept the fortunes (and misfortunes) of the labor market and learn to live within them, or escape by becoming themselves capitalists. Those laborers who did not choose to become capitalists could, at best, hope to make the most of their earnings (whilst employed, that is) by practising thrift and sound economic management. And so, says the *Short Address to Workmen,*

> When labour offered for sale is plentiful its price [i.e., wages] will be low, when it is scarce it will be high. This is a law of nature against which it is vain to contend; only "forbearance, management, and economy" could alleviate the inevitable lot of human life, as revealed in the iron law of wages. Active protest was out of place. Your complaints [labourers were informed] are sometimes exaggerated and were they better informed than they are, you would not have chosen [protest and combination as] the remedy to remove them. (Simon, 1960: 162)

The working class press was severe in its treatment both of overt and covert approaches to ideological domination by such organizations as the SPCK and the SDUK. In penetrating and revealing these strategies, working class writers exposed attempts on behalf of opposing class interests to foist an improper literacy onto workers. In the same process by developing their critiques they helped positively to enhance proper literacy among their readers. Against the political economy of the SDUK, such working class writers as William Longson, Bronterre O'Brien, William Carpenter, and numerous economic commentators for the *Poor Man's Guardian,* produced compelling yet entirely accessible rebuttals (see Hollis, 1973: 64–69). Some of the most scathing comment, however, was reserved for the exponents of diversion—for those who would neutralize workers' critical potential by channeling their reading energies into safe literature and, thereby, turn hard-won skills against the interests of those who had managed to acquire them.

The SPCK was denounced for aiming "to prop up the 'present cannibal order of things' by reconciling the poor to poverty." O'Brien referred to those who circulated the Society's tracts as "canting vagabonds" with "hypocritical pretensions to religion," lamenting that the hold they had over weak minds made it even "more difficult to break through their slimy meshes" than to overcome the persuasive powers of the stamped press (cited in Hollis, 1973: 144). In a more moderate vein, Cobbett exposed the SPCK as hoping to prevent the people from reading and thinking politics.

The working class press similarly denounced "useful knowledge" as patronizing, hypocritical, and hostile to the people's interests. The kind of knowledge truly required by the people—that is, the content of a proper literacy—had nothing to do with the number of humps on the back of a dromedary, the number of transmigrations in the life of a caterpillar from chrysalis to butterfly, or with how a kangaroo jumps. It had, instead, to do with their rights as citizens; with why the class that actually produced wealth was the most degraded, while that which produced nothing was elevated; with why working people were denied a vote and any say whatsoever in legislation, while the "idle and mischievous" exercised complete power in political and legal matters; with why those whose acts revealed that they were really without religious conviction had control of the nation's religion (Hollis, 1970: 20–21).

Critiques published in the working class press of the literacy fostered by the SCPK and the SDUK came from rank and file readers as well as from the editors and other established writers. Hollis cites a laborer's assessment (published in the *Poor Man's Guardian)* of the "useful knowledge" purveyed by the *Penny Magazine.*

> Useful knowledge, indeed, would that be to those who live idly on our skill and industry, which would cajole us into an apathetic resignation to their iron sway, or induce us to waste the energy and skill of man for them all day, and seek re-

> laxation of an evening in the puerile stories or recreations of childhood . . . This first number of their *Penny Magazine,* insinuates that poor men are not qualified to understand the measures of government. 'Every man is deeply interested in all the questions of government. Every man, however, may not be qualified to understand them'. My fellow-countrymen, I beseech you now do be modest, do be very diffident,—Pray do distrust the evidence of your reasons—submit implicitly to the *dicta* of your betters! (Cited in Hollis, 1970: 143–144)

This writer shows a profound understanding of the distinction between proper and improper literacy; of reading and writing that promises enhanced control over and genuine understanding of one's daily life, and that which effectively negates them in the interests of others. The self-conscious aim of those who produced the working class press was to advance proper literacy among their readers. Nowhere is this expressed more clearly and directly than by O'Brien.

> Some simpletons talk of knowledge as rendering the working classes more obedient, more dutiful—better servants, better subjects, and so on, which means making them more subservient slaves, and more conducive to the wealth and gratification of idlers of all description. But such knowledge is trash; the only knowledge which is of any service to the working people is that which makes them more dissatisfied, and makes them worse slaves. This is the knowledge we shall give them . . . (cited in Hollis, 1970: 20)

The battle for the working class press between 1816 and 1836 provides an excellent illustration of distinct and competing literacies emerging as social constructions within the context of struggle between competing interest groups. The polarized conceptions of Charles Knight and Bronterre O'Brien capture this in microcosm. From a working-class standpoint, the form of literacy promoted on behalf of ruling class interests for worker consumption must be adjudged *improper*—and, in fact, was assessed as such by the working class press. Against this hegemonic literacy the working class press battled to create and transmit a proper literacy: a truly counter-hegemonic form which would focus workers' attention upon those structured inequalities of power and control within economic, political, and social life, that were the *real* causes of their condition.

Endnotes

1. An especially interesting development was apparent in the growth of secular Sunday schools, where a major concern was to free working people from the ideological influence of the church. This influence was seen by many working class leaders as "the chief means whereby the people were held back from action." The reform groups established in Royton maintained that there was no hope of a more liberal form of government while priests were able to awe the people with fears of being damned to eternity. Anti-religious literature and discussion was a feature of reform activity in Lancashire. This concern with undermining the unwanted ideological influence of religion extended to promoting a proper literacy among working-class children as well

as adults. Lancashire reformers "endeavoured to replace the religious indoctrination of children with a rational education in the Sunday schools they promoted, as part of the union movement for parliamentary reform from 1817 onwards" (Simon, 1960: 187). The underlying assumption was that instruction in an ambiguous, doubtful, and contradictory religion cramped children's understandings, and baffled their judgments. It made for uncritical, irrational, and distorted thought. When people were trained to think rationally, and to distinguish critically between right and wrong, it would be impossible for any king or government to tyrannize over them and deny them their rights (Simon, 1960: 187-188).

2. It is, for example, debatable how far Cobbett's concern for political reform was grounded in authentic commitment to working-class interests. In the *Address to Journeymen and Labourers,* for instance, we find him recommending the principle that the right to vote be extended only to those who pay direct—as opposed to indirect—taxes. This would have denied the vote to vast numbers of working people. In acknowledging this, the best Cobbett offers is the assumption that a reformed government could very easily hit upon an optimally just arrangement (compare Cole and Cole 1944: 214–215). The relationship between journalism and authentic commitment to working class interests is much less ambiguous in Hetherington and *The Poor Man's Guardian* and later, in the Chartist press.
3. By Barry Cornwall, *Penny Magazine,* 7 July 1932, cited in Hollis (1973: 53). See the reply, "The weaver's song not by Barry Cornwall," *Poor Man's Guardian.* 3 November 1832, cited in Hollis (1973: 54).

Bibliography

Allen, W. and McClure, E. (1898). *Two Hundred Years: The History of the SPCK.* London: Society for Promoting Christian Knowledge

Aspinall, A. (1949). *Politics and the Press, 1780–1850.* London: Home and Van Thal.

Bamford, S. (1984). *Passages in the Life of a Radical.* Oxford: Oxford University Press.

Barker, A. (1938). *Henry Hetherington. 1792–1849. Pioneer in the Freethought and Working Class Struggles of a Hundred Years Ago for the Freedom of the Press.* London: The Pioneer Press.

Clarke, W. (1959). *A History of the SPCK.* London: Society for Promoting Christian Knowledge.

Cole, G. and Cole, M. (eds) (1944). *The Opinions of William Cobbett.* London: The Cobbett Publishing Co.

Collett, C. (1933). *The Taxes on Knowledge.* London: Watts and Co.

Hampton, C. (ed.) (1984). *A Radical Reader.* Harmondsworth: Penguin.

Hollis, P. (1970). *The Pauper Press: A Study in Working-Class Radicalism in the 1830s.* London: Oxford University Press.

Hollis, P. (ed.) (1973). *Class and Conflict in Nineteenth Century England 1815–1850.* London: Routledge and Kegan Paul.

Lovett, W. (1920). *The Life and Struggles of William Lovett.* London: Bell and Sons.

Mills, C. Wright (1959). *The Sociological Imagination.* New York: Oxford University Press.

O'Neil, W. (1970). Properly literate. *Harvard Educational Review,* 40(2): 260–263.

Simon, B. (1960). *Studies in the History of Education 1780–1870.* London: Lawrence & Wishart.

Thompson, E. (1963). *The Making of the English Working Class.* Harmondsworth: Penguin.

Wearmouth, R. (1948). *Some Working Class Movements of the Nineteenth Century,* London: The Epworth Press.

Webb, R. (1955). *The British Working Class Reader.* London: Allen and Unwin.

Wickwar, W. (1928). *The Struggle for the Freedom of the Press.* London: Allen and Unwin.

FOUR

Simon Says See What I Say: Reader Response and the Teacher as Meaning-maker (1993)

Michele Knobel

Biography of the Text

"Simon Says" was originally published in 1993 in a special issue of the *Australian Journal of Language and Literacy* edited by Geoff Bull. This special issue focused on critical literacy and in many ways stands as a trope for the "social turn" in Australian literacy education underway in the late 1980s and early 1990s (the journal itself changed its title from the *Australian Journal of Reading* to the *Australian Journal of Language and Literacy* in 1991). While I had published articles in professional magazines prior to this, "Simon Says" was my first article in an academic journal. I wrote it while studying for a Masters in Education degree at the University of Southern Queensland (then the University College of Southern Queensland) in Toowoomba, Australia. I was incredibly fortunate to have Geoff Bull as my mentor throughout my Masters studies—Geoff very much encouraged me to focus on becoming a researcher and—along with his partner, Michèle Anstey—gave me an excellent apprenticeship in designing and conducting studies that were of genuine interest and import to me personally. This apprenticeship included working as a research assistant on their own nationally funded projects, being strongly supported in presenting my own research at local and national conferences, and getting to meet a great many seriously good academics. I really am indebted to them both.

Geoff went on to supervise my Masters research thesis, but the data drawn on in "Simon Says" grew out of an independent study with Geoff that set out to explore how conventional Reader Response theory could be informed or critiqued by critical literacy theory. This was such an eye-opening time for me, with the work of Carolyn Baker and Peter Freebody (e.g., 1987), Pam Gilbert (e.g., 1989), James Paul Gee (e.g., 1992a), among others, helping me to rethink taken-as-normal classroom language and literacy practices and to understand how teachers themselves shape and valorize what "counts" as doing literacy and being literate.

Reading back over the paper now is a little blush-inducing in places, but it remains an honor to have appeared amongst such prestigious and influential writers in this issue of the journal (e.g., Gee, Gilbert, Baker and Freebody) especially at such an early point in my career.

Introduction

Recent developments in literacy research and application of literacy theory have encompassed critical analyses of school literacies and discourses (Luke, 1993; Luke and Walton, 1994), discourse analysis of classroom talk about text via transcripts (Baker and Freebody, 1989; Baker, 1991a, 1991b), examination of response to texts and how texts position readers in specific ways (Machet, 1992), and critical self-reflection on literacy teaching practices (Morgan, 1992).

This paper, by means of reference to research literature, discourse analysis and reflection on my own teaching practices, aims to explore anomalies regarding reader response and the interconstruction of meaning discovered within my classroom reading pedagogy. My exploration is structured in three parts; first, an examination of reader response traditions and theories; second, an analysis of reader response from transcripts of my classroom reading sessions; and third, my position regarding children's response to literature and the role of the teacher as a mediator of meaning.

Part 1: "Come Sit By Me and I'll Read You a Story"

The practice of reading stories to children has attracted increased research attention in recent years (Davies, 1989; Baker and Freebody, 1989). Children's literature has been invested with enhancing literacy development (Wells, 1991), promoting particular world and social views (Baker and Luke, 1991), providing opportunity for moral development (Machet, 1992), and helping children develop a "love" of reading (Patterson, 1991). My position is that these investments are ideological. Ideology describes a set of values and beliefs that position the individual within a particular form of life (cf. Gee, 1992). It affects the construction of meanings from texts and impacts on response via those meanings. Before elaborating on ideology, meaning construction and reading response ideology, however, it is necessary to examine some of the traditions underlying reader response theory.

Simon Says, See What I Say

In the beginning was the text; to the New Critics it was wholly within the text that meaning was located, and the teacher was the privileged holder of this meaning (Thomson, 1984; Probst, 1986). Reader Response theorists reacted strongly against this notion of "authoritative interpretation" of meaning and championed the possibility of multiple readings and responses to text. This involved "transactions" of meaning between the reader, text and author's purpose (Rosenblatt, 1969, 1982). Psychoanalysts added escapism and realizing unconscious fantasies to reader response theory, and included emotions in the analysis of response (Holland and Lesser, as cited in Iser, 1976).

Structuralists maintained that textual meanings were derived from literary codes and conventions, focusing on grammar and literary concepts (Stott, 1987; Olson, 1987). Writers employing developments in post-structuralism and postmodern theory, such as Morgan (1992), Willinsky (1991) and Lankshear and McLaren (1993), have moved beyond the text to include social, political, historical, cultural, discursive and biographical conditions and experiences in the exploration of meaning making and response. A metatheoretical framework begins to emerge from synthesis of these traditions. The framework consists of "meaning transactions" between the reader, text and author's intention, and includes agents of response, such as emotion, biography and culture. It also concerns the reader's mastery of literary codes, conventions and discourses. Of course, this framework requires contextualizing to create a more accurate description of reader response in primary school classrooms. What is needed is a firm grounding in the school, social and literacy cultures in which response to literature often occurs, plus an awareness of ideologies and agendas present in the practice of reading aloud to children. This involves an examination of meaning making and the teacher as a mediator of meaning.

It Means What It Says . . .

Response to texts requires meaning making. Constructing meaning from texts involves deciding the sense or the representation of reality of the words. Meaning making is not confined to decoding text, but encompasses the meaning maker's biography and culture, as well as the wider effects of historical, political and discursive positioning (Baker and Luke, 1991; Lankshear, 1991; Gee, 1992). Deciding a text's representation of reality, therefore, actively involves readers in meaning making and response (Taxel, 1989).

When a text is read aloud to children however, the meaning-making process is consciously, or unconsciously, mediated by the teacher (Golden, 1988; Baker and Freebody, 1989; Baker, 1991a, 1991b; Wells, 1991). Mediation in this sense is more than acting as a "go-between" or interface for meaning. It is an active, interpretative role of varying degrees of influence, that directs children towards particular world views via the teacher's ideologies and agendas. The teacher's personal ideologies (viewpoints and values) act as filters, coloring the types of comments made, questions asked and answers accepted by the teacher. Teacher "agendas," or frames of reference (Anstey, 1989; McHoul, 1991) also impinge on meaning making in the reading session. Agendas describe the specific purposes driving particular patterns of thinking or behavior. These patterns are directly affected by the "filter" of ideology. The effect of teacher meaning mediation on children's responses to read-aloud texts was the driving force behind critical reflection on my own reading-aloud pedagogic practices within primary school classrooms.

Part 2: Are My Students Responding to Texts or to Me?

Discourse analysis of reading session transcripts provided a methodological framework for my self-reflection. The study was conducted within four classrooms over six months. I decided to share with my students Margaret Wild's book, *Let the Celebrations Begin!*, illustrated by Julie Vivas (1991). The agenda behind this selection consisted of three aspects. First, I wanted to read a recently published book to diminish the effects of previous readings on student response. Second, I wanted to use an Australian author and illustrator familiar to my students and third, I wanted a book that would evoke response.

Each reading session followed the same format. Whole class, teacher-led discussions were conducted prior to and following reading the story. The narrative was read continuously to minimize the impinging effects of "running metatextual commentary" on meaning making and response (de Castell, Luke and Luke, 1987, cited in Baker and Freebody, 1989: 163). Each reading session was video or audio-taped, then converted into a written transcript.

Analysis of my transcripts revealed anomalies that lay outside the metatheoretical framework synthesized from reader response theories. The framework was constructed around the reader as the locus of meaning making, yet my transcript analyses indicated that my students were by no means responding entirely from their own construction of text meaning.

Anomalies: Reading the Teacher Instead of the Text

Anomaly 1

It was clear from the transcripts that the "discussion" around the book mostly took the form of teacher question—student response—teacher feedback. Luke

(1993) and Baker (1992) also identified this pattern within classroom discussion about texts, labeling it "Initiation, Response and Evaluation" (IRE). Closer analysis of the IRE patterns revealed that the verbal student responses consisted predominantly of unison answering, or "chorusing" (see Transcript 1a and b below; in these transcript excerpts and the others that follow, "T" refers to the teacher or to me; "S" refers to a single student; and "Ss" refers to multiple students).

Transcript 1a

152T: Right. Fit more people on. So do you think it was crowded, or was it fairly spacious?
153Ss: Crowded.
154T: Crowded. Yes. Nicholas?
155S: Oh. I was just going to say it was crowded.

Transcript 1b

180T: So are they wearing fairly thin clothes, do you think?
181Ss: Thin.

Comment

Students appeared to be reading *me* through my questions rather than the text. This was evidenced by their ability to predict the "right" answer (the answer I required, the one that would please me). Prediction was based on children's understanding of the meaning I (the teacher) held personally and indicated by the framing of my questions. Of course, this type of prediction is also based on the teacher's voice inflections, facial expressions, gestures and body positioning. Such factors are difficult to transcribe, however, and have been excluded from analysis for this reason (cf. Edwards and Westgate, 1987; Baker and Freebody, 1989; Baker, 1991a).

Anomaly 2

A response pattern I call "association" was identified repeatedly within the transcripts. This patterning involved the students repeating each other's words or references and was generally in response to a teacher question (see Transcript 2a and b).

Transcript 2a

451T: So what sort of life do you think—Oh, yeah. Yeah. What sort of life do you think she had back home?

452S: A really um, good life.
453S: Comfortable.
454S: Lazy.
455T: Good one, good one. What do you think—what sort of life are they having here?
456S: Rough.
457S: Rough.
458S: Rough.
459T: Good.
460S: Rough and just hard.

Transcript 2b

48T: Oh, I see. They've got toys, haven't they? O.K., let's have a look inside (turns to title page). Now, you think the story might be about a party—does this change your ideas at all?
49S: Um. Christmas!
50T: Christmas. Why Christmas?
51S: Yes.
52S: Because it's got a star.
53S: Oh. There's a star here.
545: It's probably Christmas.
55T: Why does a star make you think of Christmas?
56S: Because it's—
57S: Because, there, there was, um that kind of star on top of Jesus.
58T: Yep.
59S: Jesus
60S: That led the way to Jesus.

Comment

Association patterns can be explained from different positions. Students may have perceived the reading session as a "ritual performance" (McLaren, 1986) instigated by me as teacher and evidenced in my repetition of my own words and the words of the students (see Transcript 2a: line 455T and Transcript 2b: line 50T). Student desire to "do well" may have prompted repeating or paraphrasing a peer's answer I had acknowledged as "correct" or "good" (see Transcript 2a: line 460S). Predicting, or trying to predict, the "correct" answers to the teacher's questions flows from student perceptions of teacher authority, teacher access to privileged knowledge and to the rites/rituals of the dominant social discourse within the classroom (Baker and Luke, 1991). Alternatively, association responses may have been attempts at student solidarity, whereby confidence and some sense of power within the classroom discourse was gained through group support. Again, students appeared to read me, rather than the text.

Anomaly 3

A further reader response anomaly was apparent when listening to the reading session tapes. I noted that many of the students' answers to my questions ended with an upward inflection; effectively restructuring student statements as questions. Baker (1991b) has also observed this interrogative intonation in children's answers in her reading session recordings.

Comment

Such patterns of inflection may occur for different reasons. Students may construct the teacher as a holder of privileged knowledge and consequently the qualifier of acceptable and non-acceptable responses. Baker's (1991b:105) interpretation was that students were "acknowledging that they could be wrong and/or asking if they (were) right." Answering in question format defers to the perceived power of the teacher, resulting in diminished student confidence regarding personal ability to make meanings from texts.

Another line of explanation focuses on the child's desire to give the "correct" answer (which involves predicting the answer wanted by the teacher) and demonstrate competence within the classroom discourse.

Anomaly 4

Critical reflection on the content and purposes underlying my questions about the book revealed obvious ideologies and agendas (see Transcripts 3a and b).

Transcript 3a

72T: Forty years ago? About 40, 40 to 45, 50 years ago. Good. Alright, and in that time, can you remember what happened? (1.0) Who was fighting who? (1.0) Do you know? Yes?

73S: British against the Nazis?

74T: Right. And the Nazis were in Germany weren't they. Remember Hitler? He wanted to take over the whole wide world. And who did he especially, especially hate? And he used to lock them up in something called (1.0) concentration camps. Who did he especially, especially hate?

75S: The Nazis?

76T: No, he was a Nazi. He loved the Nazis. Yes?

77S: The Jews.

78T: The Jews. Spot on. And he used to hate the Jews. And remember, I don't know—(1.0) oh, well you don't remember, but he used to make them wear these bright yellow stars to say that they were Jews and they were a horrible people. And were they horrible do you think?
79Ss: No.

Comment

I presented an anti-Hitler/Nazi position to students through my comments and questions; neither Jews nor Nazis were mentioned in the book. The surface transfer of my ideological position was signified by the children's ability to chorus a "correct" response (see line 79Ss).

Transcript 3b

281T: Right. Can you imagine, these children are about your age, and they don't have their parents with them, they only have one set of clothes, they don't have any toys. (1.0) How would you feel? Think about it. (2.0) You'd feel great?
282Ss: No.
283T: How'd you feel, Jamie?
284S: Um (1.0). I'd feel horrible, really.
28ST: Why would you feel horrible?
286S: Oh, lonely. 'Cause you don't have your mother anywhere.
287T: Right. What a good word—lonely.
288S: Depressed.
289T: Depressed. Why depressed Luana?
290S: Um, (1 .0). Because, like, they might have lost their good friends and their, their nice bed to sleep in.

Comment

My agenda was not clear to the students, as evidenced by their initial hesitancy in answering my questions. Once I suggested a response the children were able to predict the answer I required (see line 282Ss). My agenda shifted, however, from focusing on affective response to addressing language use when I commented on the word "lonely" and ignored the use of "horrible." This shift was identified by another student who "did language" and suggested the term "depressed."

The interaction patterns, ideologies and agendas that emerged from analysis of my reading sessions served as powerful indicators of how I as teacher construct the child within my classroom. Assumptions are made regarding children's meaning-making abilities and their current knowledge which affect my teaching agendas and questions. My students, in turn, acquire a particular interpretation of "reading" within the classroom and how they should position themselves in relation to my construction of them (Luke, 1993). When reading aloud to students and discussing the narrative, I am filtering the text through my own meanings; sifting and sorting language and concepts to produce a particular version of the story matched to a particular version of my students. I am in effect, a mediator of meaning.

Part 3: Getting Rid of "Simon Says"

Reflection on the anomalies discovered within my reading sessions increased my awareness of the impact of mediation on reader response. Mediation is not necessarily a negative interface in reading response, but it can limit children to particular ways of viewing the world. My reaction to my findings was to explore avenues for ameliorating student response and meaning making within my classroom. One avenue was modifying my questioning techniques and practices. Perceptions of "only one right answer" and the teacher as a holder of privileged knowledge are weakened by encouraging inter-student discussions where I assume a detached, yet supportive, role (Kelly, 1990; Baker, 1991a). Restrictive teacher-student discussion patterns (such as IRE) are also minimized accordingly and the possibility of multiple readings is established.

Discussion is not the only classroom activity available for responding to text. Alternatives include responding through art, drama, music, debating, and writing (Chase and Hynd, 1987; Smith, Greenlaw and Scott, 1987; Singh, 1988; Chambers, 1991; Wells, 1991; Patterson, 1991). Such processes can show students that meaning making and response to literature lie within themselves rather than within the teacher. Students will hopefully begin to develop their own strategies for making meanings from texts and come to value their personal responses to reading.

Awareness of critical literacy practices may enable students to identify and critique ideologies and agendas present in the text and in the teacher. Critical awareness, in this sense, involves not only examination of the text for gender, racial, political and discourse bias, but also includes the examination of response itself. Students are encouraged to identify the sources of their own response and determine the discourses impinging on them. Such meta-reflection may be promoted through student analysis of the origins of their responses to literature, or by having them examine reading session transcripts in much the same way as I have within this article.

Before students can be taught to be critically aware of their responses to literature (if indeed, such a thing can be "taught"), teachers themselves need to examine their own ideologies and agendas in light of how this "personal baggage" impacts students and classroom practices (Patterson, 1991; Newman; 1991). This is not to say that all ideologies and agendas are detrimental to reader response. Response to literature actively involves each reader in personal meaning making, and for this reason will never be neutral or detached.

My position regarding teacher and student response to literature in the classroom is twofold. Firstly, teachers need to be aware of how their own meanings and interpretations of reality may affect, and sometimes conflict with, their students' meaning making and reality construction in relation to literature and reader response. Secondly, teachers can become more effective "reflective practitioners" by developing a "meta" knowledge and language of reading mediation in the classroom and applying it to their pedagogy as praxis. This article has aimed at facilitating the development of these practices.

Conclusion: Celebrating Authentic Response

Responding to literature within the classroom is subject to myriad influences, and the focus of this article was on one of these: the teacher as a mediator of meaning and response. Teachers should be aware of their perceived (and therefore real) power to affect how children construct meanings and responses from texts read to them. Teachers as mediating readers filter meanings and present particular constructions of students and the world through their "personal baggage" of ideologies and agendas. Teachers should be able to recognize how much of student response to literature is simply a parroting of teacher response in an elaborate "Simon Says" pattern of interaction, and how much is "authentic" response.

I have been led to re-evaluate and modify my teaching practices and strategies through meta-knowledge of the conscious and unconscious promotion of my ideologies and agendas in my role as mediating reader. My sharing of literature with children will hopefully no longer take the form of "Simon says see what I say" but, rather, will allow my students to celebrate their freedom to respond to literature authentically. To snaffle the title of a very hopeful book; let the celebrations begin!

References

Anstey, M. (1989). Composing, comprehending and the content areas: Can integration confuse the teaching/learning agenda? Darwin: Proceedings of the 14th National ARA Conference. 76–87.

Baker, C. (1991a). Reading the texts of reading lessons, *Australian Journal of Reading*, 14(1): 5–20.

Baker, C. (1991b). Classroom literacy events. *Australian Journal of Reading*,14(2): 103–8.

Baker, C. (1991c). Literacy practices and social relations in classroom reading events. In C. Baker and A. Luke, A. (eds), *Towards a Critical Sociology of Reading Pedagogy: Papers of the XII World Congress on Reading*. Amsterdam: John Benjamins Publishing Company.161–189.

Baker, C. (1992). Classroom talk. Paper presented at the Working Conference on Critical Literacy, Brisbane, 29 June.

Baker, C. and Freebody, P. (1987). *Children's First School Books*. Oxford: Blackwell.

Baker, C. and Luke, A. (eds) (1991). *Towards a Critical Sociology of Reading Pedagogy: Papers of the XII World Congress on Reading.* Amsterdam: John Benjamins Publishing Company.
Booth, D. and Thornley-Hall, C. (eds) (1991). *The Talk Curriculum.* Carlton, Victoria: Australian Reading Association.
Bull, G. (1989). *Reflective Teaching.* Carlton, Victoria: Australian Reading Association.
Chambers, A. (1991). *The Reading Environment.* Newtown, NSW: Primary English Teachers Association in association with The Thimble Press.
Chase, N. and Hynd, C. (1987). Reader response: An alternative way to teach students about text. *Journal of Reading,* 30: 530–40.
Davies, B. (1989). *Frogs and Snails and Feminist Tales.* Melbourne: Allen and Unwin.
Edwards, A. and Westgate, D. (1987). *Investigating Classroom Talk.* London: Falmer Press.
Egan, K. (1987). The shape of the science text: The function of stories. In S. de Castell, A. Luke and C. Luke (eds), *Language, Authority and Criticism.* London: Falmer Press. 96–108.
Freebody, P, Luke, A. and Gilbert, P. (1993). Reading positions and practices in the classroom. *Curriculum Inquiry,* 21: 435–457.
Gee, J.P. (1992a). What is reading?: Literacies, discourse and domination. Paper presented at the Working Conference on Critical Literacy, Brisbane, 29 June - July 3.
Gee, J.P. (1992b). Literacies: Tuning into forms of life. *Education Australia,* 19/20: 13–14.
Gilbert, P. with Rowe, K. (1989). *Gender. Literacy and the Classroom.* Carlton, Victoria: Australian Reading Association.
Golden, J. (1988). Text and the mediation of text. *Linguistics and Education,* 1: 19–43.
Green, J. and Meyer, L. (1991). The embeddedness of reading in classroom life: Reading as a situated process. In C. Baker and A. Luke (eds), *Towards A Critical Sociology of Reading Pedagogy: Papers of the XII World Congress on Reading.* Amsterdam: John Benjamins Publishing Company. 141–160.
Harding, D. (1937). The role of the onlooker. *Scrutiny,* 6(3): 247–58.
Harding, D. (1968). Psychological processes in the reading of fiction. *British Journal of Aesthetics,* 2(2): 133–47.
Harding, D. (1967). Considered experience: The invitation of the novel. *English in Education,* 2(1): 7–15.
Iser, W. (1976). *The Act of Reading: A Theory of Aesthetic Response.* London: Routledge and Kegan Paul.
Kelly, P. (1990). Guiding young students' response to literature. *The Reading Teacher,* 43(7): 464–70.
Knobel, M. (1991). An examination of children's response to literature. Research paper presented at the University of Southern Queensland, Toowoomba, Australia.
Koeller, S. (1988). The child's voice: Literature conversations. *Children's Literature in Education,* 19(1): 3–16.
Lankshear, C. (1991). Getting it right is hard: Redressing the politics of literacy in the 1990s. In P. Cormack (ed), *Selected Papers presented to the Australian Reading Association 16th National Conference.* Adelaide: Australian Reading Association. 209–228.
Lankshear, C. (1992). Critical literacy and active citizenship. Paper presented at the Working Conference on Critical Literacy, Brisbane, 29 June 3 July.
Lankshear, C. and McLaren, P. (eds.) (1993). *Critical Literacy: Politics, Praxis and the Postmodern,* Albany: State University of New York Press.
Lehr, S. (1988). The child's developing sense of theme as a response to literature. *Reading Research Quarterly,* 23(3): 337–57.
Luke, A. (1992). The social construction of literacy in the primary school. Paper presented at the Working Conference on Critical Literacy, Brisbane, 29 June 3 July.
Luke, A. (1993). Stories of social regulation: The micropolitics of classroom narrative. In B. Green (ed.) *The Insistence of the Letter: Literacy and Curriculum Theorizing.* London: Falmer Press. 137–153.

Luke, A. and Baker, C. (1991). Towards a critical sociology of reading pedagogy: An introduction. In C. Baker and A. Luke (eds), *Towards a Critical Sociology of Reading Pedagogy: Papers of the XII World Congress on Reading*. Amsterdam: John Benjamins Publishing Co. xi-xxi.

Luke, A. and Walton, C. (1994). Teaching and assessing critical reading. In T. Husen and T. Postlethwaithe (eds), *International Encyclopedia of Education* (2nd edn). London: Pergamon Press. 1194–1198.

McHoul, A. (1991). Readings. In C. Baker and A. Luke (eds), *Towards a Critical Sociology of Reading Pedagogy: Papers of the XII World Congress on Reading*. Amsterdam: John Benjamins Publishing Co.191–210.

Machet, M. (1992). The effects of sociocultural values on adolescents' response to literature, *Journal of Reading*, 35(5): 356–362.

McLaren, P. (1986). *Schooling as a Ritual Performance*. London: Routledge and Kegan Paul.

Mey, J. L. (1991). Literacy: A social skill. In C. Baker, C. and A. Luke (eds), *Towards a Critical Sociology of Reading Pedagogy: Papers of the XII World Congress on Reading*. Amsterdam: John Benjamins Publishing Co. 83–101.

Morgan, W. (1992). *A Post-Structuralist English Classroom: The Example of Ned Kelly*. Carlton, Victoria: Victorian Association for the Teaching of English.

Morrow, L. (1988). Young children's responses to one-to-one story readings in school settings. *Reading Research Quarterly*, 23(1): 89–107.

Newman, J. (1991). Learning to teach by uncovering our assumptions. In S. de Castell, A. Luke and C. Luke (eds) *Language, Authority and Criticism.* London: Falmer Press. 107–122.

Olson, D. (1991). On the language and authority of textbooks. In S. de Castell, A. Luke and C. Luke (eds), *Language, Authority and Criticism.* London: Falmer Press. 233–244.

Patterson, A. (1991). Power, authority and reader response. In P. Cormack (ed), *Literacies—Reading the Culture. Selected Papers presented to the Australian Reading Association 16th National Conference*. Adelaide: Australian Reading Association. 245–259.

Probst, R. (1986). Mom, Wolfgang, and me: Adolescent literature, critical theory, and the English classroom. *English Journal*, 75(6): 33–39.

Rosenblatt, L. (1969). Towards a transactional theory of reading. *Journal of Reading Behavior*, 1(1): 31–47.

Rosenblatt, L. (1982). The literary transaction: Evocation and response. *Theory into Practice*, 21(4): 268–277.

Saxby, M. and Winch, G. (1987). *Give Them Wings: The Experience of Children's Literature*. Melbourne: MacMillan.

Schön, D. (1987). *Educating the Reflective Practitioner*. San Francisco: Jossey-Bass.

Singh, M. (1988). Becoming socially critical: Literacy, knowledge and counter construction. *Australian Journal of Reading*, 11,(3):155–164.

Smith, N., Greenlaw, M. and Scott, C. (1987). Making the literate environment equitable. *Australian Journal of Reading*, 40(4): 400–407.

Stott, J. (1987). The spiralled sequence story curriculum: A structuralist approach to teaching fiction in the elementary grades. *Children's Literature in Education*, 18(3): 148–62.

Sullivan, J. (1987). Read aloud sessions: Tackling sensitive issues through literature. *The Reading Teacher*, 40(9): 874–878.

Taxel, J. (1987). Children's literature: A research proposal from the perspective of the sociology of school knowledge. In S. de Castell, A. Luke and C. Luke (eds), *Language, Authority and Criticism.* London: Falmer Press. 32–45.

Thomson, J. (1984). Wolfgang Iser's "The act of reading" and the teaching of literature. *English in Australia*, 70: 18–30.

Wells, G. (1991). Talk about text: Where literacy is learned and taught. In D. Booth and C. Thornley-Hall, C. (eds), *The Talk Curriculum*. Carlton, Victoria: Australian Reading Association. 45–88.

Wild, M. and Vivas, J. (1991). *Let the Celebrations Begin!* Norwood: Omnibus Books.

Willinsky, J. (1991). Postmodern literacy: A primer. *Interchange*, 22(4): 56–76.

Part Two

FIVE

Critical Literacy and Active Citizenship (1992/94/97)

Colin Lankshear and Michele Knobel

Biography of the Text

This chapter began in 1992 as an invited keynote address at the Working Conference on Critical Literacy, convened by Peter Freebody, at Griffith University in Brisbane, Australia. The original version (Lankshear 1992) was written in New Zealand against a backdrop of issues and critical orientations based in New Zealand writing of the time. It drew on an approach to critical literacy grounded in Paulo Freire's concept of pedagogy as praxis and Chris Searle's approach to situated pedagogy as described in his landmark books *Classrooms of Resistance* and *The World in a Classroom.* It described how a material event reported in a regional New Zealand town could be taken as a situated pretext for classroom pedagogy as community-based praxis mediated by classroom work. It was subsequently re-written in light of the emphasis within a number of Australian states at the time on critical readings and re-writings of texts. One incarnation was published by the Australian Curriculum Studies Association in 1994 as a professional development resource for teachers (Lankshear, 1994). The present version was published in 1997 as a chapter for the book that eventually emerged as the formal outcome of the 1992 conference.

Introduction

In this chapter we suggest some possible components of classroom reading and writing practice that are integral to the educational goal of promoting active citizenship, and that can reasonably be identified as components of a critical literacy. We begin with some introductory remarks about literacies as discursive constructions and explain why we address the development of classroom approaches to critical literacy in the context of a larger concern of education for active citizenship in the 1990s.

We then discuss active citizenship as contested discursive terrain by outlining an account of active citizenship advanced by an Australian Senate Standing Committee and drawing on competing views to mount a critique of that account. Proceeding from this critique, we identify some important demands imposed on citizens in new times (Hall, 1991) and develop an alternative account of active citizenship as an educational goal. The chapter concludes with our attempt to identify some components of a critical classroom literacy consonant with our view of the active citizen.

Understanding of literacy has expanded dramatically since the early 1980s with the emergence of the "new literacy studies" (Gee, 1996). Developments from a range of social theory perspectives have progressively chipped away at the virtual monopoly over educational research of text-based practices previously exercised by psychologists of one type or another. Freed from the stranglehold of positivist technicism, those working from a new literacy studies perspective have come to appreciate the radically plural and discursive character of literacy.

Literacies are many, not singular. Moreover, what all cases of literacy seem to have in common—such as a basis in a technology (e.g., print, alphabetic script), or a set of techniques or competencies, or some combination of these—is now widely seen as in many ways less significant than the *differences* among literacies. We now understand literacies as socially created constitutive elements of larger human practices—discourses—that humans construct around their myriad purposes and values. It is precisely these human purposes and values that give point to developing and working with certain kinds of texts in certain kinds of ways. Equally, these socially constructed ways of working with texts contribute, dialectically, to shaping the larger practices, processes, beliefs, attitudes, and so on, by and through which human purposes and values are "lived."

This is readily apparent from a historical standpoint. Rich historical data reveals how, for example, in late 18th-century England the opposing political values and purposes of the Corresponding Societies and the Church and King Clubs, respectively, gave rise to polarized political Discourses whose associated literacies were as different from one another as chalk and cheese (Lankshear, 1989; Willinsky, 1993). Historians of literacy, sociolinguists, and ethnographers from several disciplines have documented, individually and comparatively, many cases of social practices and their related constructions of reading and writing (see, e.g., Gee, 1990; Graff, 1979, 1981; Heath, 1983; Kress, 1985; Scribner & Cole, 1981; Street, 1984; Willinsky, 1989). They have, in other words, traced in close detail important aspects of the relationship between what Gee (1991, 1992–93,1996) calls *Discourses* (with a capital D) and *discourses.*

By Discourses we mean socially constructed and recognized ways of being in the world, which integrate and regulate ways of acting, thinking, feeling, using language, believing, and valuing. By participating in Discourses we take up social roles and positions that other human beings can identify as meaningful (cf. Gee, 1996), and on the basis of which personal identities are constituted. It is in and through Discourse that *biological* human beings are constituted as ("identitied") *social* human beings.

This is not to imply that the "rules" governing a Discourse and membership "within" it are necessarily precise, still less to imply that they are settled or immutable. What constitutes being in or out of a particular Discourse is often highly contested. This, as we show later, is true of active citizenship as Discourse. Discourses are dynamic and alive. Living in and through them is very much a process of constantly renegotiating them. Discourses are profoundly dialectical. At the same time, discrete Discourses are identifiable as such, together with the sorts of identities they make possible. Discourses, then, are of widely varying types and scope. In his list of typical Discourses, Gee (1991) included:

> (enacting) being an American or a Russian, being a man or a woman, being a member of a certain socio-economic class/being a factory worker or a boardroom executive, being a doctor or a hospital patient, being a teacher, an administrator, or a student, being a member of a sewing circle, a club, a street gang, a lunchtime social gathering, or a regular at a local watering hole. (p. 4)

To these we add "being an (active) citizen."

discourses (with a small d) are the language—or the "saying/writing/listening/reading/viewing"—components of Discourses. Discourse is "always more than just language" or "discourse"; but "discourse" is always and necessarily present in Discourse (Gee, 1990, p. 142). Although Gee himself advanced a different, and much more elaborate view, we define *literacy* for our purposes here as those "language bits" (in Discourses) that involve text. Literacies, then, are the "textual components" of discourses: that is, texts and textual practices.

In their relationship, Discourse and discourse are mutually constitutive. As a locus of meaning-making a Discourse both shapes and is shaped by its discourse. This is to recognize that making meaning requires purposes and means as well as a medium. The important points for us here are that we can conceive of literacies only in conjunction with cogenerative integrated Discourses. This helps explain why we are grounding our attempt to construct elements of a critical literacy in the larger discursive context of educating for active citizenship. Practicing active citizenship, we argue, presupposes an orientation toward texts and certain capacities for meaning-making that are increasingly being identified with conceptions and practices of critical literacy.

There is more, however. Once we understand Discourses and literacies for what they are and in their relationship to each other, we can self-consciously engage in framing and constructing (new) Discoursal visions—which, as we have seen, incorporate literacies, in the same way as it has long been argued people can enter consciously into making and transforming history and culture once they understand history and culture for what they are.

The dialectic noted earlier applies here. Teaching a critical social literacy is part of what is involved in educating students for active citizenship for an anticipated multicultural republican Australia (Kalantzis, 1992/1993). Our conception of this critical social literacy is partly shaped by our vision of the active citizen. At the same time, however, our evolving vision of active citizenship is itself partly constituted out of attitudes, values, competencies, and so on, that we already recognize as aspects of "reading and writing the social" from a critical perspective. What follows is an attempt to "play out" this dialectic imaginatively on the surface texture of this text.

Active Citizenship: Contested Discursive Terrain

Right now citizenship is a high profile educational concern in the United States and Britain, as well as in Australia. Following the release of the *Report of the Senate Standing Committee on Employment, Education and Training* (Department of Employment, Education and Training, 1989. See also *Active Citizenship Revisited,* Department of Employment, Education and Training, 1991), the Australian Education Council ratified a citizenship education goal among its 10 *Common and Agreed National Goals for Schooling in Australia:* namely, to "develop knowledge, skills, attitudes and values which will enable students to participate as active and informed citizens in our democratic Australian society within an international context." All state and territory education systems in Australia have responded at some level to this goal.

Broadly speaking, citizenship marks a domain of Discourse at the intersection of state (or "political society") and civil society. More specifically, however, we can identify multiple extant competing discursive constructions of citizenship, within and between societies, as well as attempts to reframe new Discourses of citizenship that go beyond those currently in evidence.

Something of the contested nature of active citizenship as Discourse is revealed in the debate surrounding the release of the Senate Committee's 1989 report, *Active Citizenship,* and subsequent developments in Australia associated largely with the prospect of that country becoming a republic. As Trinca (1993) noted, at the level of discourse at least, monarchies (and their colonies) are about constituting subjects: the discourse of republics espouses constituting citizens. According to the Senate Committee, education for active citizenship must ensure students (a) understand how government works (at the commonwealth, state and local levels), (b) appreciate the role of community groups and nongovernment organizations, and (c) be motivated to be active citizens. Such an education was seen by the Committee as necessary to counteract a number of growing concerns, including the following:

1. Media ownership and presentation trends are restricting the range of information and viewpoints represented on matters pertaining to government, and emphasizing "personalities and image" over "issues" (Department of Employment, Education and Training, 1989: 8).

2. Many migrants have only experienced political structures and traditions quite different from Australia's. Not all "necessarily share our democratic traditions." Some may actually "fear the political process" (ibid.: 8).

3. Many young people feel alienated from community affairs, lacking skills and interest to participate in formal community processes and decision making. Many young people, failing to see "any direct link between . . . government. . . and their own lives," do not register to vote (ibid.: 9, 14).

4. Unacceptably high levels of "political ignorance," especially among young Australians, have been proclaimed, indicating "indifference and apathy towards political dimensions of experience," and an attitude of leaving it to "other people" to take care of governance. Where "accountability is weak" and power concentrates "in the hands of a few," the "quality of democracy [comes] under threat" (ibid.: 14).

In the view of the Senate Committee, for citizenship education to be effective in helping counteract such concerns certain things would have to be observed. According to the report, "active citizenship" means more than mere knowledge of political-governmental and community affairs. Education for active citizenship must not revert to "old-style 'civics' courses," merely force-feeding facts about the political system. Rather, active citizenship is about believing in the concept of a democratic society and being "willing and able to translate that belief into action." In this context, the report stresses upholding and participating in formal democratic processes, noting that for migrants this begins with taking up Australian citizenship. Beyond knowledge of (formal) political structures and processes at government and community levels, "being active" as a citizen calls for relevant skills and attitudes including, notably, respect for rights and duties that have evolved constitutionally and institutionally throughout Australia's colonial history.

The report maintains that sufficient curriculum policy already exists to underwrite the syllabus and pedagogical requirements of active citizenship education. Rather, the need is to reduce the gap between policy rhetoric and action, and to extend citizenship education beyond books alone and out into civil society.

Teacher education must prepare teachers to educate for citizenship (Department of Employment, Education and Training, 1989: Ch. 3). Appropriate curricular resources must be developed (Ch. 4, especially p. 51). Curriculum and resource developers, teacher educators, and teachers themselves, must identify and build on available exemplars of pedagogy and resources for active citizenship (Ch. 4).

Unfortunately, the report glosses complex and contested ideals and, consequently, masks the extent to which concepts like *democracy, politics, participation, public versus private sphere,* and *being active* denote what adds up to hotly contested discursive terrain. Three focal points of competition and dispute are sufficient to make the point.

- The notion of active commitment to democracy is vague and open to quite different discursive renderings. Respondents to the 1989 report distinguished, significantly, between "protectionist" and "participatory" approaches to democracy, as well as between a "party politics and representative democracy" ethos on the one hand, and a "personal and community politics" ethos on the other. John Fien and Wayne Kippin made it clear in their individual submissions to the original report that without careful attention to concepts and definition "it is impossible to sustain a consistent analysis" of the issues involved in education for active citizenship, or to develop "a coherent solution to the perceived problems of school curricula and teacher education" that arise in this area (Department of Employment, Education and Training, 1991, pp. 4–5). Moreover, the Committee's explicit and practically exclusive focus on formal public practices of democracy undermines its avowed concern that students learn to "appreciate the role of community groups and non-government organisations." Despite scattered references in the subsequent 1991 report to wider "activities and values which constitute the culture of a community," and through which "a community seeks both to preserve and transform itself" (p. 6), the Committee's overwhelming emphasis is on developing familiarity with "institutionalized aspects of parliamentary democracy." Certainly, this is the ground on which the Committee speaks most clearly and convincingly. The level of conceptual development of "community" and related concepts is very low in both reports, and illustrative examples of the possible role of community and nongovernmental organizations in citizenship education are impoverished.
- Both *Active Citizenship* and *Active Citizenship Revisited* have been charged with advancing an essentially conservative, unduly narrow, and conceptually underdeveloped position on the relationship between the "political" and the "personal" and the relative significance of the "public" and "private" spheres with regard to active citizenship (Department of Employment, Education and Training, 1991).

Some opponents of the Senate Committee's framing of active citizenship advocate "a holistic view" that is not confined to "the arena of civic or public concerns" but stresses individual and private as well as collaborative and public dimensions of being an active citizen. The wider view acknowledges elements of "the personal" as having political significance, and posits dialectical links between power, social arrangements, and personal life.

> Politics . . . is not simply a matter of who occupies the Lodge or what issues are attracting public lobbying activity, but [also of] who (for example) decides and who accepts responsibility for the household chores and why one particular type of household "agreement" on these matters is common. Personal life is undoubtedly political and any attempt to deny this must be seen as a political act in itself. (Department of Employment, Education and Training, 1991: 6)

From this standpoint, struggles within the private sphere to win a more equitable distribution of domestic work and decision-making power inside the family, and struggle by migrants to negotiate a viable and satisfying identity within their new life situation, become facets of actively constructing and practicing citizenship.

Against this, the Committee digs in and asserts the grounding of citizenship in the public sphere. "To be a citizen is to participate in the public practices which sustain, *and to a large extent define,* a community" (ibid.: 6, italics added). The reports explicitly recognize the formal processes of parliamentary democracy: those "concrete social and historical forms through which . . . participation and transformation are currently effected" (ibid.: 7). Yet matters that are emerging ever more obviously as major issues of citizenship in New Times—like the morally and politically significant dimensions of constructing and negotiating identity and loyalties across borders (Kalantzis, 1992/1993)—go unrecognized.

- In "Citizenship Education after the Monarchy," Mary Kalantzis (1992/1993) identified pressing citizenship issues emerging at this juncture in Australia's national history that are quite absent in the Senate Committee's ruminations. These relate to constructing citizenship around a (shifting) dialectic between national identity and the identities of persons and groups. The current preoccupation among social theorists worldwide with discourses of "the Post"—the postmodern, postcolonial, postindustrial, poststructural, and so on—reflects a growing awareness that we have reached a historical conjuncture: a crossroads between "past" ways of doing things and opportunities to negotiate new discursive orders across a wide spectrum of human experience (Slattery, 1993). For all the fuzziness that remains around the edges, it is widely accepted that we have, in a significant sense, entered New Times.

Berman (1982) and Hall (1991) captured key features of the age that speak directly to citizenship. Hall recognized global trends toward "greater fragmentation and pluralism, the weakening of older collective solidarities and block identities and the emergence of new identities associated with greater work flexibility, the maximization of individual choices through personal consumption" (Hall, 1991: 58). Berman documented ways in which and the extent to which "modem environments and experiences cut across all boundaries of geography and ethnicity, of class and nationality, of religion and ideology." These boundaries are not so much destroyed as weakened and undermined. The overall effect is an erosion of, in Hall's (1991: 58) words, "lines of continuity which hitherto stabilised our social identities."

Kalantzis' work on citizenship in Australia resonates with these themes. She identifies two factors bearing on citizenship in Australia that demand a new Discourse. These are the extent of *local diversity* in Australia, brought about largely, although not entirely, by mass immigration and the fact that Australians are having to live with an "increasingly proximate global diversity as our economic, cultural and civic associations increasingly and more comfortably cross national borders" (1992/1993: 29). Her call for an approach to citizenship based on rights and responsibilities located in geographic space rather than on a sense of national loyalty predicated on assumptions of singularity, highlights the importance of constructing and negotiating identities and loyalties across borders.

Clearly, active citizenship is assuming a new urgency but at the same time is well and truly due for a major discursive overhaul. In its attempt to frame the Discourse of active citizenship, however, the Senate Committee has simply failed to comprehend the scope and depth, and to grasp the mood of New Times. The challenge issued by the Committee must be taken up, but in much wider discursive terms than it envisages, and beginning from a more adequate purview of the life of our age. This calls for building a Discourse of active citizenship in which developing and teaching a critical literacy has a central place.

Citizenship Discourse in New Times

Obviously there is much more to be said about New Times than is possible here. We intend only to identify three important contemporary trends that impact directly on citizenship and to trace some of their discursive implications. These are:

- social changes calling for enhanced understanding of institutions in their relationship to the lives of individuals and groups and for revitalized civic participation in institutional life;
- trends in communications media that create impediments to making meanings (from text) that promote the common good;
- increased proximity to and intensified experience of diversity, calling for new approaches to negotiating identities and loyalties and to learning to live productively and harmoniously with difference (Kalantzis, 1992/1993).

Social Change and Institutional Life

Social changes during recent decades have heralded a decline in numerous institutions, wrought change in others, and thrown the validity of still others into doubt.

Many older established means of "hooking people up" socially have been eroded, often without new means emerging to take their place.

For example, the near complete entry of women into the paid workforce in many "developed" societies has had an important impact on community life and prior institutional arrangements. To date, little institutional adjustment has occurred to ensure such things as (a) provision for the unpaid work of child care, (b) care of the elderly, (c) the maintenance of neighborhood ties, and (d) the elaborate system of voluntary community activity previously undertaken by women. Much of what continues to be done in these areas is still carried out by women. Men's work hours have not decreased. There has been little growth in compensatory community services. Not surprisingly, then, the women who continue such work (and their families) endure considerable stress. Alternatively, "the [important] community-related tasks are often abandoned" (cf. Levett and Lankshear, 1992: 45).

Social theorists working from quite different perspectives advance sobering analyses of civic life in crisis. Drawing on work by Lyotard (1984), Baudrillard (1981), and on related theoretical developments that stress the centrality of the communications revolution to the "postmodern condition," Hinkson (1991) observed the extent to which in postmodern times institutions and social groupings associated with modernity are experienced as "anachronisms," and even as "forms of dogmatic imposition." Two related aspects stand out from this so far as the institutional life of citizens is concerned.

First, the "logic of information" and image demands the absence of structural or systemic constraints ("noise") to the free flow of messages and image signifiers. Second, computer and hi-tech media communications generate and constitute "social relations which are essentially *networks of temporary or fleeting interchange*" (Hinkson, 1991: 9). One way of demarcating the social settings of modernity from their postmodern counterparts is in terms of the relative significance and foregrounding of network (social) relations. Hinkson (1991: 9–10) made the point well:

> [T]he network in modernity as a structured form of relation was significantly offset by relations which were structured in other ways—relying on forms of social bonding which held together self and other through inter-generational norms, for example. What happens in postmodern settings is the emergence of a social space which holds out the possibility—by virtue of the structuring possibilities of information and the flow of images—of reconstituting social life around the network to such a degree that the social group is experienced as an anachronism (and a form of dogmatic imposition). It is as though the network relation which was always an aspect of the more comprehensive group is filleted off . . . and under the influence of the power of the new technologies moves into the foreground as a means of structuring a society which works on principles quite different from those of other known social forms.

The implication of this logic for some important familiar institutions of modernity is obvious. They become discredited as "out of time"; as so much "noise"; as impediments to postmodern individuality, autonomy, and mobility. Add to this the construction of postmodern selves through engagement with disembodied, image-mediated Others, in accordance with intense pressure to define one's *self* over and *against* Others through individualized consumption, and the threat posed to institutional life as we have known it becomes even more apparent (Kellner, 1991).

From a rather different angle, Bellah, Madsen, Sullivan, Swidler, and Tipton (1991) reported from the United States a widespread loss of faith in institutions, and emphasized the implications of this for citizenship. They argued that modern societies have been overly dependent on technical means for solving problems. We need to balance technique with a deeper understanding of "moral ecology." This means understanding how much we live our lives within institutions, recognizing that enhancing our lives presupposes better institutions, and "surveying our present institutions . . . to discern what is healthy in them and what needs to be altered" (p. 5). The challenge of New Times, then, is to reform existing institutions, and build new ones where necessary, in response to changed and changing demands. A good society is "an open quest" calling for the active participation of all citizens, because "the common good is pursuit of the good in common." The age calls for nothing less than "a new experiment in democracy, a newly extended and enhanced set of democratic institutions, within which we citizens can better discern what we really want and what we ought to want to sustain a good life on this planet for ourselves and the generations to come" (p. 9).

This calls for resisting current trends toward ideologies and Discourses of robust individualism and opening ourselves up instead to an invigorating and fulfilling sense of social responsibility. Beyond merely calling for people to get involved—reminiscent, perhaps, of the Senate Committee's pleas to match our rhetoric of citizenship education with practice—it will be necessary to *create and maintain institutions* that enable such participation, encourage it, and make it fulfilling as well as demanding. For Bellah and colleagues (1991), the question facing citizenship education is "how to educate ourselves as citizens so that we really can 'make a difference' in the institutions that have such an impact on our lives" (p. 19).

Media, Meanings, and the Common Good

The Senate Committee's concerns about media trends in Australia resonate with growing interest among educationists in promoting critical media literacy as an integral aspect of education for active citizenship. As has often been noted, theorists ranging from Fredric Jameson to Jean-François Lyotard identify "the dominance of image, appearance [and] surface effect over depth; . . . the blurring of image and reality . . . [and] the erasure of a strong sense of history" as key features of the postmodern condition (Hall, 1991: 60; cf. Harvey, 1989). Communications media are strongly implicated—as simultaneously cause and effect—in these patterns of dominance. Not surprisingly, critical media literacy is being heralded as a counterforce to this postmodern logic that, like the growing emphases on network relations, presents serious challenges to maintaining and reconstructing Discourses of active and informed political and civic involvement.

Kellner (1991) commented on the rise of the image to dominance by reference to Postman's (1985) argument that early this century Western society began to abandon print culture and enter an "Age of Entertainment" grounded in a culture of the image. This shift heralded a "dramatic decline in literacy [and] a loss of the skills associated with rational argumentation, linear and analytical thought, and critical and public discourse—resulting in 'degeneration of public discourse and a loss of rationality in public life' " (Postman, 1985: 64; cf. Enzensberger, 1992). Kellner advocated classroom activity aimed at expanding "literacy and cognitive competencies" in ways that enable us to counteract the subject constitutive effects of the deluge of media messages, images, and spectacles we encounter daily. He identified elements in hermeneutics and postmodern theory that provide relevant insights into how we are constituted socially, and techniques by which to deconstruct texts and images. From these, he argued, we can and should move to construct and teach a critical media literacy "which will empower individuals to become autonomous agents, able to emancipate themselves from contemporary forms of domination and able to become more active citizens, eager and competent to engage in forms of social transformation" (Kellner, 1991: 64).

Identities and Loyalties in Contexts of Diversity

The need in New Times to rethink citizenship radically—that is, from the roots—is nowhere more apparent than with regard to the themes of diversity, the new complexity of personal and group identities, and the need to constitute a citizenry accepting of and comfortable with difference. Economic, telecommunications, travel, migration, and political trends during recent decades have coalesced to produce situations in many societies worldwide, including Australia, where citizens are having to live with "increasingly proximate global diversity" and heightened local diversity (Kalantzis, 1992/1993: 29). This, as Kalantzis made clear, poses three specific challenges having profound educational implications. First, societies like Australia must "leave [their] boundaries open to the negotiation of global diversity" as their associations, economic and otherwise, move inevitably across national borders. This requires on the part of citizens acceptance of and ease with ethnocultural diversity.

Second, given the complex multicultural character of modern populations, many citizens are likely to develop and experience "multiple senses of loyalty and affinity." In New Times the state must be able to recognize and accommodate multiple layers of loyalty and affinity in ways that extend to "new" citizens the full enjoyment of rights and execution of their duties as citizens. Against the backdrop of war between rival ethnic factions in places like Bosnia, Kalantzis suggested that reconstituting civic life around such an accommodation may be the best route to stability and the surest safeguard against "internecine ethnic craziness." Indeed, having many links "to diverse geographic spaces and ethnic roots" can enhance stability, "diffusing and defusing past animosities," as citizens "negotiate and live with difference" (Kalantzis, 1992/1993: 29).

Third, for these conditions to be met, citizens in New Times require particular understandings and capabilities. They must understand processes by which they have been formed morally and politically so that they can "disentangle the multiple layers of their identities and political loyalties" (Kalantzis, 1992/1993: 31). They must also be able to negotiate across boundaries of ethnicity, gender, and countries. In short they need a more elaborate, New Times equivalent of what Mills (1959)—in earlier times—called a *sociological imagination.*

Elements of Education for Active Citizenship in New Times

On the basis of the ideas just sketched, we see a need to reframe citizenship around active commitment to understanding and evaluating social institutions and practices (i.e., Discourses), in both the public and private spheres, in relation to the requirements for promoting the common good in New Times. Beyond this, active citizenship involves working to modify, build, and maintain institutions that are conducive to pursuing the common good. Moreover, as we have seen, the common good has become the good of an increasingly diverse citizenry. So far as pursuit of the common good is "the pursuit of the good in common" the discursive processes, structures, relations, and values—the *institutions*—in and through which this good is pursued must accommodate such diverse cultural, ethical, and political traditions as meet the legitimate demands of an ethical human order.

In what remains, we focus on two key implications of all this for citizenship education.

1. Education for active citizenship calls for fostering a sophisticated "sociological imagination" that incorporates (what we call) institutional imagination, political imagination, cultural imagination, and moral imagination.

2. Practicing active citizenship presupposes being competent in handling media texts of varying types (e.g., news reports/public notices, advertisements, and commentaries) in a range of ways (e.g., analysis, evaluation, synthesis, criticism). Of course, active citizens need to be competent with other sorts of texts as well:

such as policy documents, political party manifestos, voting papers, bureaucratic forms, funding applications, bylaws, and many others. We are simply highlighting media texts here.

We agree with Kalantzis that teaching a critical social literacy is integral to any such education for active citizenship. And with Kellner we recognize the necessity of a critical media literacy, as an essential dimension of critical social literacy. To put things in terms of the conceptual frame we have been working with here, we are calling for a reconstituted Discourse of active citizenship, and for classroom education to support other sites within which future adults are apprenticed to this Discourse. Critical social (and media) literacy is a key element of the discourse integral to enacting the Discourse of being an active citizen. Becoming critically literate is integral to being apprenticed to a New Times Discourse of active citizenship.

Yet, both the new Discourse of active citizenship and the critical social literacy element of the discourse of active citizens are currently underformed. They are very much still-to-be-constructed. We have clues to follow in constructing them, however. And we know that they are necessarily constructed together and dialectically. Consequently, in order to conceive potential components of a critical social literacy that simultaneously informs and is informed by the practice of active citizenship, we need to grasp the nature and requirements of a sociological imagination for New Times. Because our notion of sociological imagination is derived directly from Mills' work in earlier times, it is useful to begin from Mills' original conception. Mills (1959) addressed the recurring phenomenon of people encountering troubles in their daily lives and feeling immersed in processes they do not really understand and sensing that these things are beyond their control. The more information people have of their world, said Mills, the more powerless they often feel, and the less they can sense the meaning of their times for their own lives and well-being and how they "fit into" the world of their times. This widespread experience, Mills argued, can be seen as a result of people lacking sociological imagination.

For Mills, sociological imagination involves understanding "biography" and "history" and being able to grasp their relatedness within social life. Biography is our own private experience of life: what life is, or is like, for us. History refers to social structures and processes and to changes that occur within them. The point is that biography and history are intimately, and intricately, connected, but we often have a poor understanding of how they are connected. Unemployed youth who cannot fathom their inability to obtain paid work despite having done courses and gone through work experience schemes are a case in point, as are those who explain such cases in terms of laziness, apathy, or lack of ability.

Being able to relate biography and history through sociological imagination enables us to make sense of our lived experiences and of how and where we fit into our times. It also makes possible an informed basis from which to enter history in an active way, by throwing weight behind movements, causes, actions, and institutions with a view to enhancing one's own and other people's prospects.

These days, of course, we talk more in terms of subject identities being constituted and human subjectivities constructed, in and through Discourse, than in terms of a dialectic between the biography of more or less unified subjects and the history that is created in and through more or less impersonal structures. Even allowing for Mills' undoubted sophistication among sociologists of his day. Hall's (1991) observation that, in New Times, "we cannot settle for a language. . . which respects the old distinction between objective and subjective dimensions of change" (p. 59), applies with some force to talk of sociological imagination. Accordingly, we recommend a modified view of sociological imagination tailored to take account of insights opened up by Hall and Kalantzis and others in relation to demands for civic participation along the lines advanced by Bellah and colleagues (1991). For New Times, sociological imagination should be seen to comprise at least institutional imagination, political imagination, moral imagination, and cultural imagination.

Through institutional imagination we understand the extent to which humans are what we are as a consequence of participating in Discourses within primary and secondary institutions (Gee, 1991); and relate our own identities and subjectivities to our distinctive discursive histories within the institutions we have lived in and through; and can envisage alternative ways of (personal and collective) being in the world potentially available to us through participation in different Discourses grounded in different institutions, *including Discourses and institutions as yet inchoate or incipient.* Institutional imagination, then, presupposes what Gee called "meta-level knowledge" of institutions and their Discourses and associated discourses (Gee, 1990, 1991). Meta-level knowledge permits analysis and critique of Discourses and their institutional settings, providing a basis from which either to participate in them more fully and satisfyingly or to seek to transform them as necessary for promoting the common good. If citizens are to live productively with diversity, to recognize "the multiple layers of their identities and political loyalties" and those of other people, and to negotiate across boundaries, they need specifically "to know the processes of their moral and political becoming . . . and to . . . [understand their] location in the world as cultural" (Kalantzis, 1992/1993: 31).

Through political imagination we understand our own ways of political being, our political identities and subjectivities, and those of others by reference to the discursive orderings and representations of structured power that we and they have experienced. Through political imagination we can envisage alternative actual and potential ways of political being from those we enact. In some cases this may involve understanding how people may be constituted as apolitical or anti-political. This is the deeper point lying behind the Senate Committee's recognition that some migrants to Australia retreat from involvement in any kind of formal politics because, for example, they have endured political tyranny and have no trust of politics. At a more subtle level it calls for appreciating how the aestheticization of politics, or the cumulative effect of media emphasis on image and/or personality, have acted to depoliticize people. In other cases we relate, through political imagination, our own and others' political identities as socialists, liberals, feminists, gay activists, Greens, Democrats, Republicans, monarchists, and so on, to varying discursive histories. In a parallel way, through moral imagination we relate actual and possible beliefs, values, and behaviors pertaining to constructions of "the Good Life" and "human well-being" to actual and possible discursive histories.

Cultural imagination can refer to linking "cultural ways of being"—in the sense of cultural that is commonly linked to trappings of ethnicity, race, or even class—to their respective discursive genealogies. Alternatively, it can be given a more generic meaning. In this sense, to understand location in the world as being cultural is to recognize that people's identities and loyalties, beliefs and values, conceptions and practices are always historically shaped and contingent, rather than natural, transcendent, and necessary. Seen this way, cultural imagination informs us that we fit into the world on the same terms and in the same way as Others do: as beings constituted discursively, and having the same rights to recognition and correlative duties of recognition as Others.

We are suggesting, then, that a sociological imagination is integral to the way of life of an active citizen and that in New Times sociological imagination should be seen to comprise at least institutional, political, moral, and cultural aspects. These are interrelated dimensions of human subjectivity and identity: places from which we make and live meanings that are fundamental to the active quest of citizens in pursuit of the common good. But, how might we work with texts in classrooms to promote the kinds of imaginations integral to active citizenship?

Critical Literacy for Active Citizenship

The practical suggestions offered below are constrained by space as well as by our own finite conceptions. We hope to have marked out an area for fruitful development. What follows should be seen as an initial exploratory foray to be elaborated and improved by anyone who finds our ideas helpful. The following scheme has been confined to written texts, and even within this limited scope is at best indicative and suggestive. The activities should not be seen as a discrete unit of work, nor as a bag of tricks. They are merely typical examples of the kinds of activities possible with forethought and planning among a team of participating subject teachers who seek to "orient, enhance, and synthesize" (language) learning in a unit with a citizenship focus (Queensland Department of Education, 1991: 39). Although our particular examples apply to secondary school level, the general approach may be adapted to primary and tertiary levels as well.

Clearly, teachers need institutional, political, moral, and cultural imagination themselves if they are to understand the sorts of activities we propose for what they are and to be able to make them work with students. The approach we advance here embodies the kinds of logics and learning processes that are the stuff of mastering sociological imagination in the sense we have outlined. As indicated, our scheme presupposes a context where teachers across several subjects (e.g., English, geography, modern history, social education, environmental studies) agree to work cooperatively and in an integrated way to explore aspects of citizenship through their disciplines. Our approach has two cornerstones: the English teacher and the teacher librarian.

The English teacher helps students "break texts open" by using available procedures for critical study of texts as described and modeled by Fairclough (1989, 1992), Kress (1985), and Gilbert (1993). In English, then, texts will be broken open in ways that require three kinds of resources to inform alternative "readings":

- information associated with other disciplines or subject areas;
- texts grounded in a range of theoretical-ideological perspectives;
- texts that cross historico-cultural time and space.

The English teacher, then, creates a need for exploration of issues traversing a range of subjects and from different standpoints. The teacher librarian, in collaboration with subject teachers, coordinates resources across the subject areas. He or she locates and assembles—for each subject area involved—texts conveying information that reflects different theoretical, ideological, and cultural perspectives pertaining to issues and themes that have been identified by breaking texts open linguistically and subsequently allocated to particular subject areas.

The starting place is an everyday text—a media story—that is rich in thematic possibilities for enhancing social imagination and understanding citizenship—as it has been constructed discursively, and how it might need reconstructing. Our text is a front page story from a major Australian daily (minus its accompanying photograph).

> The face of starving Africa
>
> AFRICA today. A starving child waits to die in the dust of Somalia. Hundreds of Somalis collapse every day, unwilling and unable to live any longer in the worst drought to grip their continent for 100 years.
>
> More than 30 million Africans, from Ethiopia to Mozambique, have left their villages in search of food and water.

> [The photograph accompanying this story] was taken by Care Australia's program officer, Ms Phoebe Fraser, the daughter of former prime minister Mr Malcolm Fraser, who is president of Care. "Somalia is desperate," she says. "In a country where hundreds of bodies line streets, most of them children, you can only hope the world is watching." Yesterday, the United Nations said it would send 500 armed soldiers to protect aid supplies to Somalia, which has been ravaged by drought and civil war, following agreement by warring factions to allow the safe delivery of aid.
>
> The Australian's readers have helped raise thousands of dollars to buy food and medicine by sending donations to the address of aid agencies alongside articles on Africa's heartbreaking story. Aid should be sent to: Care Australia, GPO Box 9977, in your capital city; World Vision, GPO Box 9944, in your capital city; Save the Children's Fund, GPO Box 9912, in your capital city; Austcare Africa Appeal, PO Locked Bag 15, Camperdown, NSW 2050; Community Aid Abroad, GPO Box 9920, in your capital city. *The Australian,* Friday, 14 August (1992: 1)

In the Beginning: Critical Language Analysis in the English Class

Drawing on work by linguists and literacy scholars, Luke (1992) suggests a range of techniques for undertaking critical language analysis across a range of text types, including media stories. These can be used productively with a text like "The Face of Starving Africa" to explore how it constructs reality textually and positions readers. A preliminary "opening up" of the text provides a basis for exploring its practical and ideological implications for citizenship and for enhancing sociological imagination through wider subject study that is integrated with more specialized language study in English.

Activity: Preliminary Critical Language Analysis of the Text

Luke noted that texts employ devices of various kinds to construct reality textually, and to position readers. For example, there are multiple and diverse ways in which the reality in Somalia might be represented or constructed textually. A given text, such as "The Face of Starving Africa," will represent it one way (a "possible world") rather than others. At the same time, the text will, by various means, position readers to make meaning from it in a particular way. It "positions readers in relation to a particular worldview or ideology" (Luke, 1992: 6)—which in this case has implications for citizenship.

Text Analysis Exercise

Read "The Face of Starving Africa" and explore the following questions.

1. What version of events/reality is foregrounded here?

2. Whose version is this? From whose perspective is it constructed?
3. What other (possible) versions are excluded?
4. Whose/what interests are served by this representation?
5. By what means—lexical, syntactic, etc.—does this text construct (its) reality?
6. How does this text position the reader? What assumptions about readers are reflected in the text? What beliefs, assumptions, expectations (ideological baggage) do readers have to entertain in order to make meaning from the text? (adapted from Luke, 1992; Fairclough, 1989).

Objective and Rationale

As we see things, an important part of the English teacher's job is to enable students to understand what these questions (and similar ones) are asking, and to acquire a relevant meta-language of linguistic and semantic concepts and tools by which to break the text open through such questions. Such understandings are best acquired in contexts that incorporate precisely the kinds of text and activities in question here. We should note, however, that much of what is needed in order to identify versions that have been excluded, the interests served by a particular representation, and so on, is best pursued in wider subject settings. Thus, the English class breaks the text open in ways that call for elaboration from other subject areas, from whence further inquiry and critique can proceed in the English class. At this point in the inquiry, the English class might only get as far as addressing the construction of reality and aspects of reader positioning.

A Possible (Sample) Response to the Text Analysis Exercise

A. Construction of reality

1. Those people starving in Somalia—and, indeed, the 30 million Africans from Ethiopia to Mozambique—are suffering from the effects of drought.
2. This drought is the worst in 100 years.
3. Somalia is also ravaged by civil war, which contributes to starvation.
4. The situation is so desperate the only hope is international sympathy and goodwill (aid).
5. Readers have contributed money to aid agencies.
6. Aid agencies address the situation in Somalia.
7. Six agencies are named, to whom donations should be sent.

Putting these together, we find a construction of the Somalia reality in terms of a extreme drought happening in a setting where there is also a war happening. In such circumstances, aid agencies coordinate relief. The role for ordinary people is to donate aid. The action is undertaken by the aid agencies, with assistance from some (special) history-making individuals like the Frasers. Aid agencies exist to respond to calamitous happenings like droughts.

B. Reader position

1. The reader is positioned to make meaning from an emotive rather than a rational or informed standpoint. The text conveys virtually no factual detail of consequence. (In terms of the matters we address later, it should be added that the page 7 stories, like most reporting in the mass media of the famine Somalia, construct essentially the same reality.) Wordings like "Africa's heartbreaking story," "desperate," "hope," "starving child," "collapse," and so on, set the tone.

2. Outside of an emotive basis for making meaning, the reader is presumed to respond to a propaganda device that we here term *testimonial.* A former prime minister and his daughter are invoked here as authoritative support for the particular construction of reality, as well as for constructing the reader-citizen as donor.

3. To make meaning, the reader is required to operate from the following sorts of assumptions and beliefs that, it will become apparent, are profoundly ideological: "(natural) disasters happen"; "disasters are addressed by money, converted to aid or relief;" "aid and relief agencies are the officially valid mediums of action;" "some causes are genuine (hence worthy of our generosity)"; "children are always innocent victims;" "we can extend generosity by donating—their circumstances are unrelated to us;" "extending charity freely is a virtue—a human act," and so on.

Some Implications of This Analysis for Other Subject Studies

Let us assume that something like the above response might emerge from a careful probing of the text given a reasonably typical range of (English) teacher and secondary student understandings. To go further, however, and envisage other possible versions of reality, ideological and interest serving aspects, and implications for citizenship, calls for a wider curriculum base. References in the original text to drought, civil war, the creation of a continent wide refugee population, and the operation of aid agencies, create openings for further exploration across several subject areas.

Activity: Subject Teachers with Teacher Librarian

1. Locate for each subject area a range of texts relevant to situations like that in Somalia *(viz.* disasters).
2. Ensure that among them the texts reflect different perspectives.
3. Identify and describe key differences in the perspectives—in terms of their underlying theories, the questions or issues with which they are most concerned, their key assumptions, whose standpoint they most reflect, where you would locate them on a continuum.

Four Possible Sample Texts

1. " . . . a whole variety of things [makes] people susceptible to drought. [A] factor that contributes to vulnerability is history. Many developing countries have inherited inappropriate economic structures and relatively weak positions in the global economy. In particular, colonization fostered the orientation of whole national economies towards exportation of one or a few products. . . . Such exclusive focus on one or two sources of export revenue increases vulnerability in many respects. Notably, it deprives the country of food crops for domestic consumption and leaves it at the mercy of international markets. Also, due to scarce resources, only enough staple crop is grown to satisfy immediate needs, leaving the population without a buffer of stored grain to carry it through periods of drought." (Castellino, 1992: 8–9)

2. "'The extensive plunder-culture [of plantations under colonization] meant not only the death of the forest but also, in the long run, of . . . fertility. With forests surrendered to the flames, erosion soon did its work on the defenceless soil and thousands of streams dried up.' This process is continuing in many parts of the world. The desert in West Africa is spreading. With the growth of refrigeration the number of food crops that can be exported for luxury consumption in the developed countries has increased. In Upper Volta peasants . . . organized themselves into unions to demand the right to grow crops for themselves rather than vegetables to export to France." (Hayter, 1982: 54; The quotation is from Galeano, 1973)

3. "The fundamental causes [of Africa's food crisis] are varied and complex. . . . The cumulative effects over many years have resulted in decrease in per capita food production, deforestation, desertification, increasing reliance on imported foodstuffs, and a mounting foreign debt. Such consequences are rooted, however, not in impersonal "market forces" or climactic conditions, but in political decisions involving the allocation of resources . . . [For example] when resources were allocated to infrastructure and extension services in the agrarian sector, the bias was toward revenue-generating export crops, rather than food production . . . African states were colonial creations with a fragile history of internal cohesion. That which existed has been rent by harsh economic realities. . . . These divisions have been played upon by outside interests for their own ends. In Angola the CIA-backed UNITA rebels fought a protracted civil war against the Marxist government. The South African government gave military support to RENAMO bandits in an effort to destabilise Mozambique. . . . The militarisation of Africa, frequently funded by loans under the guise of development aid is yet another drain on scarce resources. The ending of the Cold War threatens to release even greater quantities of surplus arms onto Third World markets." (Dorward, 1992: 5–10)

4. "[In Ethiopia] massive deforestation contributed to a decline in precipitation, which . . . with other factors has been responsible for the country losing [millions of] tons of topsoil a year.

Apart from adverse environmental factors, existing social institutions combined with inappropriate government policies were also instrumental in engendering famine. During Haile Selassie's reign, large segments of the agricultural sector were under the tutelage of the royal household, the military, feudal lords, or the church. Heavily taxed peasants lacked motives to innovate and increase production. . . . Ethiopian food production was among the world's lowest. . . . Following . . . the 1975 land reform program [of the new Marxist government] production began to fall markedly [and] the government arrogated to itself exclusive rights to market peasant produce, for which it paid little, on occasion less than cost. Coffee was so heavily taxed that . . . production halved. With steady rises in prices and limited availabilities of consumer goods, farmers scaled down food deliveries . . . the government . . . countered by requisitioning crops. Immediately prior to the 1984 famine, soldiers . . . confiscated surplus grain at gunpoint, and in so doing deprived peasants of the very reserves needed to ride out the drought." (Stein, 1988: 10)

Authorial Intrusion: What These Texts Suggest

We see such texts suggesting a number of things that relate back to the construction of reality in "The Face of Starving Africa" and help focus possible activities in other subject areas. They suggest, for example:

- Droughts are often not purely natural disasters.
- Droughts do not "act on their own" to cause famines and death by starvation.
- Famines are to an important extent created socially.
- Certain policies and institutions have contributed to causing starvation.
- Some people/countries benefit from activities that contribute to causing droughts, floods, famines, and so forth.
- People and countries across the entire world are implicated directly and indirectly in processes, relationships, and structures that contribute to famine.
- These same processes, structures, and relationships are integral to the social construction of refugees.

Activity: Subject Teachers Working as a Team

Around the texts located, develop complementary exercises in each subject area to reveal aspects that may have been left out or distorted in the original media story and initial (orienting) exercises. (Space limits us to considering just a few of the potential subject areas here and a limited range of orienting exercises.)

Geography Exercise 1

1. In which countries do famines mainly occur?
2. Which social groups do they affect most?
3. Are famines natural disasters?
4. Find texts that provide significantly different accounts of particular famines.

Environmental Studies Exercise 1

1. Identify ecological variables (e.g., precipitation, soil) that are important to food production.
2. In what ways have human activities impacted on ecological variables?
3. What consequences can this impact have on food production?
4. Are there different views about the nature and the extent of human impact on ecology? If so, how do they differ?

Modern History Exercise 1

1. Identify some countries that have been badly affected by famine.
2. In which cases was "political instability" a factor?
3. In those instances where it was a factor, what different accounts are provided of the causes of the "instability."
4. In the different accounts, who is seen to "be behind," or benefit from, the factors causing instability? How do these interests differ according to the accounts provided?

Back to English

In the light of work done around the exercises in other subject areas, students can return to questions posed at the outset and elaborations of them.

1. From whose perspective is "The Face of Starving Africa" presented?
2. What other possible (and actual) versions have been excluded or elided here?
3. Whose interests are served (and whose undermined) by the original representation?
4. Comment on the effect of the reader position evident in "The Face of Starving Africa." What other reader positions do you now see as possible for different texts on the same subject?
5. Analyze linguistically the role played by the references to "drought" and "civil war" in the original text. Note: For example, they might be investigated as possible instances of the syntactic device of *transformation* (Clark, 1992: 121; see also Fowler, Hodge, Kress and Trew, 1979); e.g., nominalizations or some other linguistic means of rendering content in a passive form (as described by Fairclough, 1989).

Where to Now?

At this point the ideological and practical implications for citizenship specifically can be explored. If there is a curriculum slot for social education, this might be the place to do it. Otherwise, the investigation could proceed in geography, history, or even in English. We cannot address these implications here in the detail they merit. Our thinking, however, is as follows, and we invite readers to consider what kind of activities and exercises might be appropriate. We envisage four main activities. These are related, and each comprises structured exercises or questions.

The first activity would explore Aid as Discourse (although it might be easier to construct the activity minus the jargon!). Within this, aid agencies could be investigated as a type of institution belonging to a particular *type of* Discourse. Structuring questions here might address such matters as:

1. What is the operating logic behind the aid approach to disasters, development, and so on (e.g., mopping up serves powerful interests better than preventing the spill)?
2. What roles do the various participants in the aid discourse play? How are they constructed? What are they constructed as?
3. What are they *not* constructed as? Insofar as aid activity is an example of activity (in the public sphere) intended to enhance the common good—albeit by merely ameliorating the worst harms—we are partially constituted as (global) citizens by the aid discourse. But as *active* citizens?

The second activity we envisage investigates local aid as Discourse, and its implications for citizenship construction on the home front. We might consider the extent to which local homelessness and hunger are symptoms of the same processes of power that elsewhere, and more dramatically, issue in famine. And is the approach to amelioration through appeals and charities simply a local extension of the larger logic we have observed? Who benefits from our construction as passive citizens within the local sphere? (Anyone with a markedly better than average income and standard of living?)

The third activity addresses the issue of whether the local and international aid discourses and their discursive construction of (passive) citizens are symptomatic of a larger Discoursal logic. To precis what we have in mind here, try this exercise.

Reader Exercise. Read the following public notice, which appeared in a New Zealand newspaper, and consider the extent to which it intimates a Discourse that channels power, maintains institutional models, and constructs citizens in ways that serve particular interests. Which interests might be thus served? And how? How does this (intimated) Discourse compare with the aid discourse? More generally, might representative parliamentary democracy provide a further parallel?; another insistence of the same "logic"? In what respects?

> OFFICE OF THE OMBUDSMAN
> PROBLEMS WITH BUREAUCRACY?
> TUESDAY, March 30, 1993, 11.30 am - 3.30 pm, REAP Centre, 61 Peel Street, WESTPORT. Ph. 789–7659.
> WEDNESDAY, March 31, 1993, 9 am - 5 pm. REAP Centre, 17 Sewell Street, HOKITIKA. Ph. 755–8700.
> THURSDAY, April 1, 1993, 8.30 am - 3.30 pm. Community Law Centre, Waterfront Building, Richmond Quay, GREYMONTH. Ph. 768–0584.
> The Ombudsmen may be able to help you if you are having problems with a government department, organisation or board, whether central or local. This includes local councils and boards, such as school boards of trustees, also state owned enterprises and some national boards as well as government departments. If you think that one of these have behaved unfairly or unreasonably towards you, whether by doing something or by failing to do something, or you are having difficulty in getting information from them and would like to discuss the problem, you are welcome to call on my offices at the times and places mentioned. An appointment is not essential, but appointments may be made by telephoning my officers, either on the days of their visit at the above telephone numbers, or in advance by telephoning my Christchurch Office (03) 366–8556. Collect calls will be accepted.
> Sir Brian Elwood
> OMBUDSMAN
> The (Christchurch) Press, 29 March 1993.

Finally, students need exposure through texts (as well as in lived practice, although that is beyond us here) to alternative Discourses of citizenship and to exemplars of varying constructions of the active citizen. Examples that range across time and space are helpful because they reveal the historical contingency of discursive practices and forms. At the same time, by revealing the concerns and practical issues that shape specific constructions of the active citizen, such examples remind us that for all the due emphasis being given at present to "multiplicity," "difference," "flux," and the like, humans share in the final analysis a remarkably common set of needs. It is from these that the quest for the common good as the quest for the good in common necessarily proceeds.

Interesting and challenging counterpoints to prevailing conceptions of citizenship within countries like our own can readily be found in descriptions of community development and defense organizations in a number of revolutionary—frequently single party—societies. As just one example of many that can be found, we offer *this fragment* from an account of a community development committee during the Nicaraguan Revolution. (Note: CDSs were development and defense committees.)

> Georgino Andrade in Managua is a community of factory workers, artisans, market and street vendors, domestics, and the unemployed. Located in the belt of neighborhoods, industries, and commercial centers that sprang up after the

1972 earthquake destroyed the city's center, it is a sprawling settlement of some six thousand people. Until 1981, the land on which the barrio is located was vacant. In May of that year a handful of families from a few of Managua's other overcrowded barrios decided to build homes there. News of the developing squatter's community spread by word of mouth, and within a month scores of other families joined the pioneers . . .

The residents of Georgino Andrade began to construct a neighborhood preschool in 1984. The Ministry of Education pledged sufficient funds to cover most of the cost, but the intensification of the contra war prevented the ministry from paying for the full costs. Consequently, the Sandinista Barrio Committee launched a drive to secure foreign donations. In March the Committee sent [an appeal] letter to sympathetic groups and individuals in the United States . . .

The residents of Georgino Andrade continued their efforts to develop the neighborhood throughout the next year. The government installed an electric transformer and lights. The CDSs conducted successful inoculation campaigns against polio, measles, and tetanus, and completed the third year of adult education classes. With material assistance from CEPAD, a Protestant development agency, the neighborhood committee sponsored a sewing class for more than 25 neighborhood women. The neighborhood's major achievement of the year took place in November, when [following community representations] the Ministry of Housing delivered land titles to 70% of the barrio's residents.

In January 1986, activists in Georgino Andrade's CDSs participated in a reevaluation of their organization. The reevaluation resulted in the following recommendations:

Neighborhood committees should try to involve everyone in the CDS of each block by encouraging residents to participate in the various CDS commissions.

CDS leaders should try to limit all meetings to 1 hour. Good records should be kept of all CDS meetings.

The CDSs must continue to act as an educational vehicle for barrio residents. 'Our principal task is to educate ourselves to express how we feel. To encourage people to participate in debates in order to further enrich ourselves.'

Georgino Andrade inaugurated its preschool in March 1986 and began to build a classroom for the neighborhood's first grade students. CDS leaders also planned the construction of a health center and a new communal house. (extracted from Ruchwarger, 1987: 180–186)

Postscript

We have left a great deal unsaid and untouched here. Our scope has been too narrow to address adequately Kalantzis' important concern that we understand the processes of our moral and political formation such that we can "disentangle the multiple layers of our identities and political loyalties" and be able to negotiate across the increasingly diverse boundaries we encounter daily. A fuller treatment must await another occasion. Our hope, however, is that the approach we have taken here, although far from complete, represents at least a small step in the right direction.

Finally, in no way do we want to suggest that donating aid is a mistake. At present, for many circumstances, aid is the only discursive mechanism immediately available, and participation is morally binding on those who have resources to share when called upon. The long-term point, however, is to work at building more active Discourses that can address causes rather than simply respond to symptoms and discourses that address existing configurations of power and privilege and "speak" them differently. This chapter has at most begun to explore what this might involve.

References

Baudrillard, J. (1981). *For a Critique of the Political Economy of the Sign*. St. Louis, MO: Telos Press.

Bellah, R. Madsen, R., Sullivan, W., Swidler, A. and Tipton, S. (1991). *The Good Society*. New York: Alfred Knopf.

Berman, M. (1982). *All That Is Solid Melts into Air: The Experience of Modernity*. New York: Simon & Schuster.

Castellino, R. (1992, September). Drought in Africa—again. *Red Cross Red Crescent*. 8–10.

Clark, R. (1992). Principles and practice of CLA in the classroom. In N. Fairclough (Ed.), *Critical Language Awareness*. London: Longman. 117–140.

Department of Employment, Education and Training / Senate Standing Committee on Employment, Education and Training (1989). *Active Citizenship*. Canberra: Author.

Department of Employment, Education and Training / Senate Standing Committee on Employment, Education and Training (1991). *Active Citizenship Revisited*. Canberra: Author.

Dorward, D. (1992, December). Famine: The economics and politics of the food crisis in Africa. *Current Affairs Bulletin*. 5–10.

Enzensberger, H. (1992, December 20). In defense of illiteracy. *The Australian*. 11.

Fairclough, N. (1989). *Language and Power*. London: Longman.

Fairclough, N. (Ed.). (1992). *Critical Language Awareness*. London: Longman.

Fowler, R., Lodge, B., Kress, G. and Trew, T. (eds). (1979). *Language and Control*. London: Routledge & Kegan Paul.

Galeano, E. (1973). *The Open Veins of Latin America*. New York: Monthly Review Press.

Gee, J. (1991) What is literacy? In C. Mitchell & K. Weiler (Eds.), *Rewriting Literacy: Culture and the Discourse of Other*. New York: Bergin and Garvey. 3–11.

Gee, J. (1992/1993). Tuning into forms of life. *Education Australia*, 19/20: 13–14.

Gee, J. (1996). *Social Linguistics and Literacies: Ideology in Discourses* (2nd ed). London: Taylor & Francis.

Gilbert, P. (1993). (Sub)versions: Using sexist language practices to explore critical literacy. *Australian Journal of Language and Literacy*, 16(4): 323–331.

Graff, H. (1979). *The Literacy Myth: Literacy and Social Structure in the Nineteenth Century City*. New York: Academic Press.

Graff, H. (ed.) (1981). *Literacy and Social Development in the West*. Cambridge, UK: Cambridge University Press.

Hall, S. (1991). Brave new world. *Socialist Review*, 21(1): 57–64.

Harvey, D. (1989). *The Condition of Postmodernity*. Oxford, UK: Basil Blackwell.

Hayter, T. (1982). *The Creation of World Poverty*. London: Pluto.

Heath, S. (1983). *Ways with Words: Language, Life and Work in Communities*. Cambridge, UK: Cambridge University Press.

Hinkson, J. (1991). *Postmodernity: State and Education.* Melbourne: Australia: Deakin University Press.

Kalantzis, M. (1992/1993). Citizenship education after the monarchy: Five questions for the future. *Education Australia,* 19/20: 28–31.

Kellner, D. (1991). Reading images critically: Toward a postmodern pedagogy. In H. Giroux (Ed.), *Postmodernism, Feminism and Cultural Politics: Redrawing Educational Boundaries.* Albany: State University of New York Press. 60–82.

Kress, G. (1985). *Linguistic Processes in Sociocultural Practice.* Geelong: Deakin University Press.

Lankshear, C. (1992). Critical Literacy and Active Citizenship. Invited keynote address. Working Conference on Critical Literacy, Griffith University, Brisbane, July.

Lankshear, C. (1994). *Critical Literacy.* Belconnen, ACT: Australian Curriculum Studies Association.

Lankshear, C. (1995). Afterword: Some reflections on empowerment. In P. McLaren & J. Giarelli (Eds.), *Critical Theory and Educational Research.* Albany: State University of New York Press. 301–310.

Lankshear, C. (1989). *Literacy, Schooling and Revolution.* London: Falmer Press.

Levett, A. and Lankshear, C. (1992). *Recent Global Social and Economic Trends Relevant to Education.* Wellington: New Zealand Qualifications Authority, mimeo.

Luke, A. (1992). Conference workshop materials. In S. Muspratt, A. Luke and P. Freebody (compilers), *Working conference on critical literacy.* Brisbane, Australia: Griffith University, Faculty of Education. 4–8.

Lyotard, J-F. (1984). *The Postmodern Condition: A Report on Knowledge.* Manchester, UK: Manchester University Press.

Mills, C. W. (1959). *The Sociological Imagination.* New York: Oxford University Press.

Postman, N. (1985). *Amusing Ourselves to Death.* New York: Viking.

Queensland Department of Education. (1991). *Draft Years 1 to 10 English Language Arts: Syllabus and guidelines.* Brisbane, Australia: Author.

Ruchwarger, G. (1987). *People in Power: Forging a Grassroots Democracy in Nicaragua.* South Hadley, MA: Bergin & Garvey.

Scribner, S. and Cole, M. (1981). *The Psychology of Literacy.* Cambridge, MA: Harvard University Press.

Slattery, L. (1993, May 15–16). 1 think therefore I think. *The Weekend Australian,* 20.

Stein, L. (1988). Famine in Ethiopia. *Quadrant,* 32(3): 8–13.

Street, B. (1984). *Literacy in Theory and Practice.* Cambridge, UK: Cambridge University Press.

Trinca, H. (1993, May 8–9). What is the alternative? *The Australian Magazine,* 20.

Willinsky, J. (1989). *The New Literacy: Redefining Reading and Writing in the Schools.* New York: Routledge.

Willinsky, J. (1993). Lessons from the literacy before schooling 1800–1850. In B. Green (Ed.), *The Insistence of the Letter: Literacy Studies and Curriculum Theorising.* London: Falmer Press. 58–74.

SIX

Literacy and Empowerment (1994/97)

Colin Lankshear

Biography of the Text

A preliminary version of this paper was published in the *New Zealand Journal of Educational Studies* in 1994 and was subsequently reworked as a chapter for *Changing Literacies* published in 1997. It builds on a remarkably useful analytic technique for elucidating elliptical statements involving concepts that are relational. I came across this technique in the course of investigating the concept of freedom. Within political and philosophical thought there was a long running debate about whether freedom is best understood as a negative concept—to be free is to be free from some constraint—or a positive concept—to be free is to be free to realize some desired purpose or ideal. The debate arose in part out of awareness that people can be free of various constraints or impediments but still not be able to realize what they seek. In the course of this debate, Gerald MacCallum (1967) and Joel Feinberg (1973) argued that freedom was a concept that involved multiple variables and that specific statements about freedom are only fully expressed when we know which values the people who make these statements are assigning to the variables involved in meaningful talk about freedom. They argued that any ascription or denial of freedom existing in particular cases involves a relationship between a subject—something that is either free or not free—and a constraint—which they either endure or do not endure—and an outcome or end—which they are either constrained from attaining or not constrained from attaining. This freedom could be seen as a relational concept that necessarily has both negative and positive dimensions and that rather than trying to establish once and for all what true freedom consists in—whether it is negative freedom or positive freedom—we are better advised to judge each specific ascription or denial of freedom on its merits, by establishing how the variables are understood in each case and judging where the evidence lies.

During the 1980s and 1990s talk of empowerment proliferated across many areas of daily life, and nowhere more so than in education. Claims about empowerment were often vague and were also often hotly contested. It seemed to me that empowerment was another case of a relational concept and that statements about empowerment were very often elliptical. So I attempted a similar exercise for empowerment to that which people like MacCallum and Feinberg had previously undertaken for freedom. This chapter presents the outcomes of that exercise.

Understanding "Empowerment"

Since the mid-1980s "empowerment" has become an educational buzz word *par excellence*. Unfortunately, it runs the risk at present of becoming trivialized and losing its semantic integrity and persuasive force as a result of unreflective overuse. The idea of empowerment often surfaces as a kind of "magic bullet" for fighting educational causes on behalf of disadvantaged groups. At times it seems to name a goal for educational programs and policies sought on the grounds of social justice, equity, and like ideals. Elsewhere it appears to serve more as a means or a principle for enabling learning to take place. But whether it is invoked as a goal or as a means—which itself is often unclear—"empowerment" is all too rarely given adequate conceptual or theoretical attention by those who set most store by it.

Empowerment talk often suffers from having too little meaning. This occurs whenever "empowerment" is used to "*name* the space where theoretical work is needed, rather than to *fill* that space" (Dale, 1991: 417, second emphasis added)—that is, where we simply assume the meaning and significance of empowerment to be self-evident. A common instance of this is where empowering pedagogy or empowering literacy are advanced as self-explanatory recipes for addressing educational and social needs of disadvantaged groups:

Q1: How should we assist disadvantaged learners?

A1: By empowering them.

Q2: How?

A2: By using an empowering pedagogy.

From a different angle, however, empowerment talk can be seen as having altogether too much meaning. It has connotations of positive value, which appeal to people across different and often incompatible ideological and normative positions. These connotations often substitute for substantive meaning. While all comers may agree that pedagogies and literacies should be empowering and that learners should be empowered, shared meaning and values here are typically more apparent than real. This is not a good basis from which to pursue educational advances.

There is a further problem. If people do not recognize and address the vague semantics and ambiguity of "empowerment," they may well fall for educational agendas they would otherwise have rejected on the grounds that they are in fact manipulative or oppressive. In the opening pages of *Literacy for Empowerment*, Concha Delgado-Gaitan (1990) describes how at first she resisted using empowerment as the central analytic and explanatory construct in her study of education with southern Californian migrant communities. This was because empowerment has "been used to mean the act of showing people how to work within a system from the perspective of people in power" (p. 2). Delgado-Gaitan had to address this problem to her own satisfaction before she was prepared to use "empowerment" as a frame for her work with the migrant communities.

This chapter suggests ways in which literacy work might benefit from addressing conceptual and theoretical matters surrounding "empowerment." I begin by suggesting a conceptual model of empowerment that adapts work done by Joel Feinberg in philosophy more than two decades ago on social concepts of freedom. There are some good grounds for believing that Feinberg's approach may be useful here.

First, it has long been recognized that within everyday usage "freedom" is prone to the semantic pitfalls already mentioned: sometimes being unduly vague and ambiguous and at other times connoting too much. Similarly, freedom has been used to name diverse and often incompatible educational and sociopolitical ideals and practices (MacCallum, 1967; Berlin, 1969). Second, with freedom as with empowerment, it is important to make clear the educational and ethical values held by those seeking to influence social practices by appeal to persuasive concepts; such values should be opened to scrutiny and debate. We need ways of opening up substantive debate about desirable pedagogies, literacies and the like, rather than closing them down by definitional fiat or by recourse to persuasive, trendy or emotive terminology. Finally, like freedom, empowerment is a relational concept referring to social arrangements and outcomes produced within discursive practices. This needs to be reflected in the way we approach and treat such ideals conceptually and theoretically.

Building on still earlier work by Maurice Cranston (1953) and Gerald MacCallum (1967), Joel Feinberg (1973) argues that statements about freedom are "elliptical," or abbreviated. When people appeal, for example, to economic freedom as a political or social ideal, we need to know whether they mean freedom of the economy to operate according to market forces untrammeled by government intervention or the freedom of individuals to meet their material needs by having access to adequate economic resources or something else again. Feinberg endorses

Cranston's claim that we should "call for the full version of all such abbreviated slogans" (Cranston, 1953: 12), otherwise we don't know what we are supporting, demanding or trying to bring about.

According to Feinberg's analysis, which is still as fruitful as any I know of for dealing with freedom as a social ideal, "freedom" is a relational concept involving three variables: a subject, a constraint, and an outcome. To provide the full version of statements about freedom we need to specify who or what is free (or unfree); from what they are free (or unfree); and what it is they are free (or unfree) to do, be, have, become, and so on. As Feinberg (1973: 11) puts it, we have to be able to fill in the blanks in the schema:

________ is free from ________ to do, be, have, etc. ___________.

Like "freedom," empowerment is a relational concept—although it involves at least four variables. For claims about empowerment to be clear they should spell out:

1. the *subject* of empowerment;
2. the power *structures* in relation to which, or in opposition to which, a person or group is being empowered;
3. the *processes* or *"qualities"* through or by which empowerment occurs; and
4. the sorts of *ends* or *outcomes* that can or do result from being thus empowered.

In other words, claims about empowerment should spell out the contents of A to D in the following schema:

> A (the subject) is empowered in respect of B (some aspect of the discursive structuring of power) by/through C (a process or quality) such that D (some valued ends or outcomes) may—and ideally will—result.

I present this model as a way of teasing out what I think is involved in claims about empowerment that are informed by any kind of adequate social theory. It is intended to encourage and assist educators to declare and clarify their ideas and values when they talk about empowerment—particularly in relation to literacy.

It is one thing to present a conceptual model and make claims for its usefulness in guiding educational practice; it is quite another, however, to demonstrate its usefulness by means of examples and creative applications. I will elaborate and explore this approach to empowerment by relating it to James Gee's work on literacy and Discourse. In particular, I will describe and adapt Gee's concepts of *dominant* literacies and *powerful* literacy to suggest various ways in which literacy might be said to empower.

Literacies, Dominant Literacies and Empowerment

In *Social Linguistics and Literacies: Ideology in Discourses* (1990), Gee distinguishes Discourses (with a capital "D") from discourses (p.142), and defines literacy by reference to what he calls secondary Discourses (p. 153). Building on a further distinction between dominant and non-dominant Discourses, he distinguishes dominant from non-dominant literacies (p. 153). He then identifies an ideal of *powerful* literacy, whereby we use a literacy as a "meta-language" to critique other Discourses and literacies and the way "they constitute us as persons and situate us in society" (p. 153).

For Gee, Discourses are "saying (writing)-doing-being-valuing-believing combinations:" "forms of life" or "ways of being in the world," which integrate such things as words, acts, attitudes, beliefs and identities, along with gestures, clothes, bodily expressions and positions, and so on. Through participation in Discourses individuals are identified or identifiable as members of socially meaningful groups or networks, and as players of meaningful social roles (Gee, 1990: 142–143). Language, of course, is integral to Discourse(s), but Discourse is always much more than language alone. Gee uses "discourse" (with a lower case "d") to refer to the "language bits" (Gee, 1990: 143) in Discourses: "connected stretches of language that make sense, like conversations, stories, reports, arguments, essays" (ibid.).

Outside of self-contained face-to-face communities, humans participate in both primary and secondary Discourses. We encounter and develop our primary Discourse through "face-to-face communication with intimates" (Gee, 1990: 7), or what sociologists call primary socialization. Primary Discourse is grounded in oral language, our *primary*—first—use of language. Through the process of enculturation among intimates we learn to "use language, behavior, values, and beliefs to give a . . . shape to [our] experience" (ibid.). Although each person encounters just one primary Discourse, primary Discourses and language uses vary across sociocultural groups distinguished by race, ethnicity, social class, and so on. Thus the particular shape given to experience within primary Discourse varies socioculturally.

We develop our secondary Discourses "in association with and by having access to and practice with . . . secondary institutions" beyond the family or primary socializing unit, e.g., schools, churches, workplaces, clubs, bureaucracies and professional associations (Gee, 1990: 8). Secondary institutions require us to communicate with non-intimates in ways and for purposes beyond those of our face-to-face world. Secondary uses of language are those developed and employed within our multiple secondary Discourses. Secondary uses of language include classroom talk, filling forms, interviewing, writing letters, inputting data, running stock inventories, writing policy, translating, and so on.

Gee defines being literate as having control of *secondary* language uses, that is, fluent mastery of language uses within secondary Discourses. This is a matter of using the "right" language in the "right" ways within particular discursive settings. Being literate is not a *singular* competency or attribute, however. Literacies are myriad. According to Gee, there are "as many applications of the term 'literacy' as there are secondary Discourses, which is many" (Gee, 1990: 8; see also Street, 1984).

Discourses, Gee argues, are intimately linked to "the distribution of social power and hierarchical structure in society" (Gee, 1990: 4–5). Having control over certain Discourses—and, hence, of their literacies—can result in greater acquisition of social goods (money, power, status) by those who enjoy this control; lacking control, or access to control, of these Discourses is a source of deprivation. Gee calls Discourses that are avenues to social goods "dominant Discourses," and identifies dominant *literacies* as secondary uses of language that can (help) provide privileged access to social goods (inside existing discursive arrangements) for those who have control of them. Subordinate literacies, by contrast, offer little or nothing in the way of privileged access to social goods.

In this context, the cases of 5S and 5M in Alison Jones's (1986, 1991) study of two different "ability streams" within a New Zealand secondary school are relevant. One was a low-to-middle stream composed almost entirely of students from Pacific Island migrant working-class families: 5M. The other was a top stream composed overwhelmingly of students from white middle-class (especially professional and business families: 5S. Jones observed closely these students in their classrooms during two years, the second being the year both classes sat their first national school examination (i.e., School Certificate). Within "prestige" subjects like English and science, exam success is largely a function of the ability and disposition to reason and argue in particular ways and to extrapolate from and interpret what is given in texts, as well as (to some extent) to absorb, recall and reproduce information, including that provided by teachers in class.

Both groups believed strongly that school success was the route to good life chances and that success in academic exams involves a combination of ability and hard work. Both groups wanted to succeed in School Certificate and both expressed commitment to working hard. Indeed, Jones found that both groups *did* work hard. What she also found, however, is that the two groups had very different views of the work to be done and that these corresponded to very different views of how to operate language within school-based learning.

Whereas the kinds of understandings, styles and dispositions required for exam success were more or less universal among the 5S students, the girls in 5M did not command the required style of thinking, the detached and analytic relationship to subject-matter, or the appropriate writing genre. Jones's data suggests that this is because the individual and collective discursive and experiential histories of the 5M students were far removed from the way academic examination writing was constituted as a dominant (school) literacy.

It seems that the enculturation of most 5M students steeped them in values of uncritical reverence for the teacher as authority and "font of knowledge." This underpinned their view of school literacy as being a matter largely of transcribing the teacher's words as notes and "learning them up" later. They did not seem to realize that claims advanced in texts or in classroom lectures should be checked and cross-checked for accuracy and meaning against (a range of) "authoritative" sources. Yet operating such checks is an essential aspect of the Discourse of taking academic exams, of being an academic or scholarly student. It is one essential move in the discursive process of effecting the "right" saying-writing-doing-believing-valuing, etc. combinations presupposed in being exam literate. As it happened, 5M students frequently copied information inaccurately. With regard to the function of hemoglobin, for instance, Ruth wrote that it "helps blood clots," when the teacher had said "it helps the blood to clot." 5M students seemed to lack a conception of such ideas as intrinsically complex and calling for accurate understanding—which they might lack. Inevitably, they "mislearned" such material, despite considerable "hard work"—including meticulous illustration and ornate print—as they understood it to be. They failed the exam, overwhelmingly.

While this might be described and explained in numerous ways, recent developments across a range of theoretical perspectives—including postmodern social theory and adaptations from poststructuralist theories in particular—lend support to attempts that focus on aspects of the mismatches between individuals' primary Discourses and dominant secondary Discourses and literacies, and/or between dominant secondary Discourses and literacies and their other secondary Discourses and literacies. Individually and collectively, 5S and 5M students respectively practiced different literacies within the context of exam-oriented school learning despite sharing common syllabi and curricula and even, in some cases, the same teachers.

The generic practice of academic-scholastic examination is an obvious case of what Gee calls a dominant Discourse, offering as it does avenues to a range of social goods and powers, material as well as symbolic. Having control of the associated secondary language use here is to have mastery of a dominant literacy—to *command and practice* a dominant literacy. The view advanced here—although it calls for considerable further development and scrutiny—is that 5S have control of the dominant literacy (i.e., they are exam literate) and 5M do not. As a dominant literacy, academic examination literacy might be distinguished from such other secondary language uses as reading teen romances, writing shopping lists, browsing Sunday papers, or reading bedtime stories to children, *all of which were doubtless controlled by 5M students*. The discursive processes of producing and "allocating" power can be seen here to bestow advantage in the acquisition of social goods—mediated by literacy—on one group at the expense of another.

Gee himself offers a further example of a dominant literacy by reference to literary criticism as a dominant Discourse. His example refers explicitly to talk of empowerment:

> [D]iscourses are intimately related to the distribution of social power and hierarchical structure in society. Control over certain discourses can lead to the acquisition of social goods (money, power, status) in a society. These discourses empower those groups who have the fewest conflicts with their other discourses when they use them . . . [T]o take [an] example, the discourse of literary criticism was a standard route to success as a professor of literature. Since it conflicted less with the other discourses of white, middle-class men than it did with those of women, men were empowered by it. Women were not, as they were often at cross-purposes when engaging in it. Let us call discourses that lead to social goods in a society "dominant discourses" and let us refer to those groups that have the fewest conflicts when using them as "dominant groups." (Gee, 1991: 5)

Let us accept that literacy criticism is a dominant Discourse par excellence, and that to practice its associated language uses successfully is to have control of a dominant literacy. The matter to be clarified is how dominant literacies are related to power and, to that extent, empower those with command of them.

There are two important things to note here about Discourses and their literacies. The first is that a given Discourse—any Discourse, dominant or otherwise—is constituted around specific objects and purposes. It privileges certain concepts, viewpoints, perspectives and values, at the expense of others. It consequently "marginalize[s] viewpoints and values central to other Discourses" (p. 5). Successful participation by individuals in a given Discourse and its characteristic conception and practice of reading and writing can, then, be impeded in at least two ways: by a "lack of fit" between the given Discourse and language use and their other Discourses—and especially their primary Discourse; and/or by value conflict between their other Discourses and the Discourse in question (e.g., peer group and class-cultural discourses that make it "uncool" to succeed academically).

The second facet of a Discourse is its place within the discursive ordering of access to social goods: that is, its status within the hierarchy of dominant and subordinate Discourses. In this respect the literacies associated with literary criticism and scholastic exams are inherently different from those associated with Discourses central to the being of many women, ethnic minority and migrant groups, and working-class populations. They stand in opposite relations to the power to access social goods. The literacies of formal criticism and exams provide avenues to acquiring social goods. The others do not. Nor, moreover, do they cohere (or "fit") well with dominant Discourses.

What sense can we make of literacy and empowerment against this background of concepts and examples and in terms of the conceptual model of "empowerment" outlined earlier? What would it mean to say that 5S students are empowered by "exam literacy," whereas 5M students are not? In what sense do dominant Discourses and their associated language uses "de-power" groups and individuals by marginalizing their discursive histories and allegiances? What conception and theory of power is implied?

The view of social power assumed here is that power is produced in the process of certain "qualities" being related to social "goods" (or means to "goods") through Discourse. To have access to power is to possess qualities that have been related positively to goods or means to accessing them. To actually exercise power is to draw on these qualities, to "cash them in," as it were. To be empowered is to have the qualities one possesses (or has available) made discursively—that is, through Discourse—into "currency" for acquiring goods and benefits or for having them bestowed.

Students from the kinds of social groups represented in streams or classes like 5S have social power by virtue of certain of their discursively acquired qualities (e.g., a disposition and ability to interpret, argue, debate, abstract, adopt a position) having been positively related—discursively—to exam success. 5M students—and others more or less like them—lack power as a result of being positioned discursively in the opposite relationship to exam success. Students like those in 5S are empowered by secondary Discourses centered around school having made these qualities—which are available to them on terms denied to the social groups most represented in 5M—into "legal tender" for exam success. As a dominant Discourse, scholastic examination "de-powers" those social groups whose discursive histories and allegiances impede their attaining control of the requisite language uses and related dispositions, attitudes and performances.

To employ the conceptual schema outlined earlier, we can say that:

> as individuals or as a group, *students like those in 5S* are empowered in respect of *school exams* (as a dominant discourse which affords access to social goods), *by their facility with appropriate language use* (which itself reflects a coherence between their larger discursive universe and the language demands of exams and scholastic work) such that under prevailing conditions they are able to *enjoy exam success* and access to associated social goods and powers.

Within existing discursive orders, dominant literacies empower certain groups (and depotentiate others), by making what they *already* have (or have privileged access to) into currency for acquiring social goods and benefits. These groups are empowered as a "natural" discursive consequence of who and what they already are, or will, in the normal course of events, become.

Powerful Literacies: Empowerment and Critique

While dominant literacies can be seen to empower groups who are, so to speak, "on the right side of Discourse," they are not the same as *powerful literacy*, and they empower in a different way. To develop this argument I will begin from Gee's account of powerful literacy and adapt and modify his central idea for my own purposes.

Gee defines powerful literacy as control of a secondary use of language used in a secondary Discourse, that affords us a meta-language with which to understand, analyze and critique primary Discourses and/or other secondary Discourses—notably, dominant Discourses—and the ways they constitute us as persons and situate us in society (see also Gee, 1990: 153, and 1991: 8–9). By a meta-language he means "a set of meta-words, meta-values [and] metabeliefs" (Gee, 1990: 153). Practicing a powerful literacy, so defined, can provide the basis for reconstituting our selves/identities and resituating ourselves within society. We should note here that a powerful literacy is a *particular use* of a literacy (or discourse/secondary language use) and not a particular *literacy*.

The key to Gee's conception of powerful literacies is the notion of operating secondary language uses as *meta-languages* of analysis, critique and transformation. He argues that understanding and critiquing a Discourse by means of a powerful literacy grounded in some other Discourse presupposes meta-level knowledge in both Discourses. The point here is that we can operate entire Discourses with fluent mastery and control of their literacies—including dominant varieties—*without* having (good) meta-level understanding of these Discourses, of how they constitute us and position us in relation to other individuals and groups, with what implications and consequences for them and us, and so on. How could this be?

The explanation rests on a distinction framed by Stephen Krashen (1982, 1985) between "acquisition" and "learning." Very briefly, "acquisition" is defined as a process of more or less unconscious attainment resulting from exposure to models and to trial and error within meaningful and functional natural settings. Most people come to control their first language through acquisition. "Learning" is defined as a process of gaining conscious knowledge via explanation, analysis, and similar *teaching* processes. Learning, then, "inherently involves attaining, along with the matter being taught, some degree of meta-knowledge about the matter" (Gee, 1991: 5). Where our proficiency has simply been *acquired*—as distinct from being underpinned with "meta-level" knowledges that only become available through processes of *learning*—we may indeed enjoy benefits of dominance and of distinctive forms of power and modes of empowerment accessed by dominance. We will, however, remain cut off from access to other forms of power and modes of being empowered that depend on having learned as well as having acquired. These are what mark the difference between a powerful literacy and a dominant literacy.

The link between literacy (as secondary language use), meta-level knowledge of Discourses, and operating a powerful literacy is quite straightforward. Viewed from one perspective, secondary language uses are integral aspects of their respective Discourses and must be controlled in order to participate coherently and successfully in these Discourses. From a different perspective, however, secondary language uses can be seen as providing the means for understanding Discourses themselves for what they are generically and in distinction from one another. This is a matter of grasping the relationship between "the language bits of Discourse" (as signifiers, bearers, and "enactment vehicles" of the various values, beliefs, norms, theories, etc. that constitute given Discourses), and the Discourses themselves.

Meta-level knowledge, then, is knowledge *about* what is involved in participating within some Discourse(s). It is more than merely knowing how (i.e., being able) to engage successfully in a particular discursive practice. Rather, meta-level knowledge is knowing about the nature of that practice, its constitutive values and beliefs, its meaning and significance, how it relates to other practices, what it is about successful performance that makes it successful, and so on. Meta-level knowledge of Discourses impacts in at least three ways so far as accessing power through literacy is concerned. I will refer to these as three modes of empowerment.

First, to have meta-level knowledge of a dominant Discourse and its literacy enhances our prospects of mastery and high level performance within that Discourse as it is currently constituted, thereby increasing our (potential) access to social goods and their associated forms of power and privilege. Let us call this Mode A.

Second, meta-level knowledge enhances possibilities for analysis and for applications of analysis. Being able to control secondary language uses as a means for analyzing a Discourse enables us to see how skills and knowledges can be used in new ways and new directions within that Discourse. This is Mode B.

Third—and this is Mode C—meta-level knowledge of Discourse(s) is required in order to critique a Discourse and seek to change it, or at least to change its identity constitutive and social positioning effects on oneself and others.

Meta-level knowledge of Discourses is necessary for their critique because all Discourses, necessarily, are largely immune to internal critique. To critique a Discourse and envisage possibilities for resisting or transforming it, we need a critical place to stand outside that Discourse. Critique presupposes both meta-knowledge of the Discourse itself—knowing that we are in it, and what it is we are in—and meta-knowledge of some other Discourse against which we understand the first and from which we compare and evaluate it. As Gee puts it: "One cannot critique one discourse with another (which is the only way to seriously criticize and thus change a discourse) unless one has meta-level knowledge in both discourses" (1991: 9). Hence (for instance), for women to critique patriarchal Discourses and pursue their transformation, they need(ed) meta-knowledge of what patriarchal Discourses are and do. This presupposes a perspective from the standpoint of (some) nonpatriarchal Discourse—such as a feminist Discourse—and meta-level knowledge of that Discourse, as well as meta-level knowledge of and a meta-language for understanding, describing and analyzing experience of being constituted and situated by patriarchal Discourses.

By identifying typical examples we can note some of the ways in which "powerful" literacies, construed as meta-languages of analysis and critique, differ from dominant literacies in terms of how they *empower*. The following examples are by no means exhaustive. They are intended only to suggest some ways in which we might clarify and extend our conception and practice of empowering learners through constructing and teaching powerful literacies.

Empowerment: Mode A

Let us begin with the notion of meta-level knowledge enhancing prospects of mastery and high-level performance within an established dominant Discourse by enabling the operation of a *powerful* literacy.

I can recall an instance of this from my own experience, where I completed a successful MA thesis in Analytical Philosophy of Education without having a clear understanding of why what I had done was "right." I had seemingly controlled the appropriate literacy without experiencing *being in control* of it. This produced considerable perplexity and insecurity. It was not until I became a tutor in an introductory course (and followed the excellent instructions and exercises covering the skills of analyzing claims into different categories, making conceptual distinctions and analyzing concepts, and evaluating arguments in terms of their validity and soundness) that I began to bring to consciousness what I'd been doing (by imitation, intuitive "feel," and with a good deal of supervisory help) in the thesis, and to understand why it had been given a good grade. It seems that in the act of moving from being a student apprentice of the Discourse to being a teacher my hitherto absent and tacit understandings of the Discourse became (more) explicit: largely, I believe, because the teaching guides spelled out important elements of procedure, purpose, integral theory and so on, which had previously escaped my conscious knowledge.

The process of attaining a meta-level knowledge of educational philosophy as a discursive field and a literate practice continued slowly during my doctoral work, and more rapidly after it. I felt increasingly in control of the Discourse and literacy of analytical philosophy of education and able to make them work for me, rather than experiencing them controlling my work (and my *being*) as I struggled to understand their demands. It became progressively easier to conceive and produce publishable work—to operate (knowingly) the literacy with greater fluency and confidence. With that came the usual sorts of social goods and benefits attending relative success in a Western academic arena. In this way I experienced greater power in the form of being able to operate the Discourse and to access the goods it made available discursively. Elements of social and personal power were closely intertwined here.

Three implications of this seem noteworthy.

First, so far as this example goes, there was no transformation of the Discourse involved. Nothing changed outside of one individual becoming increasingly able to operate a literacy and, to that extent, a Discourse more successfully and to access benefits ordained by the discursive structuring of power within an existing state of affairs.

Second, my "coming to power" may actually have reduced the career prospects of others whose ability to operate the Discourse in a way that "unlocked" social goods was surpassed by my growing confidence and "productivity." This marks an important difference between "having control of a dominant literacy" and "operating a powerful literacy." Those who are empowered in terms of access to social goods merely as a result of "their" acquired Discourses being constituted as dominant, are vulnerable. They are in danger of being displaced by those whose literacy within a dominant field of social practice becomes more powerful. In short, one can be dominant without being powerful in important respects and that lack of personal power may result in eventual loss of social dominance. Not all who are born to rule live and die as rulers, so to speak.

Third, to the extent that empowerment through control of dominant literacies is empowerment patterned by class, ethnicity and gender, pedagogies aimed at promoting powerful literacy have potential for democratizing access to social goods within existing power structures.

To invoke the empowerment schema by (hypothetical) reference to the example drawn from Jones's study of the girls grammar school:

> *S* (an individual who is otherwise "bound for 5M") would be empowered in respect of *school exams* (a discourse offering access to social goods), through attaining *meta-level understanding of exam demands* (reflecting the triumph of pedagogy over socially ordained deficit), such that S could *succeed in exams*, and be better positioned to access associated social goods.

This is an approach to empowerment by which (1) members of non-dominant groups (2) can access social goods (3) through personal power developed around pedagogies of powerful literacy (4) within substantially unchanged discourses and power structures. Dominant literacies largely operate a "closed shop" of empowerment. Powerful literacies can help "open the shop."

There is an important pedagogical requirement here, namely, that educators

> realize that teaching and learning [in the technical sense derived from Krashen] are connected with the development of meta-level cognitive and linguistic skills. They will work better if we explicitly realize this and build the realization into our curricula. Further, they must be carefully ordered and integrated with acquisition if they are to have any effect other than obstruction. (Gee, 1991: 9–10)

At the same time, we must avoid unrealistic expectations of explicit teaching as a means for redressing "shortfalls" in the prior acquisitions of non-mainstream students so far as dominant D/discourses are concerned. While rendering the rules, values, procedures, etc. of dominant D/discourses overt and explicit—i.e., concrete and accessible—to all students is binding on teachers at all levels, this is "not an educational panacea [for non-mainstream students] and . . . involves complex problems" (Gee et al., 1996: 12). Making "the rules of the game" overt can never be done exhaustively, since we cannot put all that is involved into *words*. And even if we could, what we could say in classroom settings would be merely the tip of the iceberg. In addition, such overt knowledge "would not ground fluent behavior [in, say, exam preparation, choice of examples, exam writing] any more than overt knowledge of dance steps can ground fluent dancing" (ibid.). This observation, however, does not detract from the ethical and vocational obligation to *teach well*, which includes rendering the implicit explicit and the abstract concrete, where it is possible and appropriate to do so.

Alternatively, if we shift the empowerment emphasis away from a strict focus on accessing social goods and attend more to the dimension of personal power, we might complete the schema in ways that stress, say, autonomous and independent competence. Within the social relations of knowledge production, degree candidates and future authors can win their own voice of competence and become independent of mentors by becoming powerfully literate in the discourse.

Empowerment: Mode B

The notion of empowerment through enhanced possibilities for analysis and applications of analysis can be dealt with quite simply. Let us take the practice of quality control in the modern workplace as an increasingly relevant example from a key site of daily life.

According to a commentator on changing work conditions, "ten years ago we saw quality control as a screening process, catching defects before they got out the door . . . Today we see quality as a process that prevents defects from occurring" (Wiggenhorn, 1990: 71–2). Quality control as Discourse changed, and enterprises are forever on the lookout for improvements to the practice that may help them win a market edge over competitors. Quality control is but one aspect of the way in which, in Wiggenhorn's words, "all the rules of manufacturing and competition [have] changed." Motorola's management learned "that line workers had to actually understand their work and their equipment," and the relationship between them (1990: 72).

This, in effect, is a call for greater meta-level understanding of manufacturing Discourses on the part of workers. There are implications here for workers' employability, career prospects and control over their own work, as well as for corporate profits. Workers who have and operate a meta-language of quality control can analyze the purposes and processes of quality control and, by application of this analysis, contribute to enhancing and transforming quality control practices—including, specifically, the literacy components of quality control. This may well presume that they have meta-level understanding of Discourses and secondary language uses of *quantity*-oriented Discourses of manufacture or other forms of production. The terms of the empowerment schema can be embodied in various ways in relation to this example. The subject variable can range over individual workers or groups (cells or teams) of workers, who in turn may vary by gender, ethnicity, prior experience, first language, and so on. The second variable could range variously over decision-making procedures, the system of wages and rewards, career ladders, and hierarchies of responsibility. The achievement or quality variable refers, of course, to command of a meta-language of quality control. The outcome variable could range over better wages, promotion prospects, control over work conditions, relative autonomy from management, and enhanced possibilities of finding (better) work in other enterprises. Hence, shop-floor workers might develop innovations on the basis of their understandings of a Discourse (e.g., quality assurance, quality control), have these recognized, and be promoted as a consequence. Alternatively, by harnessing inventiveness to meta-level understandings of markets and niches, such workers might "go entrepreneurial" and sell their innovations to high bidders, take out patents, and so on. Many contemporary work D/discourses, markets, niches, routines, etc. are sufficiently new and different to mitigate some of the differences among employees in prior experiences and acquisition, enhancing prospects of empowerment through relevant metalevel knowledge.

Empowerment: Mode C

Gee's particular emphasis in his discussion of literacy and empowerment falls squarely on the possibilities afforded by powerful literacy for critique and *transformation* of Discourses.

With regard to classroom pedagogies of powerful literacy, Gee calls for teaching to be constructed around a critical intent aimed at enabling children from all social groups to critique their primary and secondary Discourses. This entails using texts and other resources to expose children "to a variety of primary and secondary Discourses" in ways that promote meta-level knowledge of these Discourses (Gee, 1991: 10). Meta-level knowledge makes possible meta-languages of critical preference among Discourses and knowledge of how to act on critical preference. And with critical awareness of alternative Discourses and the possibilities of meaningful choice among them, comes power—including the power to resist the way particular Discourses constitute us as identitied persons and situate us within society and the social hierarchy of Discourses; and so, too, comes the power to *choose* apprenticeship to participation in other Discourses that will constitute and situate us in certain ways we have come to see as desirable or preferable—for various reasons, including ethical and political reasons.

This approach is reminiscent of successful forms of feminist pedagogy that offer women the option—albeit often a tough option secured at high cost—of refusing ways in which patriarchal discourses (would) constitute and position them as women (Bee, 1993; Rockhill, 1993). Beyond this, feminist pedagogies make it possible to envisage and embark upon a long historical march through sexist discourses and institutions, with a view to their eventual transformation in accordance with values and conceptions drawn from examples and visions of alternative discursive practices.

Variations on D/discourse analysis and critique approaches to promoting powerful literacy are apparent in pedagogical initiatives currently under development in Australia. Insights informing this approach and its broad lines of development are described by Luke and Walton:

> Written texts are . . . refractive; that is, they actively construct and represent the world. To read critically . . . requires awareness of and facility with techniques by which texts and discourses construct and position human subjects and social reality . . . Instruction based on text and discourse analysis aims to give students insights into how texts work and, more specifically, how texts situate and manipulate readers. In so doing, its purpose is to engage readers directly and actively in the politics of discourse in contemporary cultures, to open institutional sites and possibilities for alternative "readings" and "writings." (Luke and Walton, 1994: 1196)

Near the heart of this approach we find Kress's (1985) view that texts construct "subject positions" and "reader positions." Texts represent and construct *subjects* in the world and also position and construct model *readers*. Via a range of lexical, syntactic and semantic devices, texts portray a view of the world and position readers to read and interpret that portrayed world in particular ways.

The uncritical (unpowerful) reader is not aware of these processes and devices, submitting unknowingly to their ideological and "subject constitutive" effects. Critical (powerful) readers, by contrast, understand how a diverse range of linguistic techniques employed in texts position readers, defining and manipulating them, and employ an understanding of how syntax, grammar and words themselves act to shape human agency, cause and effect, and how the world is portrayed (Kress, 1985; see also Fairclough, 1989). In so reading the text they can resist, as appropriate, the meanings and subject positions it coaxes them to adopt.

Such meta-level knowledge of texts and how they operate is, precisely, knowledge of the ways power is produced and enacted through language. The powerfully literate reader can contest texts, resisting meanings and positions these would otherwise "impose." This is to be empowered as a recipient or "object" of text. As a writer of texts the powerfully literate person develops "powerful competencies" with a range of genres and techniques that may be employed in pursuit of personal, ethical and political purposes.

Such approaches to powerful literacy as described here likewise sustain meaningful conceptions and pedagogies of empowerment. They address the power of Discourses and their literacies to shape what we become as persons and what we are impeded from becoming, and thereby what is (and is not) available to us in the way of social goods, benefits and opportunities. They suggest pedagogical techniques that allow readers to understand these discursive and textual processes in relation to the politics of daily life in general as well as of particular spheres, like home, school and work. In so doing, they address elements of ideological-symbolic domination and control as well as more material forms of regulation—since Discourses involve both. Discourse critique promotes meta-level understanding of Discourse(s) as well as familiarity with the substance of multiple Discourses, in respect of what they make available and withhold. Learners, then, are enabled to see how they are, or have been, constituted by Discourses. They are enabled also to envisage embarking upon new and different possibilities for "self-construction" or identity formation by engaging in alternative Discourses.

Final Remarks

There are advantages in taking the conceptual demands of empowerment seriously along the lines sketched here. The more closely and clearly we specify particular learner-subjects, the more likely we are to identify forms and processes of power that limit their options within everyday settings and to envisage or negotiate learning processes and outcomes in line with the ends we seek. Similarly, the better our theory is of how power is produced and distributed, the more we will understand the varying circumstances and aspirations of different learners, and the better informed will be the means and processes employed toward (negotiated) learning outcomes.

Moreover, the approach adopted eschews chauvinism and pre-empts certain dangers associated with restrictive forms of political correctness. Rather than usurp the ideal of empowerment for a sole (typically unargued) pedagogical or ideological position, it leaves the way open for proponents of differing values to state their case clearly and to develop their pedagogical models in an informed and directed way. We can then engage in focused substantive debate, knowing with clarity and a good deal of precision what it is we are agreeing with or disputing. At present we can all agree on empowerment as an educational goal and proclaim the virtue of an empowering literacy, since "empowerment" means all things to all people: which is to say that it means nothing clear. And "nothing clear" is a poor pedagogical guide.

In the final analysis, if we believe literacy for empowerment marks an educational ideal worth striving for, having concern for clarity, coherence, and critique makes a good starting point.

References

Bee, B. (1993). Critical literacy and gender, in C. Lankshear and P. McLaren (eds.) *Critical Literacy: Politics, Praxis and the Postmodern*. Albany: State University of New York Press. 105–132.

Berlin, I. (1969). *Four Essays on Liberty*. London: Oxford University Press.

Cranston, M. (1953). *Freedom: A New Analysis*. London: Longmans Green.

Dale, R. (1991). Review of Andy Green, *Education and State Formation. Journal of Education policy*, 6(4): 418–418.

Delgado-Gaitan, C. (1990). *Literacy for Empowerment: The Role of Parents in Children's Education*. London: Falmer Press.

Fairclough, L. (1989). *Language and Power*. London: Longman.

Feinberg, J. (1973). *Social Philosophy*. Englewood Cliffs, NJ: Prentice Hall.

Gee, J. (1990). *Social Linguistics and Literacies: Ideology in Discourses*. London: Falmer Press.

Gee, J. (1991). What is literacy? In C. Mitchell and K. Weiler (eds), *Rewriting Literacy: Culture and the Discourse of the Other*. New York: Bergin and Garvey. 3–11.

Gee, J., Hull, G. and Lankshear, C. (1996). *The New Work Order: Behind the Language of the New Capitalism*. Boulder, CO: Westview Press.

Jones, A. (1986). "At school I've got a chance:" Ideology and social reproduction in a secondary school, unpublished PhD thesis, University of Auckland.

Jones, A. (1991). *"At School I've Got a Chance"—Culture/Privilege: Pacific Islands and Pakeha Girls at School*. Palmerston North, NZ: Dunmore Press.

Krashen, S. (1982). *Principles and Practice in Second Language Acquisition*. Hayward, CA: Alemany Press.

Krashen, S. (1985). *Inquiries and Insights*. Hayward, CA: Alemany Press.

Kress, G. (1985). *Linguistic Processes in Sociocultural Practice*. Geelong: Deakin University Press.

Luke, A. and Walton, C. (1994). Teaching and assessing critical reading, in T. Husén and T. Postlethwaite (eds) *International Encyclopedia of Education*, 2nd edition. Oxford: Pergamon Press, 1194–1198.

MacCallum, G. (1967). Negative and positive freedom. *Philosophical Review*, 76: 312–34.

Rockhill, K. (1993). (Dis)connecting literacy and sexuality: Speaking the unspeakable in the classroom, in C. Lankshear and P. McLaren (eds.) *Critical Literacy: Politics, Praxis and the Postmodern*. Albany: State University of New York Press. 335–66.
Street, B. (1984). *Literacy in Theory and Practice*. Cambridge: Cambridge University Press.
Wiggenhorn, W. (1990). 'Motorola U': When training becomes an education. *Harvard Business Review*, August, 71–88.

Seven

Language and the New Capitalism (1997)

Colin Lankshear

Biography of the Text

Between 1989 and 1992 I worked closely with a freelance sociologist, Allan Levett, in New Zealand. Our collaboration arose out of an invitation to do some research in a school in Auckland that served a substantial indigenous and Pacific Island migrant community. Under the leadership of a principal resolutely committed to tackling educational disadvantage by addressing social esteem, the school was wrestling with the question of what kind of education would best serve students from groups that disproportionately ended up on the wrong side of academic achievement. Allan had long been interested in large-scale change and its implications for key social institutions, organizations, and values—including schools and educational values. Part of our focus homed in on issues concerning the relationship between formal education and the changing world of work. Allan suggested I should augment my reading in the "critical sociology of work" by exploring works in the genre that Jim Gee subsequently called "fast capitalist texts." He reminded me that I did not have to agree with what they said, but that I had to know them in order to be better informed about forces of contemporary change and the challenges that schools would increasingly face. This was some of the best intellectual advice and direction I have ever received.

When I began working as an academic in Australia one of my first tasks was to lead a team charged with developing an undergraduate degree in adult and workplace education. The program had to include exactly the kind of textual engagement that Allan Levett had recommended: understanding and critical reflection across a range of perspectives on contemporary work. This period coincided with meeting Jim Gee and beginning a conversation about a "new work order." We wrote an exploratory paper called "The new work order: Critical language awareness and 'fast capitalist' texts" for the Australian journal *Discourse: Studies in the Cultural Politics of Education*, which was published in 1995. Discourse's editors, Bob Lingard and Fazal Rizvi, also had a book series and contracted us to write, with Glynda Hull, a book—*The New Work Order: Behind the Language of the New Capitalism*. Following the publication of this book I began thinking about how the new work order seemed to be associated with an emerging order of language and literacy proficiency: a new word order. My first attempt to articulate this idea and its association with work was in a paper for *The International Journal of Inclusive Education*. "Language and the New Capitalism" was published in 1997 and is reproduced here.

Introduction

In 1976, Samuel Bowles and Herbert Gintis published their landmark work *Schooling in Capitalist America*, in which they argued that the education system "helps integrate youth into the economic system . . . through a structural correspondence between its social relations and those of [capitalist] production" (Bowles and Gintis, 1976:131). According to Bowles and Gintis, the social relations of schooling replicate the hierarchical division of labor under capitalism and develop forms of personal demeanor, self-image and social-class identification in parallel with capitalist constructions of job adequacy. Likewise, alienated labor can be seen as reflected in the lack of control students have over their education, their estrangement from the curriculum, and so on.

The "correspondence principle" was subsequently challenged by many critics in respect of its alleged explanatory power and what was widely perceived as its reductionist character, its undue structuralist determinism and its theoretical rigidity. Of course, hindsight reveals that *Schooling in Capitalist America* and the debate it stimulated made an enduring contribution to our understanding of the extent to which, and ways in which, social relations, practices and outcomes of formal education are enmeshed with the (re)production of economic life under capitalism. In a period when we increasingly hear talk of "the new global economy," "information and services economies," and "post-capitalist society" (Drucker, 1993: see all), we do well to remember that the organization of productive life in societies like our own remains, implacably, capitalist—albeit in new, restless, complex and profoundly re-invented ways.

Drawing on examples from the U.S. and Australia, but which I believe to be much more widely applicable, this chapter invites readers to ponder how far conceptions and practices of language and literacy in school are currently undergoing change in conjunction with an emerging "new capitalism" (cf. Gee et al., 1996), and what implications these changes may have for inclusive education and inclusive literacy. The argument describes some key features of the new capitalism and what we have elsewhere referred to as "the new work order" (Gee et al., 1996). It then describes some trends apparent in language and literacy education at school and adult-vocational levels. These include the apparent endorsement within current policy direction of a new word order which, if realized in practice, may mediate access by individuals and groups to places and rewards within the new work order. Of course, any such relationship will prove to be complex, and any further empirical exploration and analysis must build on the many theoretical advances made within critical social theory since the days of the correspondence principle.

Some Comments on "Capital" and "Capitalism"

To appreciate what, if anything, is "new" about the new capitalism, and to get angles on the significance of the relationship between language and the new capitalism, it is helpful to begin with some general comments about capitalism per se.

In broad terms, capitalism may be understood as a system that uses wage labor to produce commodities for sale and exchange, and for generating profit, rather than for meeting the immediate needs of the producers. As such, the distinction between use value (X's value comes from using it) and exchange value (its value is for exchange and what we can get for it) is fundamental.

Capital is seen as one of the four main production factors, the others being land, labor and enterprise. Capital consists of such things as machinery, infrastructure/plant, tools and technologies, and other human creations (from ideas to exchange media like money, synthetics, etc.) that are applied to the production process. Capital is used to purchase "commodities"—raw materials and labor, mainly—in order to produce other commodities for sale at a profit—which profit is turned back into capital: the process of capital accumulation.

Of course, this highly general notion of capitalism can accommodate many different specific forms of activity, as well as many debates about what is central to and distinctively characteristic of capitalism (Marshall, 1994: 38–40). For Marx, the emphasis was on labor as the engine of value creation—i.e., it was adding labor to the other productive forces that was the key; by generating surplus value from the worker's labor, the capitalist could accumulate. This presupposed exploitation of the worker and, in Marx's view, class conflict was thereby structured into capitalism as a contradiction that would ultimately result in the historical transcendence of capitalism. Weber, by contrast, focused more centrally on markets and various institutions that enable market exchange as being the key to capitalism—notably, such institutions as private property, market networks, monetary systems and appropriate "socializing mechanisms" by which to shape attitudes conducive to capital accumulation.

Historically, a range of capitalist forms and a range of scales and grounds have been evident (cf. Marshall, 1994: 38–40). These include agrarian, industrial, financial and post-industrial or informationalist forms. Scales and grounds of operation range over small-scale/private, entrepreneurial, corporate, monopoly and transnational variants.

Some of these variations in type and scope have been around for a very long time—for instance, who was the first farmer to grow corn for sale at a profit?; who were the early entrepreneurs who drove a system(at)ic wedge between owners of capital and wage laborers?—and they are all still in evidence today. They are all part of the larger "scene" of capitalism, so to speak. And at some time or another, each could have been regarded as the "new" capitalism.

A Brief Account of Key Features of the "New" Capitalism

So, what is "new" in today's new capitalism? What features is "new" credited with drawing out? Much has been written on this theme, with many accounts dealing with specific aspects, while others attempt to gain a larger overview. The most satisfactory, succinct synthesis I have found to date on themes discussed at length in the literature is provided by Manuel Castells (1993). Castells identifies five systematically related features of what may be called the new capitalism, which have been emerging and aligning during the past half-century.

(1) Sources of productivity depend increasingly on the application of science and technology and the quality of information and management in the production process—applied knowledge and information: "The greater the complexity and productivity of an economy, the greater its informational component and the greater the role played by new knowledge (as compared with the mere addition of such production factors as capital or labor) in the growth of productivity" (Castells, 1993: 16–17). Producers are forced to build their activities around "higher value-added production," which depends on increased use of high technology and abstract thinking—or what Reich (1992) refers to as the work of symbolic analysts. Major innovations during the past thirty years, which have underwritten new spheres of production and vastly enhanced productivity, are all the results of "applying theoretical knowledge to the processes of innovation and diffusion" (Levett and Lankshear, 1994: 31).

(2) An increasing proportion of GNP is shifting from material production to information-processing activities. The same holds for the workforce: whether "foot soldiers of the information economy . . . stationed in 'back offices' at computer terminals linked to world wide information banks" (Reich, 1992: 175), or as "symbolic analysts" involved in the high-order "problem solving, problem iden-

tifying and strategic brokering activities" performed by research scientists, design and software engineers, management consultants, writers and editors, architects and architectural consultants, marketing strategists and many others besides (cf. Reich, 1992: 175). "An ever-growing role is played by the manipulation of symbols in the organization of production and in the enhancement of productivity" (Castells, 1993: 17).

(3) Major changes in the organization of production have occurred along two axes. First, goods production has shifted from standardized mass production to flexible specialization and increased innovation and adaptability. This allows for optimal customization and diversification of products and enables quick shifts to be made between different product lines—reflecting the postmodern predilection for "difference" (that makes no difference) and diversity. Second, a change has occurred in the social relationships of work. The "vertically integrated large-scale organizations" of "old," standardized mass production capitalism have given way to "vertical disintegration and horizontal networks between economic units" (Castells, 1993: 18). This is partly a matter of flatter hierarchies and increased devolution of responsibility to individual employees and the creation of quality circles, multi-skilled work teams with interchangeable tasks, and enlarged scope for workers to participate in decision-making (within definite parameters). It is also a matter of horizontal relationships of cooperation, consultation and coordination, in the interests of flexibility, decentralization and adaptability in production, which extend beyond the confines of a specific business or firm to include other partners within an integrated productive enterprise, such as collaborative arrangements between manufacturers and supplier, which help keep overheads and stock inventories down, allowing competitive pricing, which can undercut opponents.

(4) The new capitalism is global in real time. National economies no longer comprise the unit of analysis or strategic frame of reference for companies and workers. For enterprises and workers alike, work is increasingly about playing on the whole world stage. For many individual workers, their competition comes from all over the world. And, of course, many companies are "all over the world and all at once" (Reich, 1992: 172). Robert Reich says with respect to individual American workers that their prospects are now relative to the global market: "Individual American workers whose contributions to the global economy are more highly valued in world markets will succeed, while others, whose contributions are deemed far less valuable, fail" (Reich, 1992: 172).

(5) The context of this change—which reflexively spearheads and responds to it—is the information technologies revolution. The new capitalism is dynamically and inseparably linked to the current technological revolution, especially its information–communications dimension. In addition to informatics, microelectronic and telecommunications, this revolution encompasses also scientific discoveries

and applications in biotechnology, new materials, lasers, renewable energy, and the like (Castells, 1993: 19). The dynamism of the relationship is such that demands generated by the kinds of economic and organizational changes already identified stimulate ongoing developments in information and communications technologies. These technologies (in their earlier manifestations), however, themselves provided many of the material conditions needed for the emergence of the global economy in the first place. Set in train, as they are, the dynamics continue apace, creating a situation where a crucial factor—if not the fundamental source—of wealth generation resides in the ability "to create new knowledge and apply it to every realm of human activity by means of enhanced technological and organizational procedures of information processing" (Castells, 1993: 20).

To these features identified by Castells, I would add that the new capitalism is unfolding in the context of a powerful, intrusive, highly regulatory techno-rationalist business world view, which—as manifested in education reform, as well as in wider changes at the level of the state—has impacted powerfully on language processes and practices.

This world view is an assemblage of values, purposes, beliefs and ways of doing things that originated in the world of business. It has now been embraced by many governments as the appropriate modus operandi for public sector institutions, including those of compulsory and post-compulsory education and training. The logic of this world view is now powerfully inscribed on how literacy is conceived and taught within publicly funded and maintained educational institutions.

The concept of a techno-rationalist business world view is an amalgam of several ideas. The "techno" component refers to privileging technicist approaches to realizing social purposes. It captures what critical social theorists call the triumph of technocratic or instrumental rationality within the everyday conduct of human affairs (Aronowitz and Giroux, 1993). This is the idea of reducing human goals and values to constructs, which can be broken down into material tasks, steps, categories, processes, etc., and tackled in systematic ways using appropriate tools, and techniques applied in a means-to-ends fashion. It includes such procedures as operationalizing qualities (e.g., competence) into measurable and observable behavioral objectives and outcomes; defining values in terms of commodities, which can be produced technologically; framing goals in terms of programs, packages and recipes, which can be delivered as means to attainment; and the like.

In the sense intended here, "rationalist" refers to the currently pervasive tendency to analyze and measure institutional processes and provisions in cost-benefits terms, with a view to rationalizing them accordingly. This involves quantifying, measuring and comparing different options for producing particular outcomes, benefits and performances, and exercising (rational) preferences in the light of the costs or inputs incurred in producing various levels of result. Having performed the calculation, the individual or organization exercises preference in the manner of a profit or benefit maximizer.

"Business" refers to a gamut of values and characteristics associated with the preferred institutional style of (so-called) leading-edge, profit-driven organizations. These include such ideas and qualities as being "cost-effective," "lean and mean," "quality-controlled," "quality-assuring," "focused on the bottom line," "value-adding," "competitive-edged," "efficient," "rationalized" and committed to "uniform standards across all sites of activity." Organizations of this type value "transferability" (of knowledge, skills and expertise), emphasize "accountability," privilege "competence" over time on the job and insist on "audit trails" as means of verifying "performance." They are oriented towards quantifiable outcomes, subscribe to a "portfolio and project" approach to life and generally prefer individual enterprise agreements to collective awards and bargaining procedures at the point of hiring.

New Capitalism and Language: Some Macro-social Processes

A New Word Order?

Themes addressed in literature on the new capitalism resonate in current educational reform discourse. At the level of language learning, this is apparent in the emphasis on four broad "types" of literacy. I call these the "lingering basics," the "new basics," "elite literacies" and "foreign language literacy." An overarching emphasis on standard English literacy is presupposed in the first three types.

At the school level, "lingering basics" refers to mastery of generalizable techniques and concepts of decoding and encoding print, presumed to be building-blocks for subsequent education in subject content and "higher-order skills." At the adult level, they refer to functional capacities with everyday texts enabling citizens to meet basic print needs for being incorporated into the economic and civic "mainstream." These conceptions "linger" from an earlier period.

The "new basics" reflect recognition that major shifts have occurred in social practices with the transitions from: an agro-industrial economy to a post-industrial information/services economy; "Fordism" to "post-Fordism;" personal face-to-face communities to impersonal metropolitan and "virtual" communities; and from a paternal (welfare) state to a more devolved state requiring greater self-sufficiency. These shifts are seen to demand on the part of all individuals qualitatively more sophisticated ("smart"), abstract, symbolic-logical capacities than were needed in the past. Hence, "the percentage of all students who demonstrate ability to reason, solve problems, apply knowledge and write and communicate effectively will increase substantially" (U.S. Congress, 1994: Goal 3 B (ii)).

"Elite literacies" refer to higher-order scientific, technological and symbolic practices grounded in excellence in academic learning. Here "literacy" denotes advanced understanding of the logics and processes of inquiry within disciplinary fields, together with command of state-of-the-art work in these fields. This allegedly permits high-level critique, innovation, diversification, refinement, etc. through application of theory and research. The focus here is "knowledge work" (Drucker, 1993), construed as the real "value-adding" work within modem economies (Reich, 1992).

"Foreign language literacy" is seen ultimately in terms of proficiency with visual and spoken texts integral to global dealings within the new economic and strategic world order, thereby serving "the Nation's needs in commerce, diplomacy, defense, and education" (NCEE, 1983: 26), genuflections towards more "humanist" rationales notwithstanding. This calls, minimally, for communicative competence allowing functional cross-cultural access to a range of discursive practices and, optimally, for levels of fluency and cultural awareness equal to being persuasive, diplomatic and strategically effective within sensitive high-risk/high-gain contexts.

An unsettling harmony exists between these broad literacy types and trends within "the new work order" (Gee et al., 1996). Increasingly, work is becoming polarized between providing "symbolic analytic services" at one extreme, and "routine production" and "in-person" services at the other (Reich, 1992). At the same time, modern enterprises seek to infuse a sense of responsibility for the success of the enterprise throughout the entire organization and to push decision-making, problem-solving and productive innovation as far down towards front-line workers as possible.

Symbolic analytic work is seen as "substantial value-adding" work (Reich, 1992: 177) and is well paid. It provides services delivering data, words and visual and non-visual representations. As noted above, this is the work of research scientists, all manner of engineers (from civil to sound), management consultants, investment bankers, systems analysts, authors, editors, art directors, video and film producers, and the like. It involves high-level problem-identifying, problem-solving and strategic brokering activities (Reich, 1992).

Routine production and in-person service work, by comparison, are construed as "low value-adding" and are poorly paid. Beyond demands for basic numeracy and the ability to read, "routine" work often calls primarily for reliability, loyalty and the capacity to take direction and, in the case of in-person service workers, "a pleasant demeanor" (Reich, 1992: 178). The gulf between this and symbolic analytic work marks the difference between "elite literacies" and the "lingering basics."

Between these extremes, work is impacted by the "changed rules of manufacturing and competition" (Wiggenhorn, 1990: 76), whereby front-line workers must increasingly solve problems as they arise, operate self-directed work teams, understand and apply concepts and procedures of quality assurance and control, and assume responsibility for many tasks previously performed by lower-level management. Such work is widely agreed to require a "higher-level basics" than previously. Yet, this work also is often not well paid. To this extent, both the "lingering basics" and the "new basics" sanction systematic exploitation in the workforce. From this perspective, it becomes very important that we explore the complex interplays between developments in the economy, education reform policies, and their uptake in literacy education emphases and practices within specific sites.

The Gross Instrumentalization of Literacy: Economized Language

A brazen instrumentalism is never far from the surface in the policy pronouncements and supportive rhetoric of educational reform pertaining to literacy. The emphasis and value attached to these elite literacies is identified most explicitly in terms of the fact that high-impact innovation comes from the application of theoretical knowledge. Whereas the new industries of the last century, such as "electricity, steel, the telephone and automobile . . . were invented by 'talented tinkerers' (Bell, 1974), rather than through the application of scientific theory" (Levett and Lankshear, 1990: 4), the big-impact inventions of this century, like "the computer, jet aircraft, laser surgery, the birth control pill, the social survey . . . and their many derivations and applications, come from theory-driven scientific laboratories" (Levett and Lankshear, 1990: 4). Symbolic analysts manipulate, modify, refine, combine and in other ways employ symbols contained in or derived from the language and literature of their disciplines to produce new knowledge, innovative designs, new applications of theory, and so on. These can be drawn on to "add maximum value" to raw materials and labor in the process of producing goods and services. Increasingly, the critical dimension of knowledge work is valued mainly, if not solely, in terms of value-adding economic potential. It is critical analysis and critical judgment directed towards innovation and improvement within the parameters of a field or enterprise, rather than criticism in larger terms, which might hold the field and its applications and effects or an enterprise and its goals up to scrutiny.

Much the same is true of foreign language and literacy proficiency. Justifications for increased emphasis on foreign language proficiency advanced in policy documents and supporting texts often foreground "humanist" considerations in support of foreign language proficiency and bilingualism, whether by increasing foreign language enrollments or maintaining community languages and ensuring ESL proficiency among linguistic minority groups. Sooner or later, however, economic motives generally emerge as the "real" reasons behind efforts to promote foreign language proficiency. Australia's Language gives as its first reason the fact that foreign language proficiency enriches our community intellectually, educationally and culturally, and second, that it contributes to economic, diplomatic, strategic, scientific and technological development (DEET, 1991: 14–15). However, Australia's location in the Asia-Pacific region and its patterns of overseas trade are the only relevant factors explicitly mentioned with respect to developing a strategy which "[strikes] a balance between the diversity of languages which could be taught and the limits of resources that are available" (DEET, 1991: 15).

Elsewhere, influential statements are direct and unambiguous, for example, U.S. Senator Paul Simon's reference to tongue-tied Americans trying to do business across the globe, in a world where there are 10,000 leading Japanese business persons speaking English to fewer than 1000 Americans speaking Japanese and where "you can buy in any language, but sell only in the customer's" (Kearns and Doyle, 1991: 87).

Two main factors have generated the emergence of second-language literacy education as a new (and pressing) capitalist instrumentality. First, trading partners have changed greatly for Anglophone economies, and many of our new partners have not been exposed to decades (or centuries) of colonial or neo-colonial English language hegemony. Second, trade competition has become intense. Many countries now produce commodities previously produced by relatively few. Within this context of intensified competition, the capacity to market, sell, inform and provide after-sales support in the customer's language becomes a crucial element of competitive edge.

Individualization and Commodification of Language and Literacy

In the grip of the techno-rationalist business world view, literacy performance is measured and reported ad nauseam and compiled into personal portfolios. At a time when individuals must be prepared to move around to find employment, "portable certified literacy competence" assumes functional significance.

This is a facet of "possessive individualism" a key operating principle of current reform discourse and grounded in a liberal conception of people and society, according to which: "society is composed of free, equal individuals who are related to each other as proprietors of their own capabilities. Their successes and acquisitions are the products of their own initiatives, and it is the role of institutions to foster and support their personal developments—not least because national revitalization (economic, cultural and civic) will 'result from the good works of individuals'" (Popkewitz, 1991: 150).

At the same rime, literacy is profoundly commodified within the current reform agenda, in respect of assessment mania, evaluation packages, validation packages, remedial teaching packages, packaged standards, profiles, curriculum guidelines, textbook packages and teacher professional development packages promising recipes and resources for securing the required performance outcomes. Sometimes this commodification reaches bizarre levels, as in a model promulgated recently in Australia (NBEET, 1996), where it is proposed that industry sectors build literacy competencies into their respective competency standards. The idea behind local competency standards and associated competency-based training is to make Australian industry as competitive as possible by creating a "smart" workforce of high-quality and efficient performers by means of up-to-date training programs that prepare workers cost-effectively for the kinds of tasks they will be doing on the job. Competency standards comprise so many "units of competence"—e.g., "participates in daily team meetings and discussions." These are broken down into "elements of competence"—e.g., "participates in daily team meetings and discussions," contains as one of its elements of competence "reads team meeting documents." An element of competence has associated "performance criteria," as well as a specified "range of variables" that plots various dimensions along which performance will have to be demonstrated. Finally, an "evidence guide" for assessing competent performance has been produced for each unit of competence.

The proposed model provides a "painting by numbers" guide as to how literacy competencies can be framed and incorporated within a set of industry standards. Options include: adding literacy units of competence to the industry standards; adding literacy elements of competence to existing units of competence; including literacy aspects within the performance criteria and/or in range of variables statements; and including literacy in the evidence guide for units of competence. Following extensive "analysis of the workplace," these various options can be exercised—drawing on the information gathered to determine the best combinations of options to meet the literacy performance requirements of work at the different competency levels. Once this is done, vocational education and training programs can provide courses, modules, materials and resources for teaching and assessing literacy as a component of competency-based training initiatives. This is, indeed, commodified literacy according to a "painting by numbers" design.

The Domestication of Language as Critical Practice

While educational reform discourse emphasizes critical forms of literate practice couched in terms of a "critical thinking" component of effective literacy or as text-mediated acts of problem-solving, it is important to recognize the nature and limits of the critical literacies proposed. They are typically practices that permit subjecting means to critique but take ends as given. References to critical literacy, critical analysis, critical thinking, problem-solving, and the like, have "in the current climate . . . a mixture of references to functional or useful knowledge that relates to demands of the economy and labor formation, as well as more general claims about social inquiry and innovation" (Popkewitz, 1991: 128). The nearer "critical" literacy approaches the world beyond school, the more functional and instrumental critique becomes, with emphasis on finding new and better ways of meeting institutional targets (of quality, productivity, innovation and improvement) but where these targets are themselves beyond question. The logic here parallels that described by Delgado-Gaitan (1990: 2) as operating in notions of empowerment construed as "the act of showing people how to work within a system from the perspective of people in power." The fact that standards are specified so tightly and rigidly within the current reform agenda reveals that the ends driving these standards are to be taken as beyond critique.

A New Doublespeak?

In *The New Work Order* (1996), Jim Gee, Glynda Hull and I look at some of the language behind the new capitalism. A new genre of "fast capitalist texts" heralds the new capitalism and its new work order and revamped workplaces, using language in ways very often not borne out on the ground. These texts are replete with talk of "enchanted workplaces," "self-directed work teams," "empowered workers" and other equally positive and attractive terms. Empirical investigation, however, regularly betrays a less expansive reality. Self-direction and empowerment often amount to little more than the right of workers to discharge accountability for finding (the most) efficient and effective ways of meeting goals, performance levels, quality schedules, etc. laid down by the real decision-makers within so-called flat hierarchies. Workers are "empowered" to accept and enact such liberatory notions as that of "the working week," defined as "however long it takes to get the job done." Glynda Hull's graphic accounts of migrant workers in a Silicon Valley electronics company falling behind their schedules working to faulty specifications the work team did not believe they were at liberty to challenge or overrule—despite knowing the specifications were wrong and despite having recently been through a workplace education program intended to enhance "self-directed teamness"—is a clear case of language that has as much relationship to the particular workplace reality as the notion of educational reform has to empirical learning conditions under current policies.

The Clamor to Technologize Literacy

Escalating dependence of work and other daily tasks and processes on computer-mediated texts is associated with prominent references to technological literacy and technologized curricula in education reform pronouncements, Indeed, according to Aronowitz and Giroux (1993: 63), "The whole task set by contemporary education policy is to keep up with rapidly shifting developments in technology." A National Science Board publication, *Educating Americans for the 21st*

Century (1983; see Toch, 1991: 16), claimed that "alarming numbers of young Americans are ill-equipped to work in, contribute to, profit from, and enjoy our increasingly technological society." The "Technology Literacy Challenge" package of February, 1996, voted U.S. $2 billion over five years to mobilize "the private sector, schools, teachers, parents, students, community groups, state and local governments, and the federal government" to meet the goal of making all U.S. children "technologically literate" by "the dawn of the 21st century." The strategy aims to ensure all teachers receive the necessary training and support "to help students learn via computers and the information superhighway"; to develop "effective and engaging software and on-line learning resources" as integral elements of school curricula; to provide all teachers and students with access to modern computers; and to connect every US classroom to the internet (Winters, 15 February, 1996: no pages).

Promoting technological literacies in tune with labor-market needs is only part of the story. New electronic technologies directly and indirectly comprise key products of new capitalist economies. As "direct products," they consist in all manner of hardware and software, for which worldwide markets need to be generated and sustained. As "indirect products," new technologies consist in information and communications services, such as internet access provision, on-line ordering and purchasing facilities, manuals and guides, networking and repair services, web page design, and so on. Educational reform agendas serve crucially here as a means to creating and maintaining enlarged markets for products of the information economy—extending beyond curricular exhortations to advocate also the extensive use of new technologies within administrative tasks of restructured schools (Kearns and Doyle, 1991).

Ending

Apologists for the new capitalism, like apologists for the magical educational powers of new technologies, are currently surfing the tide of history with seemingly unbounded confidence. They have assumed the right to define the role and purposes of education in terms of service to the unfolding new work order. Their confidence is backed with the power of educational policies decreed, enforced and policed by administrators high on the waft of the techno-rationalist business world view. The choice facing educators who are committed to alternative educational visions is clear-cut. Either we "put up and shut up," or we struggle to live out the belief that education is not the servant of any single end or purpose—recognizing that:

> In the new capitalism words are taking on new meaning, language and communication are being recruited for new ends . . . and multiple literacies are being

> distributed in new ways . . . [This new capitalism] makes us confront directly, at a fundamental level, the issue of goals and ends, of culture and core values, of the nature of language, learning and literacy in and out of schools. (Gee, Hull and Lankshear, 1996: 158)

My aim has been to provide some focus for discussion about desirable and defensible relationships between classroom-based language and literacy education and the world beyond the classroom. This is not (yet) a closed issue, and the stakes are high. The new correspondences indicated here between work and literacies as promulgated in a raft of education reform policies seem likely to diminish prospects for inclusive education, inclusive literacy and, indeed, for an inclusive society. The question of the range of social purposes to be served by language and literacy education needs to be kept open and current tendencies to limit them contested. Educators committed to the principle of inclusive education must engage actively in the struggle to keep this issue alive, and be prepared to debate it long and hard from informed standpoints.

Bibliography

Aronowitz, S. and Giroux, H. (1993). *Education Still under Siege*. Westport, CT: Bergin and Garvey.

Bell, D. (1974). *The Coming of Post Industrial Society: A Venture in Social Forecasting*. London: Heinemann.

Bowles, S. and Gintis, H. (1976). *Schooling in Capitalist America*. New York: Basic Books.

Castells, M. (1993). The informational economy and the new international division of labor. In M. Carnoy, M. Castells, S. Cohen, and F. Cardoso (eds), *The New Global Economy in the Information Age: Reflections on Our Changing World*. University Park, PA: Pennsylvania State University Press. 15–43.

DEET (Department of Employment, Education and Training. Australia) (1991). *Australia's Language: The Australian Language and Literacy Policy—the Policy Paper*. Canberra: Australian Government Publishing Service.

Delgado-Gaitan. C. (1990). *Literacy for Empowerment*. London: Falmer Press.

Drucker, P. (1993). *Post-Capitalist Society*. New York: Harper.

Gee, J. P.., Hull, G. and Lankshear, C. (1996). *The New Work Order: Behind the Language of the New Capitalism*. Boulder, CO: Westview Press.

Kearns, D. and Doyle, D. (1991). *Winning the Brain Race: A Bold Plan to Make Our Schools Competitive*. San Francisco, CA: ICS Press.

Levett, A. and Lankshear, C. (1990). *Going for Gold: Priorities for Schooling in the Nineties*. Wellington: Daphne Brasell Associates Press.

Levett, A. and Lankshear, C. (1994). Literacies, workplaces, and the demands of new times. In M. Brown (ed.) *Literacies and the Workplace: A Collection of Original Essays*. Geelong: UNSW Press.

NCEE (National Commission on Excellence in Education) (1983). *A Nation at Risk: The Imperative for Educational Reform*. Washington, DC: U.S. Department of Education.

Popkewitz, T. (1991). *A Political Sociology of Educational Reform: Power/Knowledge in Teaching, Teacher Education, and Research*. New York: Teachers College Press.

Reich, R. (1992). *The Work of Nations*. New York: Vintage Books.

Toch, T. (1991). *In the Name of Excellence*. New York: Oxford University Press.

United States Congress (1994). *Goals 2000: Educate America Act*. Washington, DC: Author.

Wiggenhom, W. (1990). Motorola U: When training becomes an education. *Harvard Business Review*, July-August, 71– 83.

Winters, K. (1996). America's technology literacy challenge. US Department of Education, Office of the Under Secretary, Washington, DC, <k.winters@inet.ed.gov> posted on <acw-t@ unicorn. acs.ttu.edu> 17 February.

EIGHT

Ways with Windows: What Different People Do with the "Same" Technology (1997)

Michele Knobel and Colin Lankshear

Biography of the Text

This chapter is based on a keynote address we presented to the First Joint National Conference of the Australian Association for the Teaching of English, the Australian Literacy Educators' Association, and the Australian School Library Association, in Darwin, Australia. It draws in part on data collected within the context of an Australian government-funded Children's Literacy National Project that investigated "technology and language and literacy learning." It also draws on unfunded community-based research we were engaged in at the time, that involved fieldwork and interviewing with young people in home and youthspace settings. While it has subsequently become common to resist hard and fast distinctions between "in-school" and "out-of-school" settings, during the early period of the mass introduction of computers into Australian classrooms it was instructive to compare the kinds of "taking up" of digital tools within formal educational settings with typical kinds of "taking up" in non-formal settings where young people had access to "insider" knowledge and ways of doing things. At the time the research was done the great majority of teachers in Australian state schools had relatively little personal experience with computing applications, and many of those who participated in the project expressed their unease with the pressure they felt to integrate new technologies into their teaching, often seeing themselves as, at best, muddling through. We were struck, however, by the commitment they typically demonstrated to doing the best they could. The following paper, along with many others we presented and published during this period, was intended to provide teachers with a framework for understanding technology in terms of social practices and variations among practices, as well as a sense of the relationship between "proficiency" and "playing around."

Introduction

The clamor to integrate computer-based technologies into curriculum generally, and language and literacy education specifically, is a hallmark of contemporary education policy in countries like our own. This, of course, is merely the most visible manifestation of a wider, pervasive "technologizing" of education that has intensified and become brashly explicit in recent years; the latest incarnation of the perennial dream of enhanced human progress courtesy of refined technique. Indeed, the "techno" emphasis is so strong at present that Stanley Aronowitz and Henry Giroux (1993: 63) do not exaggerate when they claim that "the whole task set by contemporary education policy is to keep up with rapidly shifting developments in technology."

In North America, policy statements and reform initiatives from the time of *A Nation at Risk* (1983), through to President Clinton's *Technology Literacy Challenge* (Winters, 1996) have consistently emphasized the need for a technologically literate population. President Clinton's most recent State of the Union Address (1997) spells out strategies for ensuring that "all Americans have the best education in the world (ibid.: 4), including—as a key strategic plank—ensuring that over the next four years all 12 year old students and above are connected to the internet. By the year 2000, "children in the most isolated towns, the most comfortable suburbs, the poorest inner-city schools will have the same access to the same universe of knowledge" (ibid: 7; see also, Winters, 1996). Physical access to computer-mediated communications technologies is heralded as the "modern birthright of every citizen," with Clinton rallying the U.S. to take action to "bring the power of the information age into all our schools" (1997: n.p.). The State of the Union speech clearly establishes a fast track agenda for technologizing classrooms, and language and literacy education in particular, during the next four years.

Here, in Australia, the recently released draft of the new National Literacy Policy interweaves literacy, technology, and economic wellbeing and growth, claiming:

> At a time of rapid technological change and pervasive internationalization, literacy skills contribute to the increased competitiveness and productivity that the national economy demand . . . [The policy aims to extend] an active critical, productive and engaging literacy in the complex and mixed modes in which

> literacy is embedded in Australia's rapidly changing technological, cultural, and economic circumstances. (Kemp, 1997: 6–7)

A perceived need for education to keep pace and in "sync" with labor market needs is obviously an important part of the story. Claims about the escalating dependence of work and other daily tasks and processes on computer-mediated text production, transferral, and reception are foregrounded in prominent references to "technological literacy" and "technologized curricula" (cf. Bigum and Green, 1992; Green and Bigum, 1996; Lankshear and Knobel, 1997) within educational reform statements here and abroad. There is more to it, of course, than simply turning out suitably prepared workers. Of at least equal if not greater importance, however, is the need to constitute vast masses of consumers (see Montgomery, 1996, for a timely account of children as targeted consumers within marketing cultures of cyberspace). New electronic technologies directly and indirectly comprise key products of postindustrial information and service economies. As direct products, they consist in all manner of hardware and software, for which worldwide markets need to be generated and sustained. As indirect products, new technologies consist in information and communications services, such as internet access provision, online ordering and purchasing facilities, manuals and guides, networking and repair services, web page design, and so on. Educational reform agendas serve crucially here as a means to creating and maintaining enlarged markets for products of the information economy—extending beyond curricular exhortations to advocate also the extensive use of new technologies within administrative tasks of restructured schools (Kearns and Doyle, 1991).

Technology and progress have become indissolubly linked in the minds of many parents, students, educators, and policy makers. Schools are investing heavily in hardware, software, internet connections, local area networks, and so on. Increasingly, we hear of parents choosing schools for their children on the basis of internet access. Such practices and mind sets evince a "widely held discourse which associates computers in classrooms with technological progress, future employment opportunities of students, as well as enhanced learning in the classroom" (Bigum and Kenway, 1997: 2).

Policy initiatives, commercial and civic strategies, and ideological investments are working in concert to facilitate—although "coerce" might be a better word—the intensified educational uptake of new technologies. Chip Bruce (1996) identifies three widely held interlocking beliefs underlying contemporary faith in the educational and social efficacy of new technologies. These beliefs simultaneously suggest a straightforward agenda for realizing the alleged potential of new technologies. They are:

1. Education has an essential role to play in meeting major challenges and concerns facing human beings. Accordingly, improving education is seen as being

a large part of the answer to current ills. There are two sides to this conviction. On one side, when things go wrong—such as economic recessions—education bears much, if not most, of the blame: the corollary being, "fix education and you fix the rest." On the other side, educationists are constituted—and otherwise feel a need, especially in periods of fiscal strain—to promote the view that education is integral to improvement, and that the role of the teacher is complex. The obligation educators widely feel to take on social improvement/amelioration roles beyond the strict confines of classroom teaching and routine administration of learning tasks is captured in Illich's (1971: 37) conception of school defining teachers institutionally as custodians, moralists, and therapists.

2. If education is a/the key to social and cultural wellbeing and advancement, computer-based technology is, in turn, the key to educational improvement. Bruce (1996: 52) notes that "computer technology is [seen as] a tool that will in and of itself improve education, and ultimately ameliorate social ills."

3. Computing technology is simply a tool—and a benign tool at that. Hence, it is believed, there are really only two major issues to be addressed, or conditions to be met, as bases for realizing the benefits of new technologies: (a) ensuring universal access to computer technology, and (b) providing adequate training in how to use it.

This is especially apparent in Australia right now, where we encounter, on an everyday basis, expressions of pressure to equip classrooms and of the need for more teacher professional development to enable them to use the technologies.

As exemplified in President Clinton's address, access is for the most part framed in physical and quantitative terms: viz., availability of appropriate hardware, software, and wiring. Policy is often reduced to issues of cost and strategies for addressing provision. This is reflected in the reductionist view of "information rich" and "information poor"—defined in terms of whether there's a computer in the classroom and/or at home. We do not want to imply that these are unimportant matters. Rather, we want to use empirical cases to explore and document some of the ways in which they are radically incomplete. At the same time, we need to demystify the "magic bullet," "quick fix" mentality bound up with prevailing views of access and, in its place, develop and adopt informed and principled stances on the role and place of computing technologies in education. This need was reinforced on the very day of drafting this section, when an Education Adviser at a Brisbane School Support Centre phoned in search of references to theory and research that would address her concern about proliferating requests for funding under the disadvantaged schools component of National Equity Program Scheme for projects, which amounted to putting Pentiums on desks in classrooms.

Similarly, problems emerge around the concern for adequate training: that is, the idea that teachers and students have to "learn how to use it" [viz., computer technology]. So far as teacher competence goes, the training concern is often shamelessly reduced to aspects of technical knowledge, and producers of professional development packages are urged to provide content for "technologically illiterate teachers" as a first priority. Accompanying this, we find lists of skills and competencies to be mastered by students at various year levels being plugged into curriculum and syllabus statements, frameworks and profiles, which is yet another instance of applied technocratic rationality (Lankshear, 1997). Such notions of "learning how to use it" imply some kind of essence and autonomy to the technology: that it is somehow self-contained, with its own independent integrity, and that to unlock its potential and power is a matter of particular kinds of learning (uncovering its secrets). Here, too, we want to complicate matters a little by drawing on some cases we have observed within school and out of school settings.

Orienting Questions

This paper addresses three main questions:

1. What kinds of differences can be found in different people's ways with similar computing hardware and software as they pursue cultural purposes in a range of settings?
2. Can these differences usefully inform education policy and practice?
3. Do they have significant implications for language and literacy education specifically?

Four Sites of Practice

These questions are taken up by reference to four "snapshots" drawn from larger case studies of computer-mediated practices in different sites. One draws on a Year 5 (i.e., Grade 5) classroom in a country town. The school is currently classified as a Band B1 disadvantaged school (qualifying it for limited additional funding support), although in the recent past it has been classified Band A. Our account distills elements of the characteristic ways by which the class undertook a unit of work based on the theme of inventions. The second and third snapshots depict contrasting ways of producing computer-generated slide show presentations. The final snapshot focuses on some characteristic ways of a group of students from an inner-city state school who were participating in activities at a nearby community and youth space with learning assistance from the site-based program directors. These activities used cyberspaces and online imagery to explore identity, culture, and language.

The various participants in these four sites were using more or less the same electronic infrastructure. They were all working at or above a hardware baseline of 486 capacity (or Apple equivalent). Only the home-based participant was working with a Pentium-powered (or equivalent) machine at the time. All were using comparatively up-to-the-minute word processing packages. Scanners were used in each case and digital cameras in three of the four. The participants who were producing slide shows were using equivalent Apple HyperCard and Microsoft PowerPoint software, respectively. Three of the four were using the internet/World Wide Web on a regular basis in their activities, and this was not a crucial variable in the case that wasn't. Hence, to all intents and purposes, the four snapshots draw on equivalent infrastructure, although by no means equivalent access to that infrastructure, or, as we will see, ways of using it.

While we can hold infrastructure roughly constant, there were some notable differences among the four cases. Two took place in school settings, the third in a private home (albeit with input and feedback from a peer group in and out of school), and the fourth in a community setting. The school-based cases themselves differed from each other. One was a full-blown curricular program. The other was an extra-curricular project, loosely coupled to prior learning within formal classroom subjects, with the student participants all coming from Year 8 to 11 classes taught by the teacher who was coordinating the project. Finally, the two slide show productions differ in that one was an entirely self-directed individual pursuit, which amounted to an open-ended personal hobby interest. The other was a "moderately coerced" team effort oriented toward producing a functional outcome (of limited intrinsic interest to the student participants) under severe constraints of time and availability of equipment.

In Search of "Ways"

We have taken the idea of "ways" as a framing device for this paper from two sources. The more obvious, of course, is Shirley Brice Heath's classic work, *Ways with Words* (1983). Heath's research provided invaluable insights into varying cultural productions of language within specific home and community sites, and the relationship between these varying cultural productions and patterns of success and failure within formal institutional (school) settings. The second is Ursula Franklin's (1990) concept of technology as practice, where "technology" is construed generically as a shorthand for ways of doing things or getting things done, socially and culturally.

From these perspectives, "ways" is not a simple concept. "Ways" involve routines and customs characteristic of particular communities of practice. They are related to goals and purposes and draw on funds of knowledge (Moll, 1992) available to participants. Different groups/communities of practice may develop different ways around what appear to be much the same social practice. These differences, however, take on important lives of their own. For example, different ways of engaging in what seems to be the same broad practice may be associated with very different meanings and values. At one level here we might consider some of the very different ways associated with tattooing. At a somewhat different level we might think of some of the very different ways involved in using the telephone. Different ways often become associated with different locations within social hierarchies or systems of status, recognition and reward—as was evident, for example, in Heath's study. Beyond a certain point or set of conditions, what appear to be the same ways may actually turn out to have become quite different practices. Furthermore, different ways transform what appear at first to be the same tools/equipment into very different tools, so far as social and cultural meanings and values are concerned. And so on.

Ways, of course, are never completely static. And they are not "given" or predetermined. Ways are brought into being, and they evolve over time. One especially interesting facet of the present moment in the history of computing concerns the extent to which ways are being "invented"—often on the run—within educational and wider social contexts. These "inventions" may vary greatly. In many cases we find funds of knowledge from long-established pedagogical ways being brought to bear on the incorporation of new technologies into classroom learning—often resulting in some "domestication" or accommodation—of the new technology within older "logics." Elsewhere, we find attempts to create quite new and different pedagogies around appropriations of computers, in accordance with notions of changed conditions and purposes seen as themselves reflected or inherent in new technologies themselves. In between we find all manner of combinations and permutations.

The following accounts are early attempts to capture these sorts of things and to comment on aspects that appear to us interesting and illuminating from the point of view of language and literacy education. The ways we describe here are not, of course, ways in the sense of long established routines or settled cultures as, for example, many of the ways identified by anthropologists are. Some appear somewhat more established than others, and some may endure longer than others. They should, rather, be seen as moments of cultural (re)production that are related to and participate in much larger individual and collective cultural histories and patterns. We have aimed to produce "snapshots" of ways at specific points in time; ways which have capacities to be more or less enduring. Our interest in them is in terms of what they might illuminate, illustrate, and suggest about (computer-mediated) cultural practices more generally: as so many "windows" on cultural practice, as it were. We are not concerned with their comparative prospects or efficacies, and we are certainly not implying any normative judgments about or among them.

Take #1: Learning in Year 5 at Abbotsdale

From the outside, the Year 5 classroom at Abbotsdale School, which draws on an officially designated disadvantaged catchment on the outskirts of a rural town, looked like a relic from an earlier era: a single-room one-teacher school, which over the decades, has seen successive classroom and administration blocks spring up around it to accommodate a growing population. Stepping through the door, however, one immediately faced visible trappings of the present. Ranged along the back of the classroom were three revamped computers with processing speeds of 486 and 586 CPUs. Two computers were fitted with quad-speed CD-ROM players, one of them also being wired to a hand-held scanner. The third computer was networked to the internet via a local public provider and was equipped with Netscape web browsing software. All three computers were linked to the same color enabled DeskJet printer. In the sequences described below, the computers were being used in four main capacities: for using movie making software; engaging in problem solving activities; accessing the internet; and producing artifacts using desktop publishing software.

The teacher, Robert, pursued integrated approaches to learning and employed a theme-based, cross-curriculum approach to planning. The theme we observed in operation over several weeks was "inventions," and was handled as a unit of work.

Robert used the pedagogical device of activity rotations to handle such themes. Large chunks of time were set aside each week during which small groups (3 to 6 students) moved through a cycle of activities and tasks in different spatial locations. Rotation-based work involved two 90 minute segments of time divided by a break. Each 90 minute segment was broken into three 30 minute blocks. Each block was devoted to a different kind of activity, typically drawing on different communications technologies. Reading theme-related materials (sometimes aloud to a teacher aide) for practice as well as for getting information relevant to their projects, accounted for one block. Working in groups with pen, paper, worksheets, pre-set tasks, and discussion, comprised a second block—and was often concerned with preparing ideas and components to be implemented at the computers. Work at computers made up the third block. Robert's plan was that during rotations the class would move through two complete sequences of activities, to maintain a rate of focused progress, ensure continuity, and provide integration of reading, writing, discussing, and computing activities. Following rotation sequences the class typically came together to discuss issues, problems, discoveries, and so on.

Our description focuses on a typical three hour period during which the class moved through two complete rotations of activities. Typical episodes have been selected for description.

In one, Samantha, Kate, and Emma were sitting at one computer working on their animated movie. They were using Microsoft's 3D Movie Maker software to "produce" a movie featuring an invention—in this case, a jet-propelled device for personal aerial transport. As they worked on each scene, they consulted their script overview and discussed character selection, placement, actions and speech, background music and sound effects.

Meanwhile, a group of four students sat at another computer, engrossed in a software program challenging them to construct on-screen a "working" apparatus that enables a ball to travel from point A to point B. They discussed possibilities, tested out their ideas, and cheered when they added a successful component to their design.

Mark, Brendan, and Liam sat at the computer that had a connection to the internet, using a search engine to locate invention-related sites. This was their first experience of the internet, and Robert had provided a task sheet requiring them to fill in particular information about the web page (e.g., its location or URL, the invention showcased, etc.). The group located a comprehensive and well designed Japanese web site presenting a range of wacky inventions, including dusters for a cat's feet so that the cat can clean your home while you're at work, a hat that incorporates a roll of toilet paper for dispensing "tissues" to people with severe colds, and the like. While reading and laughing their way through the text, they commented on some of the syntax used and discussed with Robert whether or not the writer spoke English as a second language.

During a second round of activities, a group of students were at the computer with the desktop publishing software. They were learning how to create text boxes, and insert text, graphics, and borders, in order to make posters advertising their movies, and personal invitations to attend the premiere. (As with the scripts and character development for their movies, the ideas and content for the posters and invitations had been discussed and mapped out during previous writing and discussion segments of the rotations. This conceptual work was done with assistance from structured activities provided by Robert—worksheets and question prompts—pertaining to language features of the genres involved. Robert also fielded questions as he moved about the room. Many activities involved in the unit of work required students to reflect on their work by describing the processes they used to solve a difficulty encountered in, say, using 3D Movie Maker, or to evaluate the pluses, minuses, and interesting aspects of a piece of software.) The trio were being introduced to the desktop features by a gentle and unassuming Year 7 student, Amanda. Amanda patiently demonstrated how to perform the functions, drawing on the students' existing knowledge of computing functions. Then one student sat at the computer, mastering the routines while aiming for the textual effects desired—with suggestions from the others on choice of fonts, borders, etc., and technical responses from Amanda when requested.

Meanwhile, other groups of students were variously engaged in searching through newspapers for reports on inventions, which they would then use to analyze the structure of the genre, practicing for their upcoming oral presentation of their report on an invention or inventor, reading aloud to a teacher's aide, or working on independent projects (e.g., constructing an invention from found objects that will water both the plants and the gardener during hot afternoons). Robert circulated among the groups, monitoring their progress and providing advice or feedback when asked.

Ways of Learning at Abbotsdale

Abbotsdale's ways were characterized by an emphasis on learning through technologies while learning about technologies. The pedagogy was strongly informed by theory. A mix of conventional—traditional, even—and innovative approaches to teaching and learning were employed to integrate use of computer technology into activities in a manner that was as "invisible" and seamless as possible. Robert described new technologies as providing "new contexts" in which to learn. He insisted that the technologies in his classroom not become ends in themselves. Instead, they were employed in ways designed to maximize learning in general, and the development and practice of "higher order thinking skills" in particular. Classroom activities were scaffolded in a variety of ways. Some employed questions prompting students to reflect individually or in groups on a process or tool and/or to evaluate it (e.g., a piece of software, a reference book). Others employed guide sheets assisting students to work from cognitively simple knowledge (e.g., through literal content questions) to more complex understandings (e.g., though questions requiring students to evaluate, extrapolate, analyze and synthesize content and processes). These ways enacted Robert's constructivist theories of learning, and presented opportunities to experiment, explore, play, take risks, and solve problems using resources of more conventional and new technologies.

Year 5 operated as a community of learners, enacting a culture of collaboration within which the students exercised a lot of initiative. During small group and whole class sessions students regularly turned to each other for assistance, feedback, and advice: turning to Robert only when a problem or question proved beyond their own means. It was common during rotations to see a student break away from his or her own group/activity at the request of another and, for example, demonstrate how to access a given file or background scene within 3D Movie Maker, or help with identifying the genre of a particular text. Students were actively encouraged to display and share their expertise for mutual benefit. This was especially evident in peer tutoring sessions run by Amanda to introduce students to new software or hardware, and new applications of familiar software. Robert also actively encouraged collaborative approaches to problem solving through the kinds of activities he structured for students (e.g., pairs searching newspapers for reports; group productions of animated movies), and through his own involvement in shared activities (e.g., helping a student search the internet for information on the Acropolis).

Abbotsdale's ways sought and regularly produced multiplier effects. For example, Amanda's peer tutoring sessions enhanced her self-esteem and confidence in herself as someone with expertise to share, as well as maximizing opportunities for others to become competent users of computing programs and applications, and to work independently of their teacher in achieving syllabus guidelines (e.g., mastering diverse genres). Further multiplier effects flowed from Robert's practice of integrating new technologies into activities on the principle of using computers for things that are best (or better) done on computers. Learning practices steered well clear of decontextualized and fragmented uses of technologies aimed at technical skill mastery alone, or that otherwise reduced new technologies to mere add-ons or "uses for the sake of it." He encouraged use of these technologies to minimize "busy" or needlessly time-consuming work (such as repeated handwritten drafts, or labor-intensive information searches, which produced merely equal or inferior data to what could be got using a computer search engine). By such means, students learned both how to use a range of new technology applications and processes, and when recourse to these provides the best option.

Take #2: Woodville's HyperCard Presentation for Speech Night

Woodville is a geographically isolated rural Preschool to Year/Grade 10 school (100 primary and 200 secondary students) in southern Queensland. It receives some additional state funding under the disadvantaged schools project. At the time of observation, the school was served by a Learning Technology Education Adviser based at a school support centre serving a 150 km radius. During the period observed, the Education Advisor was trialing an Apple QuickTake camera and a scanner in activities with teachers and students from three schools. Like the other two schools, Woodville had purchased its own QuickTake camera on advice from the Education Advisor. The scanner and a computer with large memory capacity used in the activity we observed were owned by the school support centre, and carried from school to school by the Education Advisor on her visits.

The group of Years 8 to 10 students and the Business Studies teacher, Rosemary, were working on their project during lunch breaks and in available timetable spaces and free periods. Their HyperCard presentation was to be used at the fast-approaching end-of-year speech night. Produced mainly on Macintosh Performas, it was designed to be integrated into and to augment the Principal's address, by presenting images of the school year—some captured statically with the QuickTake camera, and others grabbed by video. The Education Advisor trained Rosemary and students together in using the equipment. The production process itself was broken into specific tasks, delegated to individuals or pairs of students who had undertaken responsibility for completing these tasks. At times Rosemary met with the team as a group, and in between they would report to her with photos, video takes, and other artifacts to be incorporated into the presentation.

Rosemary had recently returned to teaching after years away raising a family—"When I left we were in typewriters, but when I came back it was computers." She often felt all at sea with the new technologies but believed it was crucial for Woodville's students to have opportunities to experience current technologies. Rosemary often referred in interviews to the community's geographic and social isolation from "mainstream Australia," expressing concern that the community may be isolated from ideas and that important changes and opportunities in the cities are simply passing them by. She saw activities with digital cameras, multimedia authoring software, scanners and the like, as a way of providing Woodville students with experiences enjoyed by students in the "mainstream" (cities). Hence, she was quick to take advantage of the speech night occasion as a pretext for organizing the group of students around the project and to make use of the Education Advisor's availability and expertise, and of equipment not available at school.

Seeing Power Point presentation software and HyperCard stacks as "stuff of the urban present"—a common tool in business circles, and increasingly popular in school settings—Rosemary threw herself into organizing, teaching, and learning with these students, with the Education Advisor's input, to create an effective HyperCard presentation for a specific purpose within the larger life of the school.

A lunch-time work session began with Rosemary reminding the group that "We're all in this together, otherwise it won't work." The students had been quite enthusiastic about the project at first, but interest began to wane under competition from end-of-year distractions ranging from tests to anticipation of the long summer vacation. Time and equipment constraints were urgent. The group had only two days to access the scanner, the computer, and the Education Advisor. Rosemary recapped the last group meeting, prompting them for the basic configuration of stacks she had suggested the previous week; that is, three stacks of unlimited "cards" or slides built around the outline the principal had written for her intended speech. Rosemary suggested using only three different backgrounds for the presentation—one for each stack—and allocated the task of snapping a digital shot of the school garden and administration block for the first stack to a lad sitting near her.

Rosemary talked about the audience for the presentation—parents and visitors—and emphasized how the students were actually working for the principal in preparing the presentation. Pairs and sub-groups within this group had earlier been given specific tasks to do, such as using the digital camera to photograph tuckshop staff and specialist teachers (e.g., physical education, music, etc.), grabbing image stills from video and converting them to digital images, interviewing staff and students, or using text art software to create headings and the like. Rosemary spent the remainder of the lunchtime session checking the progress of each task and working with the two students charged with setting up the "cards" on the computer. Because time was short, she simply showed and told the different students how to use the equipment and explained as economically as possible the steps needed to complete their respective tasks. The Education Advisor scanned those photos already taken while we interviewed her. (Rosemary and the Education Advisor subsequently completed the scanning and final production of the cards, and the Education Advisor burned a CD-ROM at the School Support Centre to provide a permanent record of the project. The presentation was very well received by the community.)

Woodville's Ways of Making a Presentation

Woodville's ways of producing a HyperCard presentation reflect elements of (what might be called) cultural "Taylorism" and/or "Fordism" (Watkins, 1991). These are evident in the hierarchies involved in different aspects of the production process, the extrinsic forms of motivation employed as Rosemary struggled to keep the students interested and involved, the division of the job into so many tasks, and fragmented involvement of participants as a result. For student participants, choice about content was shaped by the contours of the Principal's speech. The format for gathering and collecting data was pre-given, in the form of a chart of components (headings) and elements of components (sub headings), presented to the students, who worked to a pre-determined formula—although they had some autonomy over detail (what pose they would photograph a teacher in; where the sports field would be photographed from, etc.).

From another angle, Woodville ways exemplify values and practices of mobilizing and organizing people to make something happen, making do with what is available, sharing resources, and the like. They reflect an interesting mix of "finding pretexts," being constructively "opportunistic," and "having a go." Rosemary's starting point was a personal wish to keep students in touch with what she saw as the technological and cultural mainstream. In many ways her parameters and options were circumscribed by her own lack of knowledge, as well as by the Education Advisor's preferences and choices of projects to promote technological learning: namely, a certain range of multimedia productions based on particular applications. Given this, speech night provided a pretext—an opportunity to find a use for what was pre-given; a problem to which an extant solution could be addressed. (Many writers would see this as a specific exemplification of the whole "computers in education scene"—i.e., we have all these things, now how can we find educational uses for them? See, for example, Bigum and Kenway, 1997). Rosemary's seizing on this pretext created a context for students to gain some awareness of certain new technologies, some of their possible uses, and some of their potential links to other technologies (videos and computers; word processing and scanning, etc.); as well as opportunities to work with some of them. "Having a go" involved Rosemary committing herself and securing student commitment to seeing a task through to completion, despite the limits on equipment and immediate access to expertise. This entailed "getting by" as best they could, making use of the availability of the Education Advisor and support center equipment on terms beyond their control. That anything happened at all is a testimony here to considerable good will, hard work, and performance beyond the call of duty by the various participants.

The outcome was a particular kind of slide show type presentation. It contained text, photographs, and video snippets as illustrations of what the Principal was communicating. It was a prop more than a production in its own right. It was very much a functional and pragmatic product, as well as being closed ended. Throughout the process of constructing the presentation there were few opportunities for exploration and experimentation, risk-taking and hypothesizing, because of the nature of the task and the larger circumstances surrounding it—including reliance on the expertise of the Education Advisor. The product, and the uses of the tools that went into it, can be seen as mediating a whole range of social and power relationships between the various participants involved—including the community members who were informed and entertained by the presentation.

Take #3: PowerPoint and *The Simpsons*

Ben (15 years) was sitting at the family computer, demonstrating what had engrossed him for two weeks of his school holidays. The computer was a 133 Pentium desktop with an 8–speed CD-ROM player, color printer, an array of current software, and a public-provider connection to the internet. With a few deft clicks of the mouse Ben opened the PowerPoint program he had recently installed on this computer, selected a file from the directory, and the screen filled with the serene blue and white of a limitless sky flecked with clouds. Suddenly, the screen burst into a bright cacophony of sound, color and movement as the familiar title from an animated television series, *The Simpsons*, dropped onto the screen to the opening lines of the show's theme song. This was followed in quick succession by a selection of images of the Simpson family going about their everyday lives. The title page gave way to a parade of slides, one for each family member and major character in the show. In order of appearance these slides depicted Homer, Marge, Bart, Lisa, Maggie, Itchy and Scratchy, Mr. Burns, Ned Flanders, Krusty, Apu, Moe and Barney, as well as providing a credits page.

Each slide showed a full color image of the character, accompanied by a few pertinent statistics (e.g., age, occupation) and followed by a short text headed:

"little known facts." Images and lines or phrases of text appeared in synchronization with the melody of the song selected to accompany the particular character depicted on each slide. In all, the slide show ran for about six minutes.

"Did you use a manual to help put your presentation together?", we asked. "Oh no," Ben replied. "I just heard from my friends at school that PowerPoint is a really cool program. You can do heaps of stuff on it." The first time Ben had seen a PowerPoint presentation had been at his school's Sports Awards night a few months earlier. "It was all right, but it was just headings for the topics for the night. The backgrounds were cool, though." There were few images in the school presentation, but Ben had been impressed by the idea. He recounted how his best friend, Joe, had later described a PowerPoint presentation he was putting together at home, show casing *The Simpsons*. Ben began to realize some of the possibilities this kind of software might have for exploring his own interests in *The Simpsons*.

Joe helped Ben to get started by giving him a disk of the Simpsons presentation Joe had put together. Spending at least an hour and a half each day for two weeks of his school holiday on his project, Ben searched official and unofficial Simpsons web sites on the internet, downloading additional graphics, scanning in images, and gathering information for the biodata to accompany each character. He supplemented this with information from his Simpsons collector cards. Ben also began experimenting with the palette of functions built into the software and found he was able to attach sounds and music to each slide. This sent him on another search of the internet, where he found hundreds of sound files. Some were compatible with the PowerPoint software. Others had to be converted to the right format using software that he and his father had spent considerable time locating and downloading from the internet. Further ideas and hints for using PowerPoint effectively were gleaned from his mates at school. Ben also discovered the animate picture function tucked away in one of the menu bars. This function enables the user to control the ways images appear on the screen (e.g., "fly from right") and come to rest at a location pre-specified by Ben. He spent hours experimenting with the effects of this function—occasionally resorting to the Help function built into the program—before using it in his presentation, creating a show filled with movement and vitality.

Ben's Ways with PowerPoint

Three dimensions of Ben's ways with PowerPoint stand out. The first is that he drew on, and added continually to, a strong conceptual understanding of computer "workings," "logics," and "potentials," built on a cultural perception of the technology as "tool 'n toy." Making his PowerPoint presentation provided a context for adding to his repertoire of technical skills and understandings while drawing on what he already knew. He used his existing knowledge to predict possible functions and capacities of the PowerPoint program and tested his hypotheses. By this process he added to the range of what his presentation could accommodate. He practiced risk-taking ("I wonder what this might do") and problem-solving: eschewing manuals, and only resorting to PowerPoint's Help menu when he got stumped. This process and growing stock of expertise extended as well to myriad internet resources and functions (such as finding software to convert one kind of sound file to another which was compatible with PowerPoint).

Ben's ways contrast sharply with what might be described as characteristically school ways, which are inherently tied to what we have called "modernist spaces of enclosure" (Lankshear, Peters, and Knobel, 1996). His project was open ended, intrinsically motivated, and "uncurricular"—in that it was not subject to measurement, categorization/classification by subject or genre, reporting, grade commodification, remediation, or timetabled closure. It was quintessentially liberal, in the sense of existing for its own sake. "The Simpsons Presentation" emerged from a popular youth culture space, rather than from teacher or other school-bound directives and routines. Indeed, Ben explicitly contrasted his presentation with the one he encountered within the formal school context. He had multiple intended audiences: himself (primarily), his mates at school, and people at large who are interested in *The Simpsons*. The presentation evolved continually on the basis of experimentation (techniques and effects), exploration (hunting down new resources), and the personal/self-directed pursuit of expertise.

More generally, Ben's ways were strongly embedded in postmodern youth culture (Rushkoff, 1994, 1996). His show developed a complex intertextuality, which built on the intertextuality of the TV show itself. Beyond this, Ben created his own cross references within the way he organized and arranged the slides (e.g., cross references between characters and links within the presentation, and creating implied links to particular episodes of *The Simpsons* through his choice of music and images accompanying the text on each slide). His work reflects his cultural alliances with a particular group of mates and unfamiliars who are "into *The Simpsons*"—a world in which kids are more media savvy, more "meta," than adults, and where a fine border line is negotiated constantly between sacred and profane, conventions and anti-establishment positions, attitudes, and practices, and so on. This comes home, graphically, in what might be seen as Ben's humorously irreverent treatment of PowerPoint software itself, subverting its marketing-oriented slide show templates, which were purpose-built for business suit culture.

Take #4: GRUNT and the Virtual Valley II Project

Virtual Valley II was initiated by CONTACT Inc., a non-profit company supporting youth arts, cultural development and training. CONTACT Inc., the radical offspring of Michael and Ludmila Doneman (see, for example: michaeldoneman.

com) is part of the ongoing Making Space community-based project. This project aims to develop a safe and comfortable, large-scale, multi-purpose youth space and a place for community groups and other youth organizations to converge and interconnect. This physical space is located "between the Roxy and the police beat" (M. Doneman, in conversation) in Brisbane's "Valley," and is known affectionately as GRUNT (Unfortunately, we are no longer able to direct you to the GRUNT website; funding issues brought about its demise . . .). The emphasis is on support for enterprise and self-sufficiency. Unlike drop-in centers and similar facilities, GRUNT, with its online telecenter and multimedia laboratory, works to provide inner city youth with "training in vocational skills, in the mastery of the new information technology and in planning, management and life skills" (Stevenson, 1994: 4).

GRUNTspace consists of three main areas defined within a 100 square meter shell on the first floor level. One area is used mainly for art exhibitions and performances. A second is a general purpose meeting, "hanging," and administrative space, furnished with deep comfortable chairs and decorated with paintings and collages, plus the occasional prop from past performances. The third is GRUNT's main production area with its two adjourning offices. This has been spray painted Star Trek silver and is equipped with ten desktop computers with minimum processing speeds of 486 cpus. A local area network wires each computer to common online storage areas as well as to the internet. An urban-industrial "feel" has been given to the monitors. Metal garbage bins are placed over them, with screen-sized holes cut into the metal of each bin. The resulting flap of metal is peeled up and back like the peak of a baseball cap. The computer boxes sit on tables covered completely by synthetic green sward, leaving only each keyboard and mouse visible. Making the space as "un-school-like" as possible is an explicit operating principle. Multimedia equipment available to GRUNT users includes color flatbed scanners, current sound, text and image authoring software, internet browser software and HTML editors, data panels and projectors, and the like.

Virtual Valley II ran in the last half of 1996. The original Virtual Valley project ran in 1995. Its goal was to produce "an alternative user's guide to the Valley"—an alternative to "official" tourist brochure representations. It presented work by nine young people who used the Valley for "work or recreational purposes and held strong opinions about the Valley's role in 'Australia's most liveable city,'" and about the urban renewal process underway in the area at the time. The result was a website and a booklet designed to guide visitors "through a number of interesting cultural sites using maps and postcard images," as well as a heightened sense among participants of belonging to a particular community that was the Valley. (Quotations were taken from the original, now defunct website). Virtual Valley II was "designed to encourage young people to map the Fortitude Valley Area . . . in ways that (were) culturally relevant to themselves and their peers" (ibid.). Two inner-city schools were involved. We focus here on the cultural production of students from one school that served Murri (indigenous Australians traditionally from the state of Queensland) youth. These students and their teachers were introduced to the internet by the CONTACT team and shown how they could claim spaces within it for their own purposes.

Participants met one morning a week over two months. Sessions varied in specific content and were divided broadly into two approaches to the overall task. One explored "identity" in a real life, workshopping mode. This worked with the preferred media of the participants—painting, drawing, and collage. Performance art/drama also was available, but since it was not a preferred medium it was not used. The other focused on learning technical aspects of web page development, including basic hypertext markup language and web page design principles, using digital cameras, manipulating the resulting digital images and anchoring them to web pages, and using flatbed scanners. Students also gathered material for their web pages on walks through the Valley, using digital and disposable cameras, sketch books and notepads. Students began compiling web pages by creating large-scale, annotated collages of aspects of the Valley that were significant to them. Collages comprised photocopies of digital and camera images they had taken of themselves, their friends, and the Valley area, plus drawings and found objects (e.g., food labels, ticket stubs, bingo cards, etc.). They were then pared back to key images and passages of texts as each student prepared a flowchart depicting the layout and content their web page or pages. During the last month of the project these flowcharts were actualized as web pages.

The end result is a fascinating look at the Valley through the eyes of these participants in its everyday cultural (re)creation. In a typical example, Justin launches his virtual tour of the Valley with word bites that capture poetically images and activities around him:

People allsorts
Ice-cream parlour allkinds
Timezone fun
Dragons
Temples colourful
China Town lots of people

Justin's text is printed in large green fonts (Courier and Times New Roman). Capital letters and italicized words, plus two photographic images, add further details to his pared-down text. The first image depicts the cultural diversity of the people in the Valley. The second underscores fun experiences at Timezone by showing a video game in action.

Following this timeless description of his response to the Valley, Justin shifts to a recount genre, recalling highlights of a particular stroll through the Valley with his class.

susan pretended to be opal
winney [Oprah Winfrey] and we was the audience

one group was police the others
was murries and shop owners after all
that we did some drawings . . .
(from the original, now defunct, web site)

The students arrived "with cut lunches and fireworks energy" (M. Doneman, in conversation). They commandeered the space, making it their own. GRUNT staff and Murri teachers who came with the students responded in turn. For the duration, GRUNTspace became Murrispace—in the physical and virtual domains alike. Learning agendas were set by the students themselves, and the adults became "part of an interface" between students' expressed interests and tangible products of learning. The overall context became one of "policy on the fly and curriculum on the fly" (ibid.). GRUNT staff and the Murri school teachers worked to mobilize and help focus student energies in ways that realized purposes that were meaningful to the students.

Virtual Valley Ways

Virtual Valley's ways with Windows enacted principles and goals that define GRUNT as a distinctive cultural enterprise, beginning with the principle that "this isn't school." The overt aim was to create a learning context that was as informal, unstructured, non-regulatory, and responsive as possible. A maximally open, experimental, exploratory space was created within broad tolerance parameters consistent with meeting the duty of care. The ideal was supported with adult-student ratios of around 1:3. In the company of adults, students set out into the Valley with digital cameras and freedom to frame their own subjects and worked with photocopiers and other equipment to create and reproduce images, which they collated to form visual narratives. These were duly supplemented with written texts and became bases for constructing Web presentations.

These ways reflected also the desire "to create a distinctively Murri learning space within the Valley, even if only for short periods" (M. Doneman, in conversation). Meeting aunties and uncles (older indigenous people within the community who are not necessarily blood relations) in the Valley and visiting favorite haunts associated with life outside school became an inevitable part of learning routines and were reflected thematically in what the students produced.

The Virtual Valley II project can also be seen as enacting an investment in expanding future prospects for Murri cultural, social, and political presence in cyberspace. New technologies were approached very much in terms of providing media and spaces for realizing identities as Murri youth. Elements of Murri youth cultures were explored and researched within virtual settings (e.g., by reference to now defunct Web sites like "Black Voices" and "Perfect Strangers"), as well as within real life settings (the "real" Valley). As conventional physical forms of graphic and written texts and "real life" experiences were carried into "virtual" space, students became aware of both "spaces" as being viable and important sites of practice for identity politics.

Reflections

Such truncated discussions cannot do justice to the depth, richness, and subtlety of social practices like those addressed here. Yet, even the short distance our descriptions go is enough to indicate that in our present clamor to technologize learning we are in danger of short-circuiting important issues and principles and, in the process, shortchanging teachers and learners.

Variations among the ways reported here show that access is a much more complex matter than merely putting hardware and software in schools (or homes). "The same tools" are by no means the same tools. They become very different tools in the presence of the different funds of knowledge people bring to the tools when they pick them up. Issues of physical availability aside, PowerPoint was appropriated in profoundly different ways, and for very different purposes, from the appropriation of HyperCard at Woodville. The result is markedly different literacy events and textual productions between the cases. There is nothing new here. In the hands of Heath's Maintown participants, books and pencils mediated very different educational performances from those they mediated among Roadville and Trackton participants, respectively (Heath, 1983). Having physical access to a pencil or a Pentium is a different matter from having access to funds of knowledge and acquisition histories (Gee, 1996) that enable certain practices to be engaged, and performances elicited, through that physical availability. This, in turn, is a very different matter from having power to influence what kinds of performance are attached to what kinds of further opportunities, social rewards, and life chances—and vice versa.

This is not, of course, a reason for skimping on physical provision. That some learners have greater physical access to tools (or physical access to greater tools) than others inescapably sets up conditions for unequal opportunities and outcomes—especially when the tools in question are part and parcel of esteemed and rewarded social performances. From this perspective, Rosemary's students were objectively at a disadvantage by comparison with Ben, the Abbotsford students, and the Murri students learning at GRUNT, when it came to extended opportunities for hands-on experience—different funds of knowledge and availability of expertise notwithstanding. We would argue that as formal education becomes increasingly devolved to local levels, it becomes absolutely essential to establish guarantees that limit physical access differentials as far as possible. Anything less is socially unjust.

At the same time our snapshots imply that technical proficiency accounts for rather little of the variation between the ways with Windows we observed. They suggest that even if technical training—i.e., training in applications and processes—were held constant, literacy events drawing on these technical proficiencies would vary greatly. Here again, we have known this for a long time but have failed to build the insight into inclusive and democratic educational practices. If anything, the current technicist fetish evident in language and literacy policy emphases is taking us in the opposite direction. Many current approaches to remediation, diagnosis, assessment, and reporting privilege code breaking and limited aspects of text participation over other essential dimensions of becoming successful readers (cf. Freebody, 1992). This creates contexts in which different cultural capitals and funds of knowledge can play out in ways that intensify unequal opportunities for access to social goods (Gee, 1996; Lankshear, 1997). Under such conditions, current demands for more professional development and inservicing are often under-informed and betray a magical consciousness (Freire, 1972) of the powers of training packages.

As with the issue of access, however, this does not mean holding back on demands for more and better professional development and inservice teacher education—or, for that matter, preservice teacher education. On the contrary, it means, that we need to make better informed demands and to meet these demands with better informed responses. This entails widening our focus on the issues surrounding the role and place of new technologies within education generally, and literacy education specifically. Apart from anything else, efforts to better prepare ourselves for integrating new technologies into successful and inclusive language and literacy education must include serious engagement with practices, theory, and research that identify and explain differences among ways with words and Windows, and the social, economic, and cultural legacies of these differences under present conditions.

Right now, we are caught up in policies and processes of further technologizing education without the necessary philosophical base and political commitments to give weight to policy rhetoric about new technologies contributing to making education more equitable, inclusive and empowering. What kinds of social practices—computer-mediated and otherwise—will contribute to making more equal the prospects of all human beings to live more satisfying and dignified lives? (cf. Paulo Freire's notion of humans living their humanity more fully; Freire, 1972) What principles of economic and social distribution are presupposed by this ideal? What role can and should education play in identifying and promoting these principles and practice that accords with them? On what bases should we estimate the educational worth of varying social and cultural ways (with words, Windows, whatever)? What is the significance of the fact that the contemporary technological revolution is accompanied by economic and social policies and practices that are increasing dramatically the income gap between the top 10% and the rest within societies like our own? (Gee, Hull, and Lankshear, 1996; Reich, 1992). Unless and until our conceptions, practices, and policies concerned with (language and literacy) education and new technologies—from issues of access, professional development, and teacher education, to concerns about inclusive curriculum and educational purposes—are informed by deep and protracted engagement with such issues, this latest round of "technologizing education" will merely position "ways with Windows" where we were under earlier technological regimes. This was a place where education was already an expensive and, for many, soul-destroying investment in legitimating the principle: Some ways will be honored, and others will not.

Accordingly, we conclude that whatever else we do pedagogically with our new technologies, we should aim to integrate them into informed practices of critical social literacy (Freire, 1972; Shor,1992; Muspratt, Freebody, and Luke, 1997). Unfortunately, there is no easy recipe or short cut to meeting that aim. We hope, however, that this paper, its bibliographic references, and their extended family of references to resources, research efforts, reported practices, and informing theories, provide at least a coherent starting point for our continuing growth as educators for better times.

Bibliography

Aronowitz, S. and Giroux, H. (1993). *Education Still Under Siege*. Westport, CT: Bergin & Garvey.

Bigum, C. and Green, B. (1992). Technologizing literacy: The dark side of the dream. *Discourse: The Australian Journal of Educational Studies*, 12(2): 4–28.

Bigum, C. and Kenway, J. (1997). New information technologies and the ambiguous future of schooling—Some possible scenarios. In A. Hargreaves, A. Lieberman, M. Fullan, and D. Hopkins (eds), *International Handbook of Educational Change*. Toronto: OISE. 375–395.

Bruce, C. (1996). Technology as social practice. Mimeo. Published in Educational Foundations, Fall.

Clinton, W. (1997). Text of the state of the union address. *The Washington Post*, February 5. Accessed Feb. 20, 1997: http://www.washingtonpost.com/wp-srv/national/longterm/union/uniontext.htm.

Franklin, U. (1990). *The Real World of Technology*. Montreal: CBC Enterprises.

Freebody, P. (1992). A socio-cultural approach: Resourcing four roles as a literacy learner. In A. Watson & A. Badenhop (Eds.), *Prevention of Reading Failure.* Sydney: Ashton-Scholastic, pp. 48-60.

Freire, P. (1972). *Pedagogy of the Oppressed.* Harmondsworth: Penguin.

Gee, J. (1996). *Social Linguistics and Literacies: Ideologies in Discourses.* New York: Routledge.

Gee, J., Hull, G., and Lankshear, C. (1996). *The New Work Order: Behind the Language of the New Capitalism.* Boulder, CO: Westview Press.

Green, B. and Bigum, C. (1996). Hypermedia or media hype? New technologies and the future of literacy education. In G. Bull and M. Anstey (eds), *The Literacy Lexicon.* Sydney: Prentice Hall. 193–204.

Heath, S. B. (1983). *Ways with Words: Language, Life and Work in Communities and Classrooms.* New York: Cambridge University Press.

Illich, I. (1971). *Deschooling Society.* London: Calder Boyars.

Kearns, D. and Doyle, D. (1991). *Winning the Brain Race: A Bold Plan to Make Our Schools Competitive.* San Francisco: ICS Press.

Kemp, D. (1997). *The National Literacy Policy for Australia.* Belconnen, ACT: Language Australia.

Lankshear, C. (1997). Meanings of "literacy" in educational reform. Invited paper presentation to Department of Education, University of East Anglia, and Warner Graduate School of Education, University of Rochester, March 1997.

Lankshear, C. and Knobel, M. (1997). Literacies, texts and difference in the electronic age. In C. Lankshear, *Changing Literacies.* Buckingham and Philadelphia: Open University Press. 133–63.

Lankshear, C., Peters, M. and Knobel, M. (1996). Critical pedagogy and cyberspace. In H. Giroux, C. Lankshear, P. McLaren, and M. Peters, *Counternarratives: Critical Pedagogies and Cultural Studies in Postmodern Spaces.* New York: Routledge. 149–88.

Moll, L. (1992). Literacy research in community and classrooms: A sociocultural approach. In R. Beach, J. Green, M. Kamil and T. Shanahan (eds), *Multidisciplinary Perspectives on Literacy Research.* Urbana: NCRE/NCTE. 211–44.

Montgomery, K. (1996). Children in the digital age. *The American Prospect.* 27 (July-August): 69–74

Muspratt, S., Freebody, P. and Luke, A. (1997). *Constructing Critical Literacies: Teaching and Learning Textual Practice.* Cresskill, NJ: Hampton Press.

National Commission on Excellence in Education (1983). *A Nation at Risk. The Imperative for Educational Reform.* Washington DC: U.S. Department of Education.

Reich, R. (1992). *The Work of Nations: Preparing Ourselves for 21st Century Capitalism.* New York: Vintage Books.

Rushkoff, D. (ed.) (1994). *The GenX Reader.* New York: Ballantine Books.

Rushkoff, D. (1996). *Media Virus: Hidden Agendas in Popular Culture.* New York: Ballantine Books.

Shor, I. (1992). *Empowering Education: Critical Teaching for Social Change.* Chicago: University of Chicago Press.

Stevenson, P. (1994). Making space: For those who invent tomorrow, a report on the feasibility of a multi-purpose youth facility for Brisbane. Brisbane: Queensland Department of Tourism, Sport and Racing and Queensland University of Technology Academy of the Arts.

Watkins, P. (1991). *Knowledge and Control in the Flexible Workplace.* Geelong: Deakin University Press.

Winters, K. (1996). America's Technology Literacy Challenge. Washington D.C.: U.S. Department of Education, Office of the Under Secretary, <k.winters@inet.ed.gov> posted on <acw-l@unicorn.acs.ttu.edu> 17 February.

NINE

"I'm Not a Pencil Man": How One Student Challenges Our Notions of Literacy "Failure" in School (2001)

Michele Knobel

Biography of the Text

This paper was originally published in the *Journal of Adolescent and Adult Literacy* (volume 44, issue 5). It grew directly out of my doctoral research work, which comprised a case study of four very different young people in Grade 7 in Brisbane, Australia. My doctoral thesis itself was located within the context of a new state syllabus, which was grounded in systemic functional linguistic and genre theory conceptions of language use. The syllabus made significant claims about how powerful uses of language could be taught explicitly in school; that is, by teaching particular text genres and their accompanying linguistic features, students could become socially powerful users of language. My research findings stood as a critique of such claims, showing that the Discourses in which my four very different research participants engaged outside school were far more influential in their ability to use language and texts effectively than what they were learning about language and texts in school (see Knobel, 1999). "I'm not a pencil man" takes the case of Jacques—a Grade 7 student with a history of failing literacy at school, but who is doing just fine outside school—and reframes my analysis of his case within the larger trend towards content-focused curricula, student-learning performance standards and standardized testing. The principal argument remains the same, however, that many young people's out-of-school proficiencies with language, texts and ways of being in the world are not leveraged to their benefit in schools.

Introduction

Many researchers agree that we are currently living in New Times (Castells, 1996; Hall, 1991; Lankshear, 1997). These New Times are characterized by the rise of multinational companies that increasingly seem to wrest control of nations away from governments. At the same time, economic production in the developed world is shifting from high-volume to high-value outputs and workers are expected to be multiskilled and trainable (Organization for Economic Co-operation and Development, 1998). Moreover, much of this high-value production is done in countries where labor is cheap, by people who have little hope of purchasing or enjoying the kinds of goods they produce.

Schools have a significant role to play in embracing and critiquing New Times, rather than trying to domesticate them or keep them at bay (de Alba, González, Lankshear, and Peters, 2000; Lankshear and Knobel, 1998). By coming to understand New Times, students will be better prepared to combat or resist alienation, cultural loss, identity dispersion, family fragmentation, and dependence on nonlocal corporations for livelihood rather than self-sufficiency, some of the social costs of New Times. In terms of addressing these social costs and in developing tactics for negotiating New Times, what students now need to learn is—and should be—vastly different to what was required in the not-so-distant past to maximize people's quality of life chances (cf. de Certeau, 1984; Lankshear and Knobel, 2000; Luke and Elkins, 2000).

However, many government responses to New Times around the world have focused on constraining what students learn by means of national curricula, increased national and state testing accountability checks for teachers, and mandated standardized tests. Without a doubt, literacy seems to have become a hot topic for governments everywhere—with many of them insisting that literacy is mostly to do with learning to read, write, and spell (Department for Education and Employment, 1996; Department of Employment, Education, Training and Youth Affairs, 1998; Ministry of Education, New Zealand, 1998; Secretária de Educación Pública, 1999; United States Department of Education, 1997). For example, in Australia at present the government has implemented a set of reading, writing, and spelling tests in Grades 3, 5, and 7. Test outcomes are indexed to a set of benchmarks that indicate whether a student performs at a standard, proficient, or excellent level (or doesn't meet the benchmark at all, which means the student needs remedial reading, writing, or spelling lessons). These benchmarks are in keeping with the latest national literacy goal for Australian education—that "every child leaving primary school should be numerate, and able to read, write and spell at an appropriate level" (Department of Employment, Education, Training and Youth Affairs, 1997: 1). The move to benchmarks and national testing in Australia is indicative of similar moves in other countries (e.g.,

Clinton, 1997; Department for Education and Employment, 1996; Ministry of Education, New Zealand, 1998; Secretária de Educación Pública, 1999).

The conception of literacy developed by means of national curricula, benchmarks, and standards is *school literacy*. This encompasses those literacy skills that most often lead to success in school (e.g., correct spelling, being able to write and speak abstract texts, being able to write five-part essays, being able to read and write for no obviously meaningful social purpose). Interestingly, however, a great deal of research shows that school literacy on its own—and as currently configured in most schools around the world—does not necessarily guarantee success in literacy practices for out-of-school contexts (Gee, Hull, and Lankshear, 1996: Lankshear, 1997; Lankshear, Peters, and Knobel, 2000; Luke, 1993; Mahiri, 1997; Prakash and Esteva, 1998; Prinsloo and Breier, 1996).

This brings us to the heart of the matter. Setting standards and benchmarks always brings into question what it is that students should know, to what ends, and how what they know is most effectively measured (Shepard and Bliem, 1995). As critical literacy approaches to education have long shown us, "the selection and organization of school knowledge contains dispositions and values that handicap certain groups while they benefit others" (Popkewitz, 1991: 151; see also Edelsky, 1990: Gee, 1996; Lankshear and McLaren, 1993; Macedo, 1994: New London Group, 1996).

Standard-setting and benchmarks only seem to make it easier for students to "fail" because "literacy"—or more accurately in such contexts, reading, writing, and spelling—is constrained to *school* literacy. However, focusing solely on school literacies at the expense of literacies that students practice out of school is for many students a grave injustice because it invalidates those literacies in which they are fluent and effective out of school. This flash point brings into play analyses of relationships between school and everyday-life experiences in any consideration of what counts as an effective language and literacy education for young people. Such analysis becomes troublesome when we focus on benchmarks and standards in relation to students who have a history of literacy "failure" at school yet engage successfully in a rich range of literacy practices outside school settings. The case of the 13–year-old discussed in this article is a telling example. It calls for educators to think carefully about what they assess as "literacy" in school and to reflect on how they respond to literacy "failure." My aim here is to analyze aspects of the relationship between school learning and one student's everyday life. I also consider the implications this might have for how educators think about teaching literacy in the face of increasing governmental assessment and analysis of literacy learning outcomes.

Meeting Jacques

Jacques (all names are pseudonyms) is stocky with a beaming smile and an infectious laugh. He has short, light brown hair and clear blue eyes. He likes to dress comfortably out of school and mostly wears T-shirts and shorts at home. Jacques wears a uniform to the large Australian public school he attends, but always manages to look somewhat disheveled in it. His talk is laced with hilarious witticisms and parodies of people, situations, and remembered conversations. Jacques appears to be well-liked, and everyone—including his teacher—calls him by his nickname, J.P.

Jacques' school literacy practices

In class Jacques sits at a group of desks close to the chalkboard at the front of the room. He either moves restlessly in his chair or sits motionless staring at the busy road that lies beyond the large bank of windows lining one wall of his classroom. By his own admission, he is easily distracted and often loses track of what is happening in class: "I get distracted a lot, by other things y'know, if they're doing something better, I'd rather- my attention's on them instead of on my work." His teacher rates him as "having great difficulty" with literacy. Jacques repeated Grade 1 and appears to have a history of school failure.

It was apparent right from the beginning of the two weeks I spent closely observing Jacques in this classroom that he was not enamored of school. He seemed to be patiently enduring school until the time came for him to be allowed legally to leave in Grade 10. Jacques claimed he has no intention of continuing his formal education into senior secondary levels, and he openly declared: "I don't like school very much." When I asked about his reading and writing practices Jacques closed down the conversation by declaring, "I'm like my dad. I'm not a pencil man."

The "snapshot" in Figure 1 is a typical example of Jacques' "participation" in a lesson. The lesson makes use of a proforma—in this case an "open compare and contrast" photocopied worksheet with a paragraph at the top describing a range of Balinese customs and two columns of lines for writing on beneath it. One column is headed "similarities" and the other "differences." This lesson is also typical of the lessons conducted by his teacher, Ms. Bryant. The indented paragraphs in Figure 1 indicate what Jacques and his classmate Sean were doing as part of the "underlife" of the lesson (cf. Gutierrez, Rymes, and Larson, 1995).

Jacques rarely contributes to lessons voluntarily and is called upon by his teacher infrequently to furnish answers or information for the class. Even when working in groups, he usually waits for others to make suggestions, asks them for the answers, or simply copies what they have written. Much of his work remains incomplete, or mysteriously becomes "lost." Indeed, Jacques does very little schoolwork unless constantly supervised and has developed a range of elaborate avoidance strategies. These include spending time looking for items he seems to have misplaced, delegating tasks to others (especially to his best friend, Sean),

(Monday 14 November 1994: Day 7 of observations)

Event: Language lesson

Sub-event 1: Reading information sheet (pair work)

(9:40 a.m.) Ms. Bryant tells students to take out their language books and explains that today they'll be working in pairs. She hands out a photocopied information sheet on Balinese customs to the students, along with an "Open Compare and Contrast" proforma. Ms. Bryant tells students to take turns reading.

Sean starts reading, and Jacques mutters to me, "I hate reading. It's boring." Sean keeps reading, and Jacques comments "Boring, hey." "You're not wrong," replies Sean. "C'mon. Keep going," counters Jacques and then pretends to fall into a deep sleep. They discuss Sean's recent trip to the beach, and Jacques talks about accidentally knocking over a girl while roller skating at a local rink.

Sub-event 2: Learning task (whole class)

(9:50 a.m.) Jacques and Sean have read only three of eight paragraphs when Ms. Bryant brings the class together again. She explains again that she wants them to compare and contrast Balinese and Australian customs.

Jacques pretends to give Sean electric shocks and is reprimanded by Ms. Bryant.

Ms. Bryant asks various children to read aloud consecutive paragraphs. She identifies things in common to both cultures and points out some of their differences.

Jacques and Sean share a running metatextual commentary on proceedings. Ms. Bryant talks about baptism and Jacques declares softly, "Yeah, and Sean was dropped."

Ms. Bryant directs the class to complete the compare-and-contrast proforma, using information from the sheet and what they know about Australia.

Sub-event 3: Work task (pair work)

A student asks Ms. Bryant what they're supposed to write, and she tells him with a sigh that she has already told the class twice.

Jacques turns to Sean and asks with a wide yawn, "What have we gotta write? She didn't even say." Sean doesn't seem to know, and they ask Nikki, who tells them to write something about baptism. Jacques asks Ms. Bryant to clarify the task, claiming he couldn't hear what she was saying.

(10:15 a.m.) Ms. Bryant explains again. Sean works on the proforma while Jacques sits, yawns, looks around, or fiddles with a pencil.

Figure 1: Snapshot: Compare and contrast

"helping" others instead of working (e.g., filling glue pots), claiming he hasn't heard Ms. Bryant's instructions, and spending large blocks of time "planning" what to do during student writing sessions. Frequently, these avoidance strategies seem to lampoon the school work he is set to do by his teacher (see Figure 2).

(Wednesday 16 November 1994, 3:00 p.m. Day 9 of observations)

Event: Teacher interview 2 (utterance 047)

Sub-event: Talking about Jacques assembling miniature books in which he writes 10-word stories about himself.

Ms. Bryant: I had a corner set up of ways to publish stories, and he would take-I used to fight with him, because he'd take *2 days* to get the paper cut out and stapled. He was wasting time because he didn't want to write. Yet, he got a lot of approval from the rest of the class for those books. I can remember him reading them out, and they'd be laughing . . . so he continued writing them . . .. He doesn't like to write—it's difficult because his spelling is poor. And for that reason he avoids writing; I find that he tries not to take it seriously. He tries to make a joke of his writing in all his stories. In first term all the children did general process writing, and he made these little books called "J.P.'s Stories." He made about six of them, and the kids thought they were hilarious. But there was nothing in them. Like, they might have had 10 words at most in them. They were very, very childish.

Figure 2: Avoidance strategies

Of course, given Jacques' comments about schooling, his subversion of Ms. Bryant's writing activities can also be interpreted as a refusal to engage in tasks that have little real-world meaning for him rather than simply as his having a problem with writing.

Jacques' Everyday Literacy Practices Out of School

Jacques' life outside school contrasts dramatically with his life in school. As practicing Jehovah's Witnesses, Jacques and his family are closely involved with church outreach work in local communities. Jacques' father owns a successful earth-moving business and is an elder in the church. Jacques' mother runs their home and is heavily involved in volunteer church work and other activities each day. Jacques and his family attend Theocratic School every Thursday night. Theocratic School has a dual purpose; it provides ministry training and acts as a forum for knowledge sharing (Watch Tower Bible and Tract Society of Pennsylvania, 1971). In Jacques' case specifically, every so often he has to write an introduction and conclusion to a Bible reading, which he then presents to the congregation (often up to 100 people). The introduction and conclusion usually tie the reading to current issues or to personal experiences in ways that help explain the meaning of the Bible text for everyday life. This reading, speaking, and exegesis is publicly evaluated by members of the Theocratic School who use checklists and criteria from the School literature to assess his performance,

Indeed, Jacques' church commitments are very much part of an adult world, and Saturdays see him dressed in a suit and carrying a briefcase as he, his family, and others, in his group go "witnessing" in their allotted "circuit." This work requires Jacques to be familiar with the literature they show to people and involves his discussing sophisticated concepts and understandings about personal values, religious beliefs, and contemporary social issues with a range of people.

Being a member of the Jehovah's Witnesses involves Jacques in a wide range of reading, writing, speaking, and listening practices. Although Jacques sometimes engages in these practices reluctantly, the amount of public and private reading, public speaking, and discussion he does in connection with his church far outstrips his application to literacy activities at school. Continual self-development is a recurrent but loving theme in Jacques' parents' talk and is not necessarily tied to academic success. In Jacques' case, for example, they actively encourage his interest in his father's earth-moving business, and he is free to stay home from school camps and the like in order to accompany his father to work.

Jacques' world of work is very much an adult world, where he is expected to assume adult responsibility. His conversations at home and elsewhere are interwoven with references to "work" and "being a worker." Thus, when talk at school turns to hit songs and trendy clothes or about being a surfer or a homeboy, Jacques either does not join the conversation or contributes comments that usually bring the discussion to an abrupt halt (e.g., "I'm nothing. I'm just a workin' man"). He definitely was a worker. He had learned to drive work vehicles at 7 years of age and operated them under supervision regularly during school breaks. Jacques understood and valued learning in relation to being able to perform well as a worker. "If you want to put gravel on a road you have to be able to work out how many meters you need."Jacques also puts his knowledge of the business world and his skills as a worker to his own personal use. For example, with the help of his mother and brother, Jacques used his father's computer to compose a flier advertising "JP's Mowing Service" (see Figure 3).

J.P.'s Mowing Service

- Efficient, reliable service.
- Grass clippings removed.
- All edging done.
- First time lawn cut FREE!
  (only regular customers)
- For free quote Ph 5551-2121

Figure 3: Jacques' promotional flier

He had posted this flier in local letterboxes and had quickly established a profitable weekend and summer-holiday business. When asked about the language used in the flier, Jacques explained he had included '"First time lawn cut FREE!" in order to entice customers; or in his words, "So they all go, 'Oh yeah, this is great' [mimes a double take] 'Whhhhaaattttt!' [grabs the flier] 'What's that number again?'" [mimes dialing frantically]. Capital letters for *free* plus the repetition of "free" in relation to a "quote" are all strategies used by business people to attract customers. The smaller font for "only regular customers" also complies with the genre of business fliers and emphasizes Jacques' sophisticated understanding of the way things are done in the business world.

Jacques' life out of school appears to have much more meaning and purpose for him than life in school. For example, as a "worker," Jacques engages in real-world tasks that are purposeful and meaningful for him. He is expected to conduct himself as an adult and is involved in various kinds of autonomous and team work. He has access to guided participation, opportunities to do things himself, and some understanding of the links between work and livelihood (Rogoff, 1995). Such opportunities and knowledge are obtained from observing and working with his father, watching him ordering machine parts and other materials during the day, and seeing him balance accounts at night. Likewise, Jacques is a fluent and mature speaker beyond his years and he participates in witnessing and in Theocratic School. Unfortunately for Jacques, the expertise he has acquired is not valued academically within his classroom, Even when opportunities arise for Jacques to demonstrate his mastery of business discourse such as his thriving car sales business established during *Earn and Learn* sessions (in a role-play simulation of a community [Vingerhoets, 1993], Jacques was a millionaire, while many other students were bankrupt)—this simulation is seen by his classmates and teachers as only a game they play on Friday afternoons, rather than as an opportunity to be successful students.

Interestingly, many current economic and social theorists agree that the nature of work and the roles of workers are changing as modes of production and consumption change in New Times (Aronowitz and DiFazio, 1994; Castells, 1996; Gee, Hull, and Lankshear, 1996; Howe and Strauss, 1993; Organization for Economic Co-operation and Development, 1998; Reich, 1992). Economic success in the future, according to these writers, will depend at the very least on the ability to identify problems and to predict future life chance by analyzing the past, present, and future. In their terms, expert performance in this kind of analytic work requires metalevel understanding of consumer and business practices. Thus, Jacques' increasing metalevel understanding of business enterprise and his ability to put these understandings to work in practical ways (such as in his lawn mowing business) will stand him in good stead with regard to economic prospects in the not-too-distant future. Indeed, despite technically failing primary school, Jacques seems confident—as do his parents—that he will be successful out of school.

Jacques' Case Is Not Atypical

It seems that current language lessons in his classroom have no bearing on the real world for Jacques. In terms of reading and writing standards and benchmarks, Jacques would most certainly be shunted into intensive remedial programs that would be likely to alienate him even further from school forms of literacy (cf. Moje, Young, Readence, and Moore, 2000). The 2 weeks spent observing Jacques for this study confirm a rupture between what Jacques is doing in school and the sets of adult practices in which he is fluent out of school.

Although Jacques' experiences with literacy in school and his literacy practices out of school cannot be generalized to all students everywhere, there is sufficient evidence to suggest that his case is not atypical (see, e.g., "Deena" in Cazden, 1988, and Gee, 1996; "Lem" in Heath, 1983; "Vinnie" in McLaren, 1994: "Murph" in Walker, 1988). These kinds of cases of young people "failing" to achieve literacy in school, but not failing in their lives outside school, have enormous implications for the role and outcomes of national testing, benchmarks, and across-the-board standards. Indeed, these cases call for educators to problematize nationally mandated testing as a measure of "literacy learning" and to ask what counts as literacy learning and failure according to these tests and standards (cf. Alloway and Gilbert, 1998; Kincheloe, Steinberg, and Gresson, 1996). Educators need to reflect critically on what learning and expertise is overlooked when pencil-and-paper tests are used to assess a student's learning, in order to be sure that they are not playing into the hands of injustice.

In addition, because of the impoverished conceptions of literacy in national curricula and benchmarks, it is important for teachers to guard against letting the tests and standards—rather than sound literacy and learning theories—direct their teaching. Indeed, time constraints and pressures on primary school teachers to produce students who are able to write, read, and spell at an "appropriate level" by the time they reach secondary school encourage transmission or banking models of teaching that focus on content alone or that promote "quick fix" approaches to learning (cf. Freire, 1972; Knobel, 1999; Lankshear and Knobel, 1998). Transmission approaches to teaching are abstracted from meaningful contexts and real world practices, so that students are often left wondering *why* they are doing a task, memorizing these words and not others, or completing work sheets and text proformas. The content of transmission approaches to teaching is usually static, grounded in the experiences and knowledges of the dominant—and often minority—group in a society, and often out of date (cf. Freire, 1972; Heath, 1983; Lankshear, Peters, and Knobel, in press; Macedo, 1994; Prakash and Esteva, 1998). None of this promotes *real*, useful learning (Heath and McLaughlin, 1991; Knobel and Lankshear, 1995; Lave and Wenger, 1991; Rogoff, 1995).

Despite education standards in most countries being linked explicitly to the health of the national economy, most standards packages don't take into account the myriad forecasts for the kinds of workers needed or even what it will mean to be a citizen in 10 years' time. Reich's (1992) predictions concerning the role of symbolic analysts in workplaces in the next millennium suggested that current students need to be taught much more than merely school-selected content and how to write or construct arguments by simply "filling in the blanks" (see also Aronowitz and DiFazio, 1994; Castells, 1996; Howe and Strauss, 1993; Lankshear, 1997, 1999). His predictions re-emphasized the problems inherent in reducing language and literacy to a list of demonstrable skills and insulating classroom language and literacy learning from "real world" practices. This situation is only underscored by the case of students like Jacques.

So What's a Teacher to Do?

The suggestion in this article is not that teachers should ignore testing schedules, benchmark evaluations, and the like; many teachers have tried to rebel but have usually found themselves marginalized or out of a job (cf. Searle, 1998). Rather, it is important to emphasize that standards tests and benchmarks should only ever be the *baseline* for a teacher. That is, effective teachers will always go beyond the standards and benchmarks to enact meaningful and richly conceived literacies in their classrooms.

There are a number of conceptions of literacy that have grown out of sociocultural theory that can serve as useful guides to planning for literacy teaching in classrooms in ways that take students beyond the skills level of standards testing and benchmarks. These include, among others, Freebody and Luke's four reading practices and Lankshear and his colleagues' three-dimensional model of literacy. I will discuss these conceptions in further detail.

Luke and Freebody (1997, 1999; also Freebody, 1992; Freebody and Luke, 1990) outlined a conception of reading that encourages a multipronged approach to classroom language and literacy planning and teaching. Although Luke and Freebody focused primarily on reading, their insights apply equally to all other forms of text processing and practice. They emphasized how in current times "the means and practices of reading had changed significantly in important ways in relation to cultural, economic, and social developments" (1997: 191). In other words, their four reader resources are grounded firmly in a sociocultural theory of literacy (i.e., collectively, they take the social and cultural dimensions of "being human" into account). Accordingly, their set of practices and resources embodies historical and relatively new perspectives on reading pedagogy and includes coding practices, text-retelling practices, pragmatic practices. and critical practices. These reader "resources" are used as elements of "successful reading" (Freebody, 1992: 49) and embody in classrooms "a richer understanding of literacy that recognizes and builds on students' prior cultural resources, experiences, and knowledges in all instruction and programs" (Luke and Freebody, 1991: 221).

In brief, Luke and Freebody's (1997) "coding practices" describe processes whereby decoding skills and habits are used to "unlock" alphabetic codes or scripts (the kinds of literacy practices most often measured by standards and benchmarking processes). Text-meaning practices inscribe the reader as a participant; someone who is able to make meaning from a text by bringing to it additional cultural and social knowledge about texts, the subject matter, and statements and discourses in the text. Pragmatic resources are brought into play in text-mediated social activities or events. That is, readers engage in using texts effectively by being able to match texts and contexts and tailor their reading accordingly. Freebody (1992: 53) warned, however, that these pragmatic resources for reading are "transmitted and developed in our society largely in instructional contexts, some of which may bear comparatively little relevance to the ways in which texts need to be used in out-of-school contexts."

Finally, critical reading practices engage the reader in analyzing probable world views and assumptions constituting a text. The reader as text analyst and critic is acutely aware that texts are also idea and value systems, and that readers often comply unquestioningly with the positions offered within these systems. Critical practices in reading require the reader to have metalevel understanding of how, why, and in whose interests texts work and be able to employ complex sociocognitive processes in interrogating representations of "how things are" in texts and how this compares with how things are in the real world (Luke and Freebody, 1997; see also Kress, 1985; Wallace, 1990).

These practices are not meant to describe neat categories of activity, rather they are one way of trying to make explicit the complexities invoked in teaching students to be effectively (or "properly") literate. They also offer insights into how we might go about teaching students like Jacques. Despite his difficulties with coding, participating in, and using aspects of literacy lessons in school, Jacques is highly literate in the codes, meaning-making, uses, and practices of the business "discourse." One way of drawing on Jacques' expertise could be to take his status as a millionaire in the role-playing game *Earn and Learn* and have him construct a guidebook based on his experiences and practices in the simulation and his expert knowledge of business discourse. Such a text would require him (a) to read or view other texts, or to draw on his knowledge of other texts; (b) draw on his participation in the world of work, as both a worker and business owner (his mowing business in the real world and his car sales business in the simulation): and (c) to draw on his use and analysis of the business discourse in terms of constructing his mowing business flier and preparing catalogues of cars he was selling in the role play. Of course, Jacques' out-of-school reluctance to write (he claimed to be no great fan of computers, either) might mean that the handbook has to be audiotaped or videotaped.

In terms of developing Jacques' (and other students') critical reading practices, the teacher could consider generating a unit of work for the class that focuses on the effects of changes in work machines on actual work and livelihood. The unit might, for example, take in texts about the Industrial Revolution in England and the Luddites' attempts to save jobs being lost to steam-powered weaving looms and lace spinners. It could embrace analyses and critiques of texts that portray men and woman at work and that discuss wage inequities, child labor, and high-cost, high-status apparel made in labor-cheap countries.

Another way of approaching how to teach students like Jacques is to use the three dimensions of literacy first sketched by Green (1988) and further developed in the national *Digital Rhetorics* project report (Lankshear et al., 1997). The three dimensions of literacy have also emerged from a sociocultural theory of literacy. In comparison with the four reading practices discussed earlier, the cultural aspect of the three dimensions of literacy encourages teachers to think more explicitly in terms of social and cultural practice, rather than focusing on the effects of texts. For example, emphasizing sociocultural practices enables us to understand how something could be simultaneously a religious artifact or ritual in one cultural context, a work of art in another, and propaganda in a third. Engaging with the sociocultural practices of literacy is crucial if we are to develop strong arguments for, and practices in, going beyond the "basic skills" of standardized tests and national curricula.

As with Luke and Freebody's four reading practices, the three dimensions of literacy proposed by Green, Lankshear, and others are interrelated and inseparable dimensions—operational, cultural, and critical—that span and integrate context, language and meaning. These dimensions are not restricted to written texts either, but include the full range of literacy practices in which people engage. The operational dimension of literacy emphasizes the language system itself and how it is used by people in "order to operate effectively in specific contexts" (Green, 1988: 160). In other words, this dimension names what people do when they are able to read, write, speak. and view effectively in terms of the socially recognized purposes of texts (e.g., to entertain, to instruct, to inform).

The cultural dimension, as we have already touched on it, focuses on the meaning in texts made by contexts and sociocultural literacy practices into which we are socialized. That is, "[t]he cultural dimension [of literacy] involves understanding texts and information in relation to contexts—real-life practices—in which they are produced, received and used" (Lankshear and O'Connor, 1999: 33). This dimension of literacy enables people to understand what makes "particular ways of reading and writing [and speaking, viewing, listening] appropriate or inappropriate, adequate or inadequate in a given situation or setting" (ibid.: 35). For example, Jacques' flier was not simply a completed cloze exercise that was to be pasted into a book and then thrown out at the end of the school year. Instead, the wording was aimed directly at people who needed their lawn mowed. It was carefully designed to alert them to Jacques' mowing business and to entice them to employ him ("first time lawn cut FREE!") on a regular basis (for "only regular customers"). The wording itself is appropriate to the task—the business-like words in an advertising context clearly convey Jacques' message.

The critical dimension of literacy "has to do with the socially constructed nature of all human practices and meaning systems" (Lankshear, 1998: 46). It is concerned with identifying and critiquing the sectional and selective meaning systems in which a text has arisen and thus, access to the principles of selection and interpretation that take them beyond being merely "socialized into the meaning system and unable to take an active part in its transformation" (Green, 1988: 162). Certainly, for proponents of the three-dimensional model of literacy, positive *transformation* for as many people as possible is a key goal.

At a practical level, and again taking Jacques as our example, the operational dimension of literacy was well and truly covered in his classroom (e.g., the compare-and-contrast proforma). In terms of the cultural dimension of literacy, his teacher did make efforts towards developing this dimension by means of a Writers' Center and its grounding in the Process Writing movement, which tried to bring professional writers' writing practices into the classroom. However, students were not apprenticed into a writing culture per se (where their writing was for a real audience or for a meaningful social purpose), nor was a wide range of genres encouraged in and by the Writers' Center—indeed, the teacher expected students to use the center only in writing narratives—which effectively excluded Jacques (and possibly other students in the class) from participating due to his lack of interest in producing this genre.

However, imagine Ms. Bryant embracing the three-dimensional model of literacy. We then see that in her classroom the Writers' Center is gone and in its place is a large pinboard full of notices from the community. These ask for help with producing animal lost-and-found notices, formatting curriculum vitae and form letters, and with compiling and publishing newsletters and advertisements for a range of organizations and services (from washing and ironing to computer tutoring and programming). In one corner a group of students is meeting with the president of a mothers' group who is proofreading a flier the students produced for the group's next summer fundraiser—an excursion to a distant rainforest. In another corner, students are putting the finishing touches on a beach scene (complete with tissue paper palm trees) for a digital photograph they're taking for a poster on discount eco-tourism holidays commissioned by a local travel agent.

A group in the middle of the room is brainstorming possible interview questions to ask the Minister for Tourism via e-mail. The students decided they wanted to interview him about the tourism policies set in place for the Sydney Olympics in order to evaluate how ecologically sound the policies were in terms of predicted numbers, destinations promoted, and guidelines to be given to the visitors. Still another group is putting the finishing touches on a booklet they have written based on interviews with 10 travel agents about the moral issues associated with trying to run a business in the tourism industry. The teacher is currently working with the final group on a song that explores the environmental and employment issues surrounding the controversial filming of *The Beach* (2000, Danny Boyle, Director) on Phi Island, off the south Thailand coast. In all of this, the teacher stresses the need for accurate spelling, grammar, and syntax on commissioned work and work destined for public consumption.

This kind of scenario is not so far-fetched. Teachers and other cultural workers are already teaching the three dimensions of literacy in effective and interesting ways (see accounts in Heath, 1983; hooks, 1998; Knobel, 1999; Mahiri, 1998; Wallace, 1999).

Going Beyond the Benchmarks

Now, six years after the case study of Jacques, I find that he did indeed leave school as soon as he was legally able, is working full time with his father, and enjoys nothing more than tinkering and fine-tuning the old car he bought with his own money before he left school. On all fronts he can be considered a successful young man—which is hardly what one would have predicted for him without knowledge of his social practices outside school contexts.

The main aim of this article was to examine national benchmarks, testing and standards in the light of one case in Australia in order to discuss the pitfalls associated with judging students on the basis of school literacy practices only. An important goal was to appeal for going beyond the literacy benchmarks to teach literacy in ways that make it culturally and sociocritically meaningful to students by embracing models of literacy teaching that are not limited to code breaking or operational literacy practices. The main argument in support of this goal is the claim that when teachers focus intentionally or unintentionally on *school* purposes and practices in language lessons, and other language-related experiences (e.g., using proformas to compare and contrast customs in different countries without discussing the social purposes for such comparisons), they may ultimately put students at a disadvantage in terms of success in school—Jacques is a case in point.

References

Alloway, N. and Gilbert, P. (1998). Reading literacy test data: Benchmarking success? *Australian Journal of Language & Literacy*. 21(3): 249–261.

Aronowitz, S. and DiFazio, W. (1994). *The Jobless Future: SciTech and the Dogma of Work*. Minneapolis: University of Minnesota Press.

Castells, M. (1996). *The Rise of the Network Society*. Oxford, England: Blackwell.

Cazden, C. (1998). *Classroom Discourse: The Language of Teaching and Learning*. Portsmouth, NH: Heinemann.

Clinton, W. (1997, February 5). Text of the State of the Union address. Original website no longer available.

de Alba, A., González, E., Lankshear, C., and Peters, M. (2000). *Curriculum in the Postmodern Condition*. New York: Peter Lang.

de Certeau, M. (1984). *The Practice of Everyday Life*. Berkeley: University of California Press.

Department for Education and Employment (1996). *General Requirements for English: Key stages 1–4*. London: Author.

Department of Employment, Education, Training and Youth Affairs (1998). *National Literacy and Numeracy Goals*. Canberra, ACT, Australia: Australian Government Publishing Service.

Department of Employment, Education, Training and Youth Affairs (1998). *Literacy for All: The Challenge for Australian Schools*. Canberra, ACT, Australia: Australian Government Publishing Service.

Edelsky, C. (1990). *With Literacy and Justice for All: Rethinking the Social in Language and Education.* London: Falmer.
Freebody, P. (1992). A socio-cultural approach: Resourcing four roles as a literacy learner. In A. Watson and A. Badenhop (eds), *Prevention of Reading Failure*. Sydney: Ashton Scholastic. 48–60.
Freebody, P., and Luke, A. (l990). Literacies programs: Debates and demands in cultural context. *Prospect*, 5(7): 7–16.
Freire, P. (1972). *Pedagogy of the Oppressed.* Harmondsworth, England: Penguin Education.
Gee, J. (1996). *Social Linguistics and Literacies: Ideology in Discourses* (2nd ed). London, Falmer.
Gee, J., Hull, G., and Lankshear, C. (1996). *The New Work Order: Behind the Language of the New Capitalism*. Sydney: Allen & Unwin.
Green, B. (1988) Subject-specific literacy and school learning: A focus on writing. *Australian Journal of Education*, 31(2):156–179.
Gutierrez, K., Rymes, B., and Larson, J. (I995). Script, counter-script, and underlife in the classroom: James Brown v. Board of Education. *Harvard Educational Review*, 65(3): 417–445.
Hall, S. (1991). Brave new world. *Socialist Review*, 21(1): 57–64.
Heath, S. (1983). *Ways with Words: Language, Life and Work in Community and Classrooms.* Cambridge, England: Cambridge University Press.
Heath, S. and McLaughlin, M. (1994). Learning for anything everyday. *Journal of Curriculum Studies*. 26(5): 471–489.
hooks, b. (l998). *Teaching to Transgress*. New York: Routledge.
Howe, N., and Strauss, N. (l993). *13th Gen: Abort, Retry, Ignore, Fail?* New York: Vintage Books.
Kincheloe, J., Steinberg, S., and Gresson, A. (Eds.). (1996). *Manufactured Lies: The Bell Curve Examined.* New York: St. Martin's Press.
Knobel, M. (1999). *Everyday Literacies: Students, Discourse and Social Practice.* New York: Peter Lang.
Knobel, M., and Lankshear, C. (1995). *Learning Genres: Prospects for Empowerment.* Brisbane, QLD, Australia: NLLIA Ltd Child/ESL Literacy Network Node.
Kress, G. (1985). *Linguistic Processes in Sociocultural Practices.* Geelong, VIC, Australia: Deakin University Press.
Lankshear, C. (1997). *Changing Literacies*. Buckingham, England: Open University Press.
Lankshear, C. (l998). Frameworks and workframes: Evaluating literacy policy. *Unicorn*, 24(2): 43–58.
Lankshear, C. (1999). Meanings of "literacy" in education reform discourse. In D. Gabbard (ed.), *International Handbook of Educational Reform.* New York: Erlbaum. 87–94.
Lankshear, C., Bigum, C., Durrant, C., Green, B., Honan, E., Morgan, W., Murray, J., Snyder, I., and Wild, M. (1997). *Digital Rhetorics: Literacy and Technologies in Education—Current Practices and Future Directions.* Canberra. ACT, Australia: Department of Employment, Education, Training and Youth Affairs.
Lankshear, C. and Knobel, M. (1998). New times!! Old ways?! In F. Christie and R. Misson (Eds.). *Literacy and Schooling*. London: Routledge. 155–177.
Lankshear, C. and Knobel, M. (2000, April). Strategies, tactics and the politics of literacy: Genres and classroom practice in a context of change. Invited presentation to the Tercer Congreso Nacional Sobre Textos Académicos, Puebla, Mexico.
Lankshear, C. and McLaren, M. (eds.) (1993). *Critical Literacy: Politics, Praxis, and the Postmodern.* Albany: State University of New York Press.
Lankshear, C. and O'Connor. P. (1999). Response to "Adult Literacy: The Next Generation." *Educational Researcher*, 28(1): 30–36.
Lankshear, C., Peters, M., and Knobel, M. (2000). Information, knowledge and learning: Some issues facing epistemology and education in a digital age. In N. Blake and P. Standish (Eds.), *Enquiries at the Interface: Philosophical Problems of On-line Education.* Oxford, England: Blackwell.
Lave, J. and Wenger, E. (1991). *Situated Learning: Legitimate Peripheral Participation.* Cambridge, England: Cambridge University Press.

Luke, A. (1993). The social construction of literacy in the classroom. In L. Unsworth (Ed.), *Language as a Social Practice in the Primary School* (pp. 3–53). Melbourne, VIC, Australia: Macmillan.

Luke, A. and Elkins, J. (2000). Re-mediating adolescent literacies. *Journal of Adolescent & Adult Literacy*. 43(5): 396–398.

Luke, A. and Freebody, P. (1997). The social practices of reading. In S. Muspratt, A. Luke, and P. Freebody (Eds.), *Constructing Critical Literacies*. Cresskill, NJ: Hampton Press. 185–226.

Luke, A. and Freebody, P. (1999). A map of possible practices: Further notes on the four resources model. *Practically Primary*, 4(2): 5–8.

Macedo, D. (1991). *Literacies of Power: What Americans Are Not Allowed to Know*. Boulder, CO: Westview Press.

Mahiri, J. (1997). Street scripts: African American youth writing about crime and violence. *Social Justice*, 24(4): 56–70.

Mahiri, J (1998). *Shooting for Excellence: African American Youth Culture in New Century Schools*. New York: National Council of Teachers of English/Teachers College Press.

McLaren, P. (1994). *Schooling as a Ritual Performance: Towards a Political Economy of Educational Symbols and Gestures* (2nd ed). New York: Routledge.

Ministry of Education, New Zealand (1998). *Literacy and Numeracy Strategy*. Wellington, New Zealand: Author.

Moje, E., Young, J., Readence, J., and Moore, D. (2000). Reinventing adolescent literacy in new times: Perennial and millennial issues. *Journal of Adolescent & Adult Literacy*. 43(5): 400–411.

New London Group (1996). A pedagogy of multiliteracies: Designing social futures. *Harvard Educational Review*, 66(1): 60–92.

Organization for Economic Co-operation and Development (1998). *The OECD Jobs Strategy: Technology, Productivity and Job Creation—Best Policy Practices*. Paris: Author.

Popkewitz, T. (1991). *A Political Sociology of Educational Reform: Power/Knowledge in Teaching, Teacher Education, and Research*. New York: Teachers College Press.

Prakash, M. and Esteva, G. (1998). *Escaping Education*. New York: Peter Lang.

Prinsloo, M. and Breier, M. (Eds.) (1996). *The Social Uses of Literacy: Theory and Practice in Contemporary South Africa*. Amsterdam: John Benjamins.

Reich, R. (1992). *The Work of Nations: Preparing Ourselves for 21st-Century Capitalism*. New York: Vintage Books.

Rogoff, B. (1995). Observing sociocultural activity on three planes: Participatory appropriation, guided participation, apprenticeship. In J. Wertsch, P. del Rio, and A. Alvarez (Eds.), *Sociocultural Studies of Mind*. New York: Cambridge University Press. 139–164.

Searle, C. (1998). Words and life: Critical literacy and cultural action. In M. Knobel and A. Healey (Eds.), *Critical Literacies in the Primary Classroom*. Newtown, NSW, Australia: Primary English Teaching Association. 73–88.

Secretária de Educación Pública (1999). *Perfil de la Educación en México*. Mexico City: Author.

Shepard, L. and Bliem, C. (1995). Parents thinking about standardized tests and performance assessments. *Educational Researcher*, 24(8): 25–31.

United States Department of Education (1997). *Strategic Plan, 1998–2000*. Washington: Author.

Vingerhoets, R. (1993). *Earn and Learn*. Mt. Waverly, SA, Australia: Dellasta.

Walker, J. (1988). *Louts and Legends: Male Youth Culture in an Inner City School*. Sydney: Allen & Unwin.

Wallace, C. (1999) Critical language awareness: Key principles for a course in critical reading. *Language Awareness*, 8(2): 98–110.

Watch Tower Bible and Tract Society of Pennsylvania (1971). *Theocratic Ministry School Guidebook*. New York: Author.

Part Three

TEN

The "New Literacy Studies" and the Study of "New" Literacies (2001-2003)

Colin Lankshear and Michele Knobel

Biography of the Text

This chapter was initially written as a keynote paper for a "by invitation" seminar organized by Jim Gee at the University of Wisconsin-Madison, shortly after his appointment to the Tashia Morgridge Chair of Literacy. The focus of the seminar was on developing a research agenda with a "new literacy studies" orientation. At the time we were struck by the extent to which studies of "new" literacies seemed radically under-represented within sociocultural approaches to literacy research. The paper we produced was intended to name this under-representation as clearly as we could and to begin to identify what we saw at the time as some of the "new" literacies that would merit being identified as such and investigated from a social practice perspective. The paper was our first attempt to start thinking about what a full length publication that began conceptualizing and documenting "new" literacies might look like for us. It eventually morphed into a framing chapter for the first edition of our *New Literacies* book, and this is the version presented here. It presents our earliest thinking about what it might sensibly mean to talk about literacies that are "new" with particular reference to formal educational practices, distinguishing between literacies that are "new" in ontological terms on account of being encoded digitally rather than typographically, and literacies that are chronologically recent or are otherwise new to being thought about in terms of literacy. We were acutely aware of the embryonic—indeed, almost entirely inchoate—nature of this early attempt to give some content and meaning to what is "new" about new literacies. (For a more recent account, see the second edition of *New Literacies*, 2006). At the same time, we believed it was important to focus on emerging trends in meaning making practices and tools and to identify areas where further theoretical and research work might be useful. Since that time the conversation about "new" literacies has grown considerably and diversely, providing one kind of umbrella for work within literacy studies at large.

Introduction

There is an interesting ambiguity and tension in the idea of "new literacy studies." From one standpoint the New Literacy Studies (NLS) refers to a new way of looking at literacy. This is sometimes referred to as a sociocultural approach to literacy or as socioliteracy studies. As such it is distinguished from "old" approaches to studying literacy typically based in some kind of psychologistic or technicist paradigm. From a second standpoint, however, "new literacy studies" can refer to studies of new forms of literacy. Rather than being about a new way of looking at literacy, this second idea is more about looking at "new forms of literacy."

This chapter begins from our observation that to date many people who "belong to" and contribute to the NLS have not been very interested in what we personally have come to regard as important and influential new literacies. On the contrary, it seems to us that the sorts of people who have provided the most useful, helpful and illuminating accounts of new literacies, and who have best modeled instructive appropriations of new literacies are, for the most part, not literacy scholars at all—whether by self-definition or in terms of formal recognition.

We believe that under contemporary conditions useful work in literacy at interfaces among schools, community, workplaces and other key sites of daily life *must* include serious engagement with a range of distinctively *new* literacies. These include literacies that are, in significant ways, new *in kind*. They also include other literacies that are *chronologically* new or that will in other ways be new to formal studies of literacy. From this perspective we find much that is valuable for informing school-based literacy education within bodies of literature and inquiry that have until now been strictly marginal to the NLS. In this chapter we will engage with some typical examples of this literature and take on board some of the challenges they present.

A small qualifying note is in order here. Any examples of "new literacies" will inevitably be selective, partial and subject to disagreement on the part of others. Some "new" literacies may come and go very quickly. Others will rise and rise. And under the fast-paced conditions of contemporary life, what appears new at a given point in time may be superseded and become "old" very quickly. There are, then, risks involved in trying to identify exemplars of new literacies. We are aware of at least some of the risks and potential disputes involved here and have done our best to select plausible candidates for new literacies. In the end, however, we think the most important point is to try and get the larger issues onto the educational agenda, since many attempts to respond to "the new" in education seem to us not to go far enough. Many simply accommodate new motifs to old ways. Others seem to attend to the new within unduly narrow scopes of concern with, say, present and foreseeable future work or citizenship. Our view is that numerous highly influential (and powerful) literacies exist that enjoy high profile places within contemporary everyday culture but that are not in any significant way accounted for in school learning. The extent to which they can and should be addressed within formal education can only be raised once we have a sense of what they are. Our aim is to identify at least some plausible candidates to be taken into account.

Concepts of New Literacies

As spelled out in the Introduction, we recognize two broad categories of new literacies. The first category, which we have identified as post-typographic new literacies, is well known, even if it is often not especially well defined or understood. These are literacies associated with new communications and information technologies or, in more general terms, the digital electronic apparatus (Ulmer, 1987). The second category is a looser and more *ad hoc* category comprising literacies that are comparatively new in chronological terms and/or that are (or will be) new to being recognized as literacies—even within the sociocultural perspective. Literacies in this second category may have little or nothing to do with use of (new) digital electronic technologies. In some cases, however, they may well comprise new technologies in their own right. Since the latter is the less intuitive and less clear category we will briefly sketch a couple of examples of what we have in mind.

Scenario Planning

Scenario planning has emerged during the past 40 to 50 years as a generic technique to stimulate thinking about the future in the context of strategic planning (Cowan et al., 1998). It was initially used in military planning and was subsequently adapted for use in business environments (Wack, 1985a, 1985b; Schwartz, 1991; van der Heijden, 1996) and, most recently, for planning political futures in such countries as post apartheid South Africa, Colombia, Japan, Canada and Cyprus (Cowan et al., 1998).

Scenarios are succinct narratives that describe possible futures and alternative paths toward the future based on plausible hypotheses and assumptions. The idea behind scenarios is to start thinking about the future now in order to be better prepared for what comes later. Proponents of scenario planning make it very clear that scenarios are *not* predictions. Rather, they aim to perceive futures in the present. In Leonie Rowan and Chris Bigum's words (1997: 73), they are

> a means for rehearsing a number of possible futures. Building scenarios is a way of asking important "what if" questions: a means of helping groups of people change the way they think about a problem. In other words, they are a means of learning.

Scenario planning is very much about challenging the kinds of mindsets that underwrite certainty and assuredness, and about "re-perceiving the world" (Rowan and Bigum, 1997: 76) and promoting more open, flexible, proactive stances toward the future. As Cowan and colleagues (1998: 8) put it, the process and activity of scenario planning is designed to facilitate conversation about what is going on and what might occur in the world around us, so that we might "make better decisions about what we ought to do or avoid doing." Developing scenarios that perceive possible futures in the present can help us "avoid situations in which events take us by surprise" (Cowan et al., 1998: 8). They encourage us to question "conventional predictions of the future," as well as helping us to recognize "signs of change" when they occur, and establish "standards" for evaluating "continued use of different strategies under different conditions" (Cowan et al., 1998: 8). Very importantly, they provide a means of organizing our knowledge and understanding of present contexts and future environments within which decisions we take today will have to be played out (Rowan and Bigum, 1997: 76).

Within typical approaches to scenario planning a key goal is to aim for making policies and decisions *now* that are likely to prove sufficiently robust if played out across several possible futures. Rather than trying to predict the future, scenario planners imaginatively construct a range of possible futures. In light of these, which may be very different from one another, policies and decisions can be framed at each point in the ongoing "present" that will optimize options regardless of which anticipated future is closest to the one that eventually plays out in reality.

Scenarios must narrate particular and credible worlds in the light of forces and influences currently evident and known to us and that seem likely to steer the future in one direction or another. A popular way of doing this is to bring together participants in a policy or decision making exercise and have them frame a focusing question about or theme relevant to the area they are concerned with. If, for instance, our concern is with designing courses in literacy education and technology for inservice teachers presently in training, we might frame the question of what learning and teaching of literacy and technology might look like within educational settings for elementary school age children 15 years hence.

Once the question is framed, participants try to identify "driving forces" they see as operating and as being important in terms of their question or theme. When these have been thought through participants identify those forces or influences that seem more or less "pre-determined" and that will play out in more or less known ways. Participants then identify less predictable influences, or uncertainties: key variables in shaping the future that could play out in quite different ways, but where we genuinely can't be confident one way or another about how they will play out. From this latter set, one or two are selected as "critical uncertainties" (Rowan and Bigum, 1997: 81). These are forces or influences that seem especially important in terms of the focusing question or theme but that are genuinely "up for grabs" and unpredictable. The "critical uncertainties" are then "dimensionalized" by plotting credible poles: between possibilities that, at one pole are "not too bland" and, at the other, not too "off the wall." These become raw materials for building scenarios: stories about which we can think in ways that suggest decisions and policy directions *now*.

A number of classic examples of scenario planning successes exist. One early and famous example concerns a petroleum company whose scenario planning entertained the possibility of a future change in the price of oil. This planning exercise occurred prior to the dramatic and world-changing oil shocks of the mid-1970s. At the time an oil price change, while possible, was practically unthinkable. Other companies certainly had not factored it into their way of thinking about the future. By dint of decisions made as a result of the scenario planning exercise, the company in question was well prepared for the unexpected and dramatically improved its business position among oil companies after the OPEC countries acted to increase oil prices.

This is not to suggest that scenario planning is good just for business and profiteering activities. Rather, we think the kind of work that goes into scenario planning is exactly the kind of work that should be built into learning activities in schools, communities and workplaces. Since it is a form of reading and writing the world, it seems to us to qualify nicely as a new literacy—one that is comparatively new chronologically, and one that would most certainly be new so far as prevailing mindsets within the formal study of literacy is concerned.

Zines

This category of new literacies relates directly to a central motif in our thinking about new literacies: namely, the issue of "mindsets" that we will address in greater detail a little later in this chapter. The rapidity and extent of change during the past 20 years has left many people who remain comparatively young in chronological terms out of touch with the tenor of the times. This manifests itself in diverse ways. One interesting but unfortunate example of this concerns the extent to which some progressive educators seem quick to dismiss or simply ignore a range of cultural forms as being apolitical—even regressive—because they fall outside the range of what progressives of their generation regard as being *political.* Zines—whether conventional print zines or electronic zines (ezines)—are a typical case in point.

"Zines" are the grunge frontier of publishing. Hard copy zines are typically hand-crafted, using found papers, card stock, typed texts, drawn or photocopied images, photographs, stickers, paper cutouts and so on. They are usually produced in small print runs and mainly distributed via word of mouth. Generally they cost only one or two U.S. dollars (or equivalent in the form of postage stamps or a "trade"—someone else's zine in exchange). Many are "defiantly personal" texts "which confidently explore their maker's passions, no matter how obscure they may seem" (Bail, 1997: 44). They often articulate strong, counter-mainstream cultural themes or social activist projects (cf. Vale, 1996, 1997). Mainstream discourses and values are subverted and pilloried in ways that can be breathtaking in their serious playfulness. In the "Space and Technology" issue of *Mavis McKenzie,*

> Mavis writes letters to Hewlett Packard asking for advice about her ancient computer. She writes to a glass company about replacing the windows in her house with Windows 95. (They write back kindly advising her that Windows 95 is a software program.) She gets tickets to *The Price is Right* [one of those afternoon TV game shows]. Mavis has far too much time on her hands (Bail, 1997: 44).

Dishwasher Pete's zine, *Dishwasher,* presents a sophisticated and trenchant critique of and resistance to the McWork world.

> I'm addicted to that feeling of quitting; walking out the door, yelling "Hurrah!" and running through the streets. Maybe I need to have jobs in order to appreciate my free leisure time or just life in general. ... Nowadays, I can't believe how *personally* employers take it when I quit. I think, "What did you expect? Did you expect me to grow old and die here in your restaurant?" There seems to be a growing obsession with job security, a feeling that if you have a job you'd better stick with it and "count your blessings" (Dishwasher Pete in conversation with Vale, 1997: 5, 6).

Generation X zinesters have evolved an array of distinctive trade-offs around the economics and politics of talent, passion and community. Bail (1997: 44) observes that

> Zines are cheap and fast. Yet their makers often struggle to pay for photocopying, stamps, even paper. I was showing a zine to a friend and coincidentally its producer was employed in her office mailroom. She'd always thought he was too talented for the job but suddenly realised why he stayed there. Zines are not easy

> to come by: producers [often] swap their publications with each other ... Basically, you need to be in the loop to get the good stuff. It's give and take—those who contribute to the culture get the most from it ... Nice to find a place with a generosity of spirit among peers ... and to find a culture that hasn't been censored, sanitized and target-marketed. (Bail, 1997: 44)

Zines are a medium for young people's opinions, thoughts, creativity, and for affirming a stance as active makers rather than passive consumers of culture.

We believe a case can be made for serious consideration of zines within progressive studies of new literacies. This would be similar to the ways critical historians of literacy have paid scholarly interest to progressive resistant textual productions of earlier times—particularly the working class underground press in the early industrial age. While we do not necessarily agree with the substantial positions advanced in some zines, the genre as a whole provides useful clues about some of the different ways young people understand and practice cultural politics. These often involve a blend of the anarchic, the edge-dweller, the intensely personal, the do-it-yourself ethic, dressed to spoof, the critique and subversion of mainstream culture and constructions of publishing. To that extent, zines reveal ideas about how many young people understand the nature and role of literacies in cultural and political practices.

The Study of New Literacies

There now exists a huge and growing body of literature that describes uses of new information and communications technologies (ICTs) within school settings, including within literacy education specifically. It is highly questionable, however, whether much of this literature actually gets at anything that is significantly *new* so far as literacies and social practices are concerned. In other words, it does not follow from the fact that so-called new technologies are being used in literacy education that *new literacies* are being engaged with. Still less does it imply that learners are developing, critiquing, analyzing, or even becoming technically proficient with new literacies.

In some of our other work (e.g., Lankshear and Bigum, 1999: 454–56; Lankshear and Snyder, 2000: Ch. 5; Goodson, Knobel, Lankshear, and Mangan, 2002: Ch. 4), we have identified recurring features of new technology-mediated literacy practices in classrooms. These reflect a marked tendency to perpetuate the old, rather than to engage with and refine or re-invent the new. Many researchers have identified the "old wine in new bottles" syndrome, whereby longstanding school literacy routines have a new technology tacked on here or there, without in any way changing the substance of the practice. Using computers to produce neat final copies or slideshow presentation software for retelling stories are obvious examples. In some cases, practices involving literacies that might reasonably be described as "new" emerge within classroom activities, but at the same time appear very odd or "unhinged" from the kinds of practices that are engaged beyond the classroom. Such practices are often "solutions" to the demand that new technologies be employed in classroom programs without the conditions existing for this to occur in sensible ways. Using email to share cryptic clues and interpretations of these clues among participating schools in an email-based competition requiring students to track down a "criminal" is a case in point. So is inventing an activity for groups of students to write "five-minute plays" taking turns to enter the text on a single laptop computer available for each group of 4 or 5 students in the class. Such examples are not so much instances of new literacies as pedagogical inventions born of necessity that infringe against the principle of efficacious learning. That is, for learning to be efficacious, what a person learns *now* must be must be connected in meaningful and motivating ways with subsequent points in a personal trajectory through a range of social practices and a range of social institutions. To put this another way, learning something *efficaciously* is to progress toward a fuller understanding and fluency with doing and being in ways that are recognized as proficient relative to socially constructed and maintained ways of "being in the world." Simply digitizing long-standing classroom practices, or casting about for things to do with digital technologies, does not ensure efficacious learning.

Such subversions of "the new" are not at all surprising when we take larger and underlying institutional characteristics of school into account. School routines are highly regular forms of practice that are intimately linked to what we call the "deep grammar" of schooling, as well as to aspects of policy development and imposition, resourcing trends, professional preparation and development, and so on. In fact, we suggest that schools and classrooms are among the last places one would expect to find "new literacies." We can begin to see why this is so by considering two key elements of the deep grammar of school, which constructs learning as teacher directed and "curricular."

First, schooling operates on the presumption that the teacher is the ultimate authority on matters of knowledge and learning. Hence, whatever is addressed and done in the classroom must fall within the teacher's competence parameters, since he or she is to *direct* learning.

Second, learning as "curricular" means that classroom learning proceeds in accordance with a formally imposed/officially sanctioned sequenced curriculum, which is founded on texts as information sources. Seymour Papert (1993: 9) observes the long-standing pervasive tendency in the education literature "to assume that reading is the principal access route to knowledge for students." The world, in other words, is accessed via texts (books; school is bookspace). This imposes a pressing and profoundly instrumental value and significance on the capacity to *read*. It also promotes and encourages a view of (school) literacy as *operational* in the first instance (which, unfortunately, is often the last instance as well): that is, reading as a matter of competence with *techniques* of decoding and encoding.

Current policies concerned with technologizing learning intersect, for example, with a teaching workforce that is largely un(der)prepared for the challenge of *directing* computer-mediated learning in the role of teacher as authority. In a climate of shortage, schools value almost any computer skills in teachers. In practice, this means that low-level operational or technical skills and knowledge predominate (Bigum, 1997: 250). Not surprisingly, teachers look for ways of fitting new technologies into classroom "business as usual." Since educational ends are directed by curriculum, and technologies are "mere" tools, the task of integrating new technologies into learning is often realized by adapting them to familiar routines. One corollary of this is that making learners "technologically literate" is largely reduced to teaching them how to "drive" the new technologies. The emphasis is very much on technical or operational aspects: how to add sound, insert a graphic, open and save files, create a HyperCard stack and so on.

This logic can be seen as a specific instance of a much larger phenomenon: the systematic separation of (school) learning from participation in "mature" (insider) versions of Discourses, which are part of our life trajectories (Knobel, 1999). School learning is learning for school; school as it always has been. The burgeoning take up of new technologies simply gives us our latest "fix" on this phenomenon. It is the "truth" that underpins many current claims that school learning is at odds with authentic ways of learning to be in the world and with social practice beyond the school gates. The reason why many school appropriations of new technologies appear "odd" in relation to "real world" practices—with which children are often familiar and comfortable—has to do with this very logic. It is precisely this "deep grammar" of schooling that cuts schools off from the new (technological) literacies and associated subjectivities that Bill Green and Chris Bigum (1993) say educators are compelled to attend to. To put it another way, new literacies and social practices associated with new technologies "are being invented on the streets" (Richard Smith, personal communication). These are the new literacies and practices that will (many of them) gradually become embedded in everyday social practice: the literacies against which the validity of school education will be assessed. But the "deep grammar" of school is in tension here with its quest for legitimation in a high tech world—which is potentially highly problematic for schools.

Accepting the Challenge of New Literacies in the New Literacy Studies

In the remainder of this chapter we lay out in a preliminary way some concepts and examples we think might help take us forward in studying new literacies in relation to school settings, as well as home, community and workplace settings. These will be presented under three headings.

- The nature and significance of "mindsets"
- Some typical examples of new literacies
- Descriptive, analytic and critical accounts of new literacies

The Nature and Significance of "Mindsets"

In recent work we have addressed some of the issues involved here by reference to a fruitful distinction advanced by John Perry Barlow (see Lankshear and Bigum, 1999; Rowan et al., 2000; Goodson, Knobel, Lankshear, and Mangan, 2002). In an interview with Nat Tunbridge (1995), Barlow spells out a distinction between two mindsets that are brought to bear on cyberspace specifically and spaces of digitized practices more generally. Barlow refers to these two mindsets as "immigrant" and "native" mindsets, respectively. We prefer to call them "outsider" (or "newcomer") and "insider" mindsets respectively, since the terminology of "immigrants" and "natives" might reasonably be seen as offensive by members of some social groupings.

Very briefly, Barlow distinguishes between those who have, as it were, "been born and grown up" in the space of "the internet, virtual concepts and the IT world generally," whom he calls "natives," and those who have, as it were, migrated to this space. The former (insiders) understand this space; the latter (outsiders/newcomers) do not. Barlow's distinction is between mindsets that relate to how this space is constructed and controlled in terms of values, morals, knowledge, competence and the like. Since "newcomers" lack the experience, history, and resources available to them that "insiders" have, they cannot—to that extent—understand the new space the way insiders do. On fundamental points and principles of cyber/information/virtual space, says Barlow, newcomers "just don't get it" (in Tunbridge, 1995: 4).

We may use "newcomers" and "insiders" as markers for two competing mindsets. One affirms the world as the same as before, only more technologized; the other affirms the world as radically different, precisely because of the operation of new technologies (Lankshear and Bigum, 1999: 458). Of course, these distinctions are not the only way of "carving up" the world, but we find them useful when talking about new technologies and education.

The "deep grammar" of school—embedded in its administrative systems, policy development, curriculum and syllabus development, systemic planning and the like, as well as in its daily enactment within classroom routines and relations—institutionalizes the privileging of the newcomer/outsider mindset over the insider mindset. Many classroom constructions of literacy involving new technologies are classic instances of outsider understandings of literacy grounded in the familiar physical world (book space) being imported into cyber/virtual/information space. This generates familiar tensions for schools: tensions that may, however, be seen as choice points—where choice about mindsets is, in principle, open and up for grabs.

For example, schools already face sizable cohorts of "insiders" largely indifferent to and bemused by the quaint practices of schooling. This is a cohort that is in tune and largely at ease with the dizzy pace of change, with the development of new technologies, and with social and economic shifts that cause pain to many outsider/newcomers (Lankshear and Bigum, 1999: 461).

Some Typical Examples of "New" Literacies

This section will provide a kind of "operational definition" of how we see some of the territory of new literacies. It comprises a broad series of examples we find illuminating. They are only a very small sample of the kinds of textual practices we identify as new literacies and the larger Discourses in which they are embedded. We have aimed to span a wide range in terms of the cultural politics of new literacies, giving particular emphasis to practices that adopt an active or critical stance.

(a) Multimediating: Michael Doneman

Michael and Ludmila Doneman are performance artists who use a range of digital technologies in their work with disadvantaged young people and Indigenous people in Australia. They identify themselves as cultural *animateurs*. In the following account of "multimediating," Michael Doneman (1997: 131) emphasizes cultural production over consumption. In so doing, he identifies what we see as an important principle of new literacies.

> In matters of definition, why spend so much time on *multimedia* as a noun when we could be looking at *multimediating* as a verb? I can have almost any number of windows open—let's say I open a chat window (or I-phone or video chat), the Web, an ftp file transfer, a usenet news reader, a telnet MUD session, a low-end graphics app, a simple word processor, net radio or streamed video and e-mail. Let's say I am mixing-and-matching my time in each environment, communicating in different ways among different communities, cutting and pasting, sending and receiving simultaneously, producing and consuming simultaneously, role playing, documenting and archiving, selecting, discarding, maintaining, filtering, reciprocating, researching, criticising, responding, arguing, judging, broadcasting ...
>
> How is this multimediating constructing my world and my response to the world? How is it constructing and responding to *community*? How is it fitting me to operate effectively in the world?
>
> Let's also say that I am doing this on-line activity from a workstation in the telecentre of a place like GRUNT [an inner city cultural youthspace located in

> a former warehouse], where there might be a rehearsal, music, informal chat, meetings and office work going on in *meatspace*. Other environments, other roles. Is the negotiation of these roles *on the fly* enabling or distracting me?

(b) E-zining: Grrrowling with the Digitarts

Digitarts (launched 1998) is an online multimedia project space originally constructed by young women for young women, exploring alternative perspectives on style, food, everyday life and commodities, and expressing different conceptions and constructions of female identity through poems, narratives, journal pages, how-to-do texts and digital images (now defunct; archived at the National Library of Australia, 2010a). To begin with, the project was dedicated to providing young women who are emerging artists and/or cultural workers with access to the knowledge and equipment necessary for the development of their arts and cultural practices in the area of new technologies. It aimed at challenging "the 'boys' toys' stigma often associated with electronic equipment," and to "provide young women with access to information technology in a non-threatening 'girls own' space, to encourage involvement in technology based artforms" (see National Library of Australia, 2010b: n.p.). In recent times, the brief of *Digitarts* has changed—possibly for funding reasons—and now includes socially and culturally disadvantaged young people of both genders. As the welcome page to the website puts it:

> Most people are happily oblivious to the inequities around them. We are not. Instead we look for the gaps, and seek to provide those missing out with access to knowledge and equipment in innovative and user-friendly ways. (webpage no longer available)

Despite changes, *Digitarts* remains a venue for emerging multimedia artists to showcase their work, a place for young people to display their burgeoning computer skills, and still seeks to attract young, traditionally disadvantaged people to new technologies by providing 6 to 8 week webpage development courses for beginners. Other training provisions in the women-oriented days have included a 12 week advanced web-development course, and a 12 week digital animation course. The collaborative production of *grrrowl* by young women around Australia (what the Digitarts—as they call themselves—refer to as a "semi-regular ezine"; see National Library of Australia, 2010c) ran for a total of six issues, all with an emphasis on women's experiences and skills. These issues are titled: #1 machines, #2 fashion, #3 action, #3.5 party, #4 simply lifeless, #5 circle/cycle.

Two projects are typical of the early *Digitarts*. *Girls in Space* (National Library of Australia, 2010b) was prompted by the low visibility of young women in public spaces and the lack of research in Australia about women's recreational and public space needs. It gathered information from young women who made use of public spaces and those who didn't, and made this information available to public policy makers. The information was also used to generate models of service and activity delivery designed to increase young women's participation in a range of public sites (e.g., recreation and public parks, sporting venues), and to promote collaboration between local government and community organizations. Spin-offs include an on-line gallery of poster art inspired in part by some *Girls in Space* participants' reflections on women and public spaces, and an online pajama party (reported in *grrrowl* # 3.5; see National Library of Australia, 2010c, which explores real and virtual spaces in participants' lives).

A second project involves the ezine *grrrowl* (National Library of Australia, 2010c). This is a collaborative publishing endeavor. One of its early issues focused on grrrls and machines. Each contributor constructed a page that is either a personal introduction—much like a conversation between newly met friends—or contains poems or anecdotes about women and technology. Hotlinks to similar web sites on the internet also define each writer's self, and her self as connected with other selves. Issues of *grrrowl* provide alternative readings of fashion trends and body image, perspectives of contemporary culture and everyday life and the like.

grrrowl #4 investigates the theme, "Simply Lifeless." It documents online the everyday lives of young women in Darwin and Brisbane. Its thesis is: "Our culture informs our everyday activity. Our everyday activity informs our culture." The issue celebrates the "everyperson" (cf. Duncombe, 1997; de Certeau, 1984), with eight young women, ranging in age from 12 years to 25 years, broadcasting by means of web pages "snapshots" of their lives—including digital videos of key elements (composing music on a much-loved guitar, a daughter feeding a pet chicken, etc.), or hypertext journals that span a day or a week of her life and that also include personal digital images (family album snaps, etc.), hand-drawn graphics, digital artwork and so on (see, especially, 12–year-old Gabrielle's page: National Library of Australia, 2010d). By documenting the "banal" and "everyday," this issue of *grrrowl* aims at "increasing the range of criteria by which our cultures are measured and defined" (National Library of Australia, 2010e: online).

Each *Digitart* project engages its participants in developing a range of "operational" technology and literacy skills needed to produce effective web pages (e.g., becoming fluent in web page design skills, HTML and VRML, scanning images, hyperlinking files, digital photography and image manipulation, developing electronic postcards and "mail to:" forms online, embedding digital video clips in web pages). Items in the *Digitarts* portfolio are steeped in cultural analyses of everyday life, as well as in processes that properly blur the relationship between effective web page construction in cyberspace and meaningful social practices in "meat space." This includes broadcast publishing of online magazine-type commentaries, the use of the internet to establish and nurture interactive networks of relations between like-minded people, and the exploration and presentation of cultural and community membership and self-identity through writing, images, and hyperlinks. *Digitarts'* work has an overt critical dimension by virtue of its keen-edged critique of "mainstream" Australian society. For instance, the editorial in the third issue of *grrrowl* explains how to override/subvert the default settings on readers' internet browser software, and encourages young women to override/subvert other socially-constructed "default settings" that may be operating in their lives. It challenges social scripts that allocate various speaking and acting roles for young women that cast them as passive social objects or as victims (e.g., "This is not about framing women as victims—mass media vehicles already do a pretty good job of that," *Girls in Space,* National Library of Australia, 2010b), and that write certain types of girls (or grrrls) out of the picture altogether. *Digitarts* offers a coherent alternative to the commodification of youth culture—i.e., youth as a market category—by making space for young women, and now young men, to become *producers,* and not merely consumers, of culture in the way it privileges the personal over the commercial (cf. Doneman, 1997: 139; Duncombe, 1997: 68, 70).

(c) Meme-ing: David Bennahum

Sender: Meme -- Information on Cyberspace <MEME@MAELSTROM.ST-JOHNS.EDU>
Poster: "David S. Bennahum" <davidsol@PANIX.COM>
Subject: MEME 5.01

----

meme: (pron. "meem") A contagious idea that replicates like a virus, passed on from mind to mind. Memes function the same way genes and viruses do, propagating through communication networks and face-to-face contact between people. Root of the word "memetics," a field of study which postulates that the meme is the basic unit of cultural evolution. Examples of memes include melodies, icons, fashion statements and phrases.

David Bennahum is the author of *Extra Life: Coming of Age in Cyberspace* (1998). We learned about his book through his regular MEME newsletter (see the MEME website, including back issues: memex.org).

MEME-ing is a powerful metalevel literacy: an enactive project of trying to project into cultural evolution by imitating the behavioral logic—replication—of genes and viruses. It involves at least two necessary conditions: susceptibility (for contagion) and conditions for replication to occur. Susceptibility is tackled by way of "hooks" and "catches"— something that is likely to catch on, that gets behind early warning systems and immunity (even well-developed critical consciousnesses can get infiltrated by the Nike icon or the Coca-Cola white swirl on red). Electronic networks provide ideal contexts for replication.

Examples of successful memes and their respective "cloners," "high profile carriers" or "taggers" include: "cyberspace" (William Gibson), "screenagers" (Douglas Rushkoff), "GenX" (Doug Coupland), "the information superhighway" (Al Gore), "global village" (Marshall McLuhan), "cyborgs" (Donna Haraway), "clock of the long now" (Stewart Brand and colleagues), "digital divide" (Clinton administration), "complexity" (the Santa Fe Group), D/discourse (James Paul Gee), and so on. The principle that lies behind Meme-ing, so far as generating active/activist literacies are concerned, is simple yet fundamental: If *we* don't like *their* contagious ideas, *we* need to produce some of our *own*.

(d) Blogging: Personal weblogs from Jane Doe to Andrew Sullivan

In a May 2002 online archive of *Wired*, a writer for the *New York Times*, Andrew Sullivan, suggested that "blogging"—the practice of publishing personal weblogs—is changing the media world. He goes so far as to suggest that blogging could "foment a revolution in the way journalism functions in our culture" (Sullivan, 2002: 1).

Weblogs are online personal diaries or journals that are added to by the owner anywhere from every now and then to multiple times a day. Sullivan traces their history from the mid 1990s, when blogging was confined to small scale "sometimes nutty, sometimes inspired writing of online diaries" (2002: 1). Today, however, the internet is home to blogs covering diverse topics and types. Sullivan notes examples ranging from tech blogs, sex blogs, drug blogs and onanistic teenage blogs to news and commentary blogs. The latter comprise sites that are "packed with links and quips and ideas and arguments" that as recently as 2001 "were the near monopoly of established news outlets" (2002: 1). While they are often still "nutty," and can be painfully banal, prejudiced, angst-ridden, or downright nasty, they can also be erudite and scale the pinnacles of sophisticated commentary and critique. Weblogs transcend traditional media categories. In the broad area of journalism, for instance, they may be as "nuanced and well sourced as traditional journalism" yet have "the immediacy of talk radio" (Sullivan, 2002: 1).

Furthermore, blogging is in tune with the tenor of the times. Blogs invoke the personal touch and put the character and temperament of the writer out front, rather than disguising it behind a facade of detached objectivity underwritten by the presumed editorial authority of the big formal newspaper or network. Sullivan, whose own blog reaches an estimated quarter of a million readers monthly and has become economically profitable, claims that Net Age readers are increasingly skeptical about the authority of big name media. Many readers know that the editors and writers of the most respected traditional news media are fallible and "no more inherently trustworthy than a lone blogger who has earned a reader's respect" (Sullivan, 2002: 1).

In general, blogs mix comment with links, in varying ratios. Although there are any number of services online that help users create the basic HTML code and "look" of their blogs (e.g., Williams, 2002), there is no content formula, and personal style and the nature of the topic appear to be key variables in whether a blog privileges commentary or links, or effects a reasonable balance. Chris Baker (2002: 1), writing in the same issue of *Wired*, notes that an index of blogs created at MIT "tracks top blogs based on how many people link to them" (see blogdex.net; now defunct). Baker identifies a number of current high profile "power bloggers," briefly describes the theme or purpose of their blogs, and reports their respective ratios of comments to links. At one pole of Baker's selection is the blog of Australian illustrator and Web designer, Claire Robertson (www.loobylu.com), which "serves up a breezy diary accompanied by exquisite illustrations," with a ratio of 20 comments to every link (Baker, 2002: 1). At the other end of the spectrum we find the blog of Jessamyn West, the "hippest ex-librarian on the Web" (Baker, 2002: 1). West is credited with "bringing the controversy and intrigue of library subculture to life" (Baker, 2002: 1) and, not surprisingly, perhaps, provides an average of five links for each commentary.

The vast majority of bloggers, however, are individuals writing for relatively small audiences on themes, topics or issues of personal interest to themselves. Chris Raettig's popular weblog drew international attention in 2001 with an enormous collection of corporate anthem sound files he had gathered (and had been sent by others) that were, as he put it, "so bad, they were good" (Raettig, 2001: 1). Raettig is a 22–year-old software programmer and developer living in London, and his blog—titled *i like cheese* (now defunct)—features livecam images of his studio apartment; links to a digital photo "warehouse" he has established and which documents everyday moments in his life; hyperlinks the technology development company he is partner in, the corporate anthem webpage, alternative rock music, and suchlike; and his journal entries. The principal theme that loosely ties his blog entries together is new technologies, with recent diary entries including commentaries on: the relaunch of his corporate anthems webpage on the zdnet.com website and the renewed international media interest this created, a trip to Cambridge to meet with owners of a technology start-up company, a review of his day, ideas for large-scale internet-based projects, and comments on the current health of the Web industry in general.

At a rather different point on the politico-cultural spectrum we find the blog of Asparagirl, a 23–year-old Web designer who works for IBM. Her blog (asparagirl.com/blog/) was recently showcased in a *Washington Post* article on weblogs (Kurtz, 2002: 1). It attracts over 1,000 daily visitors and presents self-styled "solipstic" commentaries that engage with themes, issues and events such as: the nature of the internet (e.g., issues concerning the joys of interactivity and the need for privacy), issues concerning statecraft and the Palestinian-Israeli conflict written from her perspective as a Jewish-American, meeting other bloggers at a party, the latest bubblegum collector cards that feature victims of the September 11 attacks on the U.S., meta-comments on the content of other people's blogs, and so on.

(e) Map Rapping: The Global Business Network

The Global Business network (GBN) is, among other things, an originating force behind the practice of scenario planning (www.gbn.com). Founded in 1987, GBN is a network of organizations and individuals "committed to re-perceiving the present in order to anticipate the future and better manage strategic response" (Rowan and Bigum, 1997: 76). In some ways the organization approximates to an actor network. Its services include a "WorldView program" where members are brought together via meetings, publications, and online conferences, and a training service—Learning Journeys—which introduces members and the public to the use of scenarios within their own organizations and contexts.

The GBN site contains a Scenario Planning section, which describes how scenarios are crafted and used, together with reports of several projects and examples of the kind of thinking scenarios promote and demand. Its "map rap" page by Peter Schwartz (no date; now defunct)—and originally one of the site's earlier "front pages"—exemplifies a new literacy in the way we are thinking of here. It employs new technologies to communicate new ways of reading and writing the world that challenge old/outmoded or no longer efficient mindsets.

> If you were an explorer in the early 1700s this map, by cartographer Herman Moll, might well have guided your explorations of North America. It is, for the most part, recognizable to modern eyes, except for one thing—it shows California as an island…
>
> This error was the result of good Cartesian reasoning: Spanish explorers coming from the south had encountered the tip of the Baja Peninsula; voyaging further north they sailed into the Straits of Juan de Fuca. When they connected the first point to the second they created the Gulf of California (Schwartz, no date: n.p.; no longer available).

Figure 1: Erroneous 18th Century Map of Baja California

As Schwartz puts it, this would be merely a historical curiosity were it not for the missionaries sent from Spain to convert the heathens in New Mexico. After landing in California, they prepared to cross the Gulf as their maps instructed: they packed up their boats and carried them up over the Sierra Nevada and down the other side, and found . . . not sea, but the longest, driest beach they'd ever seen.

When they wrote back, protesting that there was no Gulf of California, the mapmakers replied: "Well, the map is right, so you must be in the wrong place" (or words to that effect). This misunderstanding persisted for 50 years until one of the missionaries rose high enough in the Church to be able to persuade the King of Spain to issue a decree to change the maps.

As Schwartz reminds us, once you come to believe in a map, it's very difficult to change it, and, if your facts are wrong, then you'll be relying on a map that's wrong too. One aim of the Global Business Network is to "challenge 'mental maps' " (Schwartz, no date) that blind people to the lay of the land rather than helping them get to where they want to go. Thus, "[t]hrough the process of scenario planning and strategic conversation, GBN can help decision makers develop more subtle, flexible maps that enable people to navigate the uncharted territory of the future" (Schwartz, no date: n.p.).

(f) Culture Jamming: Adbusters

At Adbusters' Culture Jamming Headquarters (www.adbusters.org), a series of elegantly designed and technically polished pages present information about the organization and its purposes, describe an array of culture jamming campaigns, describe the Adbusters paper-based magazine, and target worthy media events and advertising, cultural practices, and overbloated corporate globalization with knife-sharp critiques in the form of parodies that act as exposés of corporate wheelings and dealings, and/or online information tours focusing on social issues. By turning media images in upon themselves or by writing texts that critique the effects of transnational companies, the Adbusters' culture jamming campaigns scribe new literacy practices for all people. An early image from a critique of a past trend to claim an "equality" ethic in the fashion world shows how combining familiar images and tweaking texts can produce bitingly honest social commentaries that everyone everywhere is able to read—a kind of global literacy.

Figure 2: Adbusters Culture Jam of Benetton Colors

The nature of culture jamming and the philosophy that underlies it, together with many practical examples of how to enact culture jamming literacies, are described in a recent book. *Culture Jam: How to Reverse America's Suicidal Consumer Binge—And Why We Must*, is written by Kalle Lasn, publisher of *Adbusters* magazine and founder of the Adbuster Media Foundation. The potential effectiveness of culture jamming was clearly demonstrated recently. An email posting in January 2002 to the Culture Jammers Network informed recipients that Adbusters' activities had come under increased scrutiny following September 11, 2001. According to the posting,

> Recently, our Corporate America Flag billboard in Times Square, New York, attracted the attention of the Federal Department of Defense, and a visit by an agent who asked a lot of pointed questions about our motivations and intent. We wondered: What gives? "Just following up a lead from a tip line," the agent admitted (Adbusters, 2002: 1).

Adbusters were clearly worrying people in high places. They observed that any campaign daring to question "U.S. economic, military or foreign policy in these delicate times," or any negative evaluation of how the US is handling its "War on terrorism," runs the risk of being cast as "a kind of enemy of the state, if not an outright terrorist" (Adbusters, 2002: 1).

Adbusters' response was to mobilize internet space to engage in a classic act of cyberactivist literacy. The posting asked other Culture Jammers whether their activities had received similar attention, noting that they had received messages from other activist organizations that had found themselves under investigation "in a political climate that's starting to take on shades of McCarthyism" (Adbusters, 2002: 1). Noting that they could live with vigilance, but that intimidation amounting to persecution is "another bucket of fish," Adbusters established its own "rat line" and invited others to publicize instances of state persecution within a medium that has immediate global reach.

> If you know of social marketing campaigns or protest actions that are being suppressed, or if you come across any other story of overzealous government "information management," please tell us your story. Go to <http://www.adbusters.org/campaigns/flag/nyc.html> (Adbusters 2002: 1).

Shortly after this occurred, Adbusters was contacted by Miramax, a Disney corporation, and informed that the corporate U.S. flag billboard was hindering filming sequences in Times Square. Miramax requested Adbusters to either take the billboard down for a few weeks, cover it up, or replace the corporate logos Adbusters had inserted in place of stars on the flag with the original stars. Adbusters decided to reject all three options. Instead, they invited internet users to produce their own spoofs, on the corporate flag theme. The result was a series of high quality spoofs which Adbusters published on their website, together with their own commentary on Miramax's intervention (www.adbusters.org/campaigns/flag/nyc; no longer archived on the site). In the context of the U.S. war on (other nations') terrorism, the corporate flag campaign lifted adbusting as a new literacy to a new plane.

(g) Communication Guerrilla Literacies

Digital hacking is one of a number of practices identified by Michael Doneman (1997) as being part of a suite of "communication guerrilla" activities. Originally, as with journal hacks who would churn out texts to deadlines, hacking meant nothing more than writing a computer program. Over time, however, "hacking" has become associated in the popular mind as the practice of breaking into computer networks in order to read or tweak data on machines to which one has no authorized access. Doneman (1997: 139) draws on Umberto Eco's concept of "communications guerillas" to describe an "emerging opposition to the pervasive and coercive use of information imagery by powerful groups." According to Doneman, communications guerillas are committed to urging people to read media and other messages in ways that open onto critical—and multiple—analyses and interpretations of these messages and, hence, multiple active responses generated by the readers and not by the message writers. As Doneman (1997: 139) explains:

> "[t]hese guerrillas are not always young people or outlaws, but often enough they are—often enough for us to consider much of the following list (downloaded from the Net) as aspects of emerging youth cultures:
>
> § **Hacking** - the infiltration and manipulation of systems.
> § **Subvertising** - the production and dissemination of anti-ads [cf. Adbusters in the previous section].
> § **Sniping** - late night raids on public places (as in the work of Australia's own BUGA UP [Billboard Utilising Graffitists Against Unhealthy Promotions www.buga-up.org]).
> § **Media hoaxing** - the hoodwinking of journalists [Dishwasher Pete once sent a friend in to be interviewed in his place on the David Letterman Show. The behind-the-scenes people twigged that Pete's friend was not Pete, but the show was about to go to air and there was nothing they could do. See Vale 1997: 18 for a detailed account]
> § **Audio agitprop** - the deconstructing of pop music and challenging of copyright laws [music file websites are a good example of this]
> § **Academy hacking** - cultural studies conducted outside university walls by insurgent intellectuals [e.g., zinesters]
> § **Subcultural bricolage** - the refunction, by societal "outsiders," of symbols associated with the dominant culture (as in the "voguing" of poor, black, urban drag queens).
> § **Slashing** - the renewing of tales told for mass consumption (as in the pornographic and often homoerotic Kirk/Spock stories published by male and female Trekkies in Star Trek fanzines [and ezines]).
> § **Transmission jamming** [e.g., Adbusters]
> § **Pirate radio and TV broadcasting**
> § **Neo-situationist demonstrations** (as in late-night dancing in ATM lobbies).
> § **Camcorder counter-surveillance** (as in the celebrated tape of the police bashing of Rodney King).
>
> All of these activities [involve] the introduction of *noise* to the signal in order to "restore a critical dimension to passive reception." All are examples of the

potential power, for good or ill, of the notion of *interactivity* (Doneman, 1997: 139).

Descriptive, Analytic and Critical Accounts of New Literacies

New literacies are complex and diverse. Within education research contexts we need to find ways of researching these literacies that do them justice, that do not water them down, or leach the color from them. We think that to be useful, the investigation and interpretation of new literacies should involve descriptive, analytic and critical accounts.

(a) Descriptive

The field needs rich descriptive sociological accounts of new literacies. Ideally these will be produced as much as possible by insiders who can "tell it like it is practiced"—avoiding the risk faced by academic literacy scholars and researchers of getting tripped up on self-conscious allocations to categories. For people like ourselves this might mean that more of our work will assume a kind of "brokerage" role—sifting through what is already available and working to find ways of projecting this work usefully into literacy education and research spaces.

Several years ago, Douglas Rushkoff published his first book, *Cyberia* (1994). While this has a degree of analysis and comment running through it, it is first and foremost an attempt to describe a world. As a description of a domain of social practice it differs markedly from, say, the kind of account Stephen Duncombe provides of zines (1997). We see both kinds of books playing important roles in informing educational practices generally and literacy education in particular. The kind of account Duncombe provides is work we believe is urgently needed on behalf of the growing range of new literacies. At the same time, we personally learned a great deal from Rushkoff's book (and his subsequent works) in the way of the knowledge it offers of emerging social practices that we think is indispensable for literacy scholars and educators. This included insights into Discourses that are increasingly central to the lives of young people. It also included insights into aspects of subjectivity, identity formation, existential significance, worldviews, and so on, as seen from the perspective of participants. This kind of material speaks directly to the sorts of issues raised by educationists who are concerned about things like the presence of aliens in our classrooms (Green and Bigum 1993). Generation X books, such as those written by Douglas Coupland (1991, 1995, 1996, 2000), also provide the kind of verité descriptions educators can learn much from.

We welcome the growing body of work done by "new literacies" insiders and see much value in bringing this work into the gamut of study of new literacies as an important "data base" for dealing with educational issues at the interfaces of work, community, schools and homes.

(b) Analytic

Different forms of analytic work are relevant to studying and documenting new literacies. Analytic tools from formal academic and scholarly work might be applied to the kinds of descriptive studies noted immediately above (remembering, of course, that such descriptive accounts will always and inescapably involve a degree of interpretation and analysis). This would involve taking the descriptions as a kind of "secondary data" and making further "sense" of it in various ways. At one level of analysis one might identify and relate the Discourse and discourse aspects of a set of social practices (i.e., the ways of speaking, acting, believing, thinking, etc. that signal one is a member of a particular Discourse, along with the "language bits" of this Discourse; Gee, 1996). At a different analytic level, the work might involve a form of sociological imagination (Mills, 1959): exploring how subjectivity and identity are related to participation in or membership of Discourses in which new literacies are developed, employed, refined and transformed.

As an example of the kind of analytic work that might be done from the standpoint of sociological imagination we can take a case like David Bennahum's book, *Extra Life* (1998) that we mentioned earlier in this chapter. Grounded in Bennahum's experiences of growing up in a digitally saturated environment, *Extra Life* has been nicely described by Douglas Rushkoff as a "*Catcher in the Rye* for the Atari Generation" (see book's back cover). Bennahum locates computers and computing practices within key phases and events in his adolescence and adult life in New York City. For example, a particular computer teacher, the computing classroom, and experiences within his computing classes emerge as pedagogical oases in the desert. There is more than a hint that Bennahum's interest in computing may have helped prevent him going down a road of delinquency and precocity that brought sticky ends to some of his adolescent peers.

Bennahum describes an array of Discourses to which he was apprenticed: for example, the Jewish faith, downtown Manhattan street culture, school and home-based computer nerd culture, teenage sex and romance and, eventually, sophisticated forms of Discourse in cyberspace. From the analytic perspective of sociological imagination, the book provides, among other things, rich data for relating the emergence of a particular kind of symbolic analyst and knowledge worker and the kinds of subjectivity that support creating and maintaining a list like "Meme" to participation in various computing culture Discourses. When the links to Bennahum"s "Meme" list and his home page are taken into account, it is possible to generate and reflect on a diverse range of ideas about how a stratum of the Atari generation builds networks, employs marketing strategies, maintains support systems and so on.

Bennahum is an author, cyber-activist, public intellectual, member of a cyberculture "elite," and a symbolic analyst and knowledge worker of high order and standing. How Bennahum became an insider to particular Discourses, and how this afforded him various "tracks" so far as access to peer groups, employment, and making his way within an emerging global order are concerned, offer some crucial insights for contemporary literacy educators.

With respect to the Discourse-discourse relationship, and a close focus on the language bits within larger social practices, the world of new literacies is rich, varied and easily accessed. We hint at some of the possibilities for investigation and reflection in our chapter on online ratings practices (see Chapter 13 in this volume).

(c) Critical-Evaluative

Two types of critical-evaluative accounts of new literacies seem especially important in relation to literacy education—bearing in mind that, as the outcomes of the practice of *critique*, "critical" accounts span the range from strongly positive or affirmative to strongly negative or condemnatory.

One type involves taking an ethical perspective toward new literacies, such that we can make *sound and fair* judgments that have educational relevance about the worth of particular new literacies and the legitimacy of their claims to places within formal literacy programs. The matter of mindsets arises again here. Engaging in critique of new literacies should also be taken as an invitation to examine our own mindset as much as an invitation to judge a practice. The educational literature on practices involving new technologies and, more generally on popular cultural practices, is replete with what seem to us to be straightforward *dismissals* of new ways of doing and being from the standpoint of unquestioned outsider perspectives. Such critiques are more likely to obstruct than enable progress in curriculum and pedagogy. On the other hand, the kind of critique of the D/discourses involved in the design and production of a particular range of computing games provided in Allucquère Rosanne Stone's chapter on "Cyberdämmerung at Wellspring Systems," in her book on *The War of Desire and Technology at the Close of the Mechanical Age* (1996), may be invaluable for literacy educators. Stone provides a stunning account of life inside an electronic games production sweatshop. The chapter disrupts concepts like "symbolic analysis," presents a new line on what lies behind the text, and takes a sober look at the theme of gender and power in the world of new technologies.

The second kind of critique we have in mind involves taking a curriculum and pedagogy perspective based on the criterion of *efficacious learning*. From a sociocultural perspective,

> the focus of learning and education is not children, nor schools, but human lives seen as *trajectories* through multiple social practices in various social institutions. If learning is to be efficacious, then what a child or adult does *now* as a learner must be connected in meaningful and motivated ways with "mature" (insider) versions of related social practices (Gee, Hull, and Lankshear, 1996: 4).

For literacy education to be soundly based, we *must* be able to demonstrate the efficacy of any and every literacy that is taught compulsorily. This, of course, immediately questions the basis of much, if not most, of what currently passes for literacy education. The criterion of efficacy applies very strongly to attempts to promote new literacies in classrooms. As we have argued elsewhere (Lankshear and Snyder, 2000; Goodson, Knobel, Lankshear, and Mangan, 2002), efforts to incorporate new ICTs into language and literacy education are often misguided from the standpoint of efficacious learning. Often they enlist learners in characteristically "schoolish" practices that have little or no present or future purchase on life outside the classroom.

For such reasons we believe that if the New Literacy Studies is to continue playing an important role in relation to literacy education in the years ahead, it will be necessary for its scholars to pay increasing attention to "the 'new' literacies."

References

Adbusters (2002). Retrieved March 21, 2002, from adbusters.org.

Bail, K. (1997). Deskbottom publishing. *The Australian Magazine*, May 3–4. 44.

Baker, C. (2002). Power bloggers. *Wired*. May 10(05). Retrieved April 20, 2002, from wired.com/wired/archive/10.05/mustread.html?pg=3.

Bennahum, D. (1998). *Extra Life: Coming of Age in Cyberspace*. New York: Basic Books.

Bigum, C. (1997). Teachers and computers: in control or being controlled? *Australian Journal of Education*, 41(3): 247–61.

Coupland, D. (1991). *Generation X: Tales for an Accelerated Culture*. New York: St. Martin's Press.

Coupland, D. (1995). *Microserfs*. New York: HarperCollins.

Coupland, D. (1996). *Polaroids from the Dead*. New York: HarperCollins.

Coupland, D. (2000). *Miss Wyoming*. New York: Pantheon.

Cowan, J. et al. (1998). Destino Colombia: A scenario process for the new millennium, *Deeper News*, 9(1): 7–31.

de Certeau, M. (1984). *The Practice of Everyday Life*. Berkeley: University of California Press.

Doneman, M. (1997). Multimediating. In C. Lankshear, C. Bigum, C. Durrant, B. Green, E. Honan, W. Morgan, J. Murray, I. Snyder, and M. Wild (investigators), *Digital Rhetorics: Literacies and Technologies in Education—Current Practices and Future Directions*, Vol. 3. Children's Literacy National Projects. Brisbane: QUT/DEETYA. 131–48.

Duncombe, S. (1997). *Notes From Underground: Zines and the Politics of Alternative Culture.* London: Verso.

Gee, J. P. (1996). *Social Linguistics and Literacies: Ideology in Discourses.* 2nd edition. London: Falmer.

Gee, J. P., Hull, G. and Lankshear, C. (1996). *The New Work Order: Behind the Language of the New Capitalism.* Sydney: Allen and Unwin.

Goodson, I., Knobel, M., Lankshear, C. and Mangan, M. (2002). *Social Technologies and Learning Cultures.* New York: Palgrave Press.

Green, B. and Bigum, C. (1993). Aliens in the classroom, *Australian Journal of Education*, 37(2): 119–41.

Knobel, M. (1999). *Everyday Literacies: Students, Discourses and Social Practice.* New York: Peter Lang.

Kurtz, H. (2002). Who cares what you think? Blog, and find out. *Washington Post.* 22 April. C01. www.washingtonpost.com/wp-dyn/articles/A25512–2002Apr21.html (accessed 22 Apr. 2002).

Lankshear, C. and Bigum, C. (1999). Literacies and new technologies in school settings. *Pedagogy, Culture and Society*, 7(3): 445–65.

Lankshear, C. and Snyder, I., with Green, B. (2000). *Teachers and Technoliteracy: Managing Literacy, Technology and Learning in Schools.* St Leonards, NSW: Allen and Unwin.

Lasn, K. (1999). *Culture Jam: How to Reverse America's Suicidal Consumer Binge—And Why We Must.* New York: Quill.

Mills, C. Wright (1959). *The Sociological Imagination.* London: Oxford University Press.

National Library of Australia (2010a). Trove: Digitarts. Retrieved August 19, 2010: trove.nla.gov.au/website/result?q&l-urlKeyDomain=10258.

National Library of Australia (2010b). Digitarts: Girls in Space. Retrieved August 19, 2010, from: pandora.nla.gov.au/nph-arch/Z1998–Dec-15/http://digitarts.va.com.au/gis/gis.html.

National Library of Australia (1020c). *grrrrowl* [ezine]. Retrieved August 19, 2010, from: pandora.nla.gov.au/tep/10258

National Library of Australia (2010d). Gabrielle: A typical day. *grrrowl #4: Simply Lifeless* [ezine]. Retrieved August 19, 2010, from: pandora.nla.gov.au/nph-arch/Z1998–Dec-15/http://digitarts.va.com.au/grrrowl4/front.html

National Library of Australia (2010e). *grrrowl #4: Simply Lifeless* [ezine]. Retrieved August 19, 2010, from: http://pandora.nla.gov.au/nph-arch/Z1998–Dec-15/http://digitarts.va.com.au/grrrowl4/index.htm

Papert, S. (1993). *The Children's Machine: Rethinking School in the Age of the Computer.* New York: Basic Books.

Raettig, C. (2001a). Analysis and thoughts regarding kpmg (warning: maximum verbosity!). *Chris Raettig—Personal Website.* Retrieved January 15, 2002, from chris.raettig.org/email/jnl00040.html (no longer available; archived here: www.xtdnet.nl/paul/kpmg/jnl00036.html).

Rowan, L. and Bigum, C. (1997). The future of technology and literacy teaching in primary learning situations and contexts. In C. Lankshear, C. Bigum et al. (investigators) *Digital Rhetorics: Literacies and Technologies in Education—Current Practices and Future Directions*, Vol. 3. Children's Literacy National Projects. Brisbane: QUT/DEETYA.

Rowan, L., Knobel, M., Bigum, C., and Lankshear, C. (2002). *Boys, Literacies and Schooling: The Dangerous Territories of Gender Based Literacy Reform.* Buckingham: Open University Press.

Rushkoff, D. (1994). *Cyberia: Life in the Trenches of Hyperspace.* San Francisco: HarperSanFrancisco.

Schwartz, P. (no date). *Origins: The Map Rap.* Retrieved August 17, 2000, from: www.gbn.org/public/gbnstory/origins/maprap.htm (no longer available).

Schwartz, P. (1991). *The Art of the Long View.* New York: Doubleday.

Stone, A. (1996). *The War of Desire and Technology at the Close of the Mechanical Age.* Cambridge, MA: MIT Press.

Sullivan, A. (2002). The blogging revolution. *Wired.* May. 10(05). Retrieved April 20, 202, from: wired.com/wired/archive/10.05/mustread.html?pg=2.

Tunbridge, N. (1995). The cyberspace cowboy. *Australian Personal Computer*. December. 2–4.
Ulmer, G. (1987). The object of post-criticism. In Hal Foster (ed.) *Postmodern Culture*. London: Pluto Press: 57–82.
Vale, V. (ed.) (1996). *Zines!* Vol. 1. San Francisco: V/Search.
Vale, V. (ed.) (1997). *Zines!* Vol. 2. San Francisco: V/Search.
van der Heijden, K. (1996). *Scenarios: The Art of Strategic Conversation*. Chichester: Wiley.
Wack, P. (1985a) .The gentle art of re-perceiving. *Harvard Business Review*. September-October. 73–89.
Wack, P. (1985b). Scenarios: Shooting the rapids. *Harvard Business Review*. November-December. 139–150.
Williams, E. (2002). *Blogger: Push Button Publishing for the People*. Retrieved April 23, 2002, from: www.blogger.com.

ELEVEN

Do We Have Your Attention? New Literacies, Digital Technologies, and the Education of Adolescents (2002)

Colin Lankshear and Michele Knobel

Biography of the Text

This chapter began as an invited keynote address at the seminar "New Literacies and Digital Technologies: A Focus on Adolescent Learners," convened by Donna Alvermann and held in the Institute for Behavioral Research at The University Georgia during January 2001. It was subsequently published in Donna's edited collection, *Adolescents and Literacies in a Digital World* (Lang, 2002).

Our interest in the concept of the attention economy was sparked during an extended period of close collaboration with Chris Bigum and Leonie Rowan when they were at Central Queensland University in Australia. They engaged us in themes that seemed to us to be prescient and of the essence. The concept of attention economics was a key recurring theme in conversations that ranged over two or three years. With more "free" time available in Mexico that we had been used to previously and with our geographic and (initially) linguistic isolation leading to long periods of time spent online, we had the opportunity to explore themes that would likely have been "off the radar" for us had we still been involved in full time work within teacher education. As time went on we increasingly saw a deep irony in this situation. It seemed to us that the phenomena we were increasingly experiencing and exploring online provided invaluable windows on where contemporary culture appeared to be headed. They provided interesting—for us, *compelling*—references points and perspectives for thinking about literacy, learning, meaning making, and values under conditions of new media. In the case of the economics of attention we were interested in the different emphases and purposes evident in the work available at the time, as well as in some of the ways digital media were being used to capture and sustain attention.

In terms of personal pleasures attached to researching, thinking and writing about literacy, the period from 1999 to 2003 was for us especially rich, fruitful and *liberal*—in the sense of having the chance to work on pretty much anything we wanted to. It was, for us, time very well spent.

Introduction

> What information consumes is rather obvious. It consumes the attention of its recipients. Hence a wealth of information creates a poverty of attention and a need to allocate attention efficiently among the overabundance of information sources that might consume it. (Simon, 1971: 40–41)

> If one is looking for a glimpse of what literacy will look like in the future, the fighter cockpit is a good place to look . . . The most interesting conversation I have had about literacy at the end of the twentieth century was with a fellow who designed avionic displays for fighters. He knew all the basic questions and a good many of the answers. (Lanham, 1994: n.p.)

This chapter is based on the idea that a new kind of economy—an attention economy—is currently emerging, which will become increasingly dominant in the future (Goldhaber, 1997). We will explore this idea in relation to new literacies and digital technologies from the standpoint of formal—school-based—learning opportunities available (or not available) to adolescents being educated in contemporary classrooms. After introducing the concept of an attention economy in some of its main variations, and considering some of its surrounding theory, we will address three key questions about the relationship between education and the attention economy.

1. What significance do digital technologies have for paying, attracting, and maintaining attention?

2. What significance do new literacies have for effective participation in an attention economy?

3. What do findings from 1 and 2 above imply for formal education?

Conceptions and Theory of an Emerging Attention Economy

While there is growing agreement that an attention economy is emerging, we nonetheless find significant differences in substance and perspective among those who employ and support the idea. These differences will result in varying implications for formal education if we decide that education should indeed attend to the nature and demands of an attention economy. To help clear the terrain, then, we will describe what seem to us to be three significantly different "takes"—albeit with broad "family resemblances"—on the concept and theory of an emerging attention economy. These are associated with work by Michael Goldhaber (1997, 1998a, 1998b), Richard Lanham (1994), the Aspen Institute (Adler, 1997), and the NCR Knowledge Lab (MacLeod, 2000). We will take these briefly in turn.

Michael Goldhaber on the "Attention Economy"

Goldhaber links the superabundance of information to the hypothesis of an emerging attention economy. He believes the fact that information in over-saturated supply is fatal to the coherence of the idea of an information economy—since "economics are governed by what is scarce" (Goldhaber, 1997). In other words, economies are based on "what is both most desirable and ultimately most scarce" (Goldhaber, 1998b: n.p.). Yet, if people in postindustrial societies will increasingly live their lives in the spaces of the internet, these lives will fall more and more under economic laws organic to this new space. Goldhaber (1997, 1998a) argues that the basis of the coming new economy will be attention and *not* information.

Attention, unlike information, is inherently scarce. This, says Goldhaber (1998b), is because "each of us has only so much of it to give, and [attention] can only come from us—not machines, computers or anywhere else" (n.p.). But like information, attention moves through the Net. Goldhaber identifies cyberspace as being where the attention economy will come into its own.

The idea of an attention economy is premised on the fact that the human capacity to produce material things outstrips the net capacity to consume the things that are produced—such are the irrational contingencies of distribution. In this context, "material needs at the level of creature comfort are fairly well satisfied for those in a position to demand them" (Goldhaber, 1997: n.p.), the great *minority* of living human beings, it should be noted. Nonetheless, for this powerful minority, the need for attention becomes increasingly important and increasingly the focus of their productive activity.

Goldhaber (1998a) argues that when our material desires are more or less satisfied such that we do not feel pressures of scarcity (such as being afraid of hunger or lack of shelter), we are driven increasingly by "desires of a less strictly material kind" (n.p.). Several such desires, he believes, converge toward a desire for attention. These include, for example, a desire for meaning in our lives—which is no longer a "luxury" once material needs are assured. Goldhaber links the search for meaning to gaining attention. "Why are we here, and how do we know that we are somehow worthwhile? If a person feels utterly ignored by those around her, she is unlikely to feel that her life has much meaning to them. Since all meaning is ultimately conferred by society, one must have the attention of others if there is to be any chance that one's life is meaningful" (Goldhaber, 1998a: n.p.). Hence, the attention economy:

> [T]he energies set free by the successes of … the money-industrial economy go more and more in the direction of obtaining attention. And that leads to growing competition for what is increasingly scarce, which is of course attention. It sets up an unending scramble, a scramble that also increases the demands on each of us to pay what scarce attention we can (Goldhaber, 1997: n.p.)

Goldhaber makes six points of particular relevance to our concerns here. First, in economically advanced societies young people during recent decades have spent a huge proportion of their waking hours within two key contexts: either in school, or engrossed in media—especially television and recordings. The experiences of these contexts involve paying great amounts of attention and, moreover, focusing attention on "a relative few" (Goldhaber, 1998a: n.p): TV personalities; stars in different fields (music, sport, films, etc.) whom we attend to via television, audio media, or contemporary multimedia; teachers; selected members of our peer group, and so on. Goldhaber notes that

> everyone who is seen on television models one common role, as do all teachers in schools, and that role is to be the object of a good deal of attention. Thus, without planning or intention, there has been a kind of cultural revolution, telling us that getting attention is a fine thing. And for many of us, having the attention of others turns out to feel very good, something we often want more of (Goldhaber, 1998a: n.p.).

Second, Goldhaber envisages two "classes" within the attention economy. These are "stars," who have large amounts of attention paid to them, and "fans," who pay their attention to the stars. Because paying attention requires effort, fans supply most of the effort in the attention economy. Unlike most workers in the industrial economy who had/have only one boss, fans will typically devote their attention-paying effort to multiple stars. While stars are the great winners in the attention economy, the losers are not necessarily the fans—who may receive sufficient "illusory attention" to meet their attention needs. The losers, says Goldhaber, are those who don't get any attention, who are simply ignored. This entails having "less of a clear identity and place in the community" (Goldhaber, 1998a: n.p.). The extreme case is a homeless person who dies in the street but is ignored for days—as happened in L.A. around the time Goldhaber was writing (Goldhaber, 1998a). "Losers" may be people who do not stand out sufficiently to attract attention, or individuals who do not effectively reward attention paid to them, or else individuals who repel others by demanding too much attention.

Third, being able to participate in the attention economy involves knowing how to pay and receive attention. A distinction between real attention and illusory attention is involved here. This is because in order to get attention one has to pay attention. Goldhaber (1997) argues that in a full-fledged attention economy the goal is simply to get enough attention, or to get as much as possible. Clearly, accumulating more than one's "share" of attention involves receiving more than one puts out. On the other hand, if one is to get attention one has to pay attention. The conundrum, so far as the attention rich are concerned, is resolved by the distinction between real and illusory attention. Stars and performers pay "illusory attention" to fans and audiences. They create the illusion that they are paying attention to each fan, to each member of their audience. Attention involves an exchange. People will withhold attention if they have no interest in the exchange. When readers lose interest in a chapter, they put the book down. To maintain interest they have to believe that the author is attending to them and their needs or desires. Creating illusory attention may be done by "pretending to flatter" the audience, "creating questions in their minds which you then 'obligingly' answer," claiming you will "help them with some real problem they have," making eye contact, gesturing, and so on (Goldhaber, 1997: n.p.). Methods of creating illusions of attention may lose worth (effect) if they become too common or well recognized.

Fourth, the emerging attention economy is creating large markets for attention technologies—technologies that allow us to get attention or that make it possible for us to go after it. The internet is a classic example (see below). But old technologies, such as stages where entertainers perform, are also important. The recent invention of digitized wearable display jackets (Kahney, 2000) is a new trend in generating attention (see below). This may involve gaining attention directly, for example, by advertising oneself. It may, however, involve a form of "three-way attention transaction" (Goldhaber, 1998a: n.p.). This is where one has attention passed to one by somebody else—as when advertisers use stars to pass attention to clients and their products, or show hosts pass attention to guests (but in turn also receive attention from fans of the stars who watch the shows). Hence, someone wearing a display jacket may screen clips of a popular star.

Fifth, the attention economy necessarily entails a new kind of privacy from the familiar kind of having private space away from the public gaze. Those who would accumulate attention have to be "out there." Attention wealth accrues from expressing oneself fully, living one's life "as openly as possible," and expressing oneself "as publicly as possible as early as possible"—e.g., putting drafts on the web, rather than keeping them under wraps until publication (Goldhaber, 1998b: n.p.). The quest for privacy under these conditions becomes one of avoiding being constrained by "would-be attention payers" and avoiding having to pay them too much attention (Goldhaber, 1998a). Whereas the old privacy was about not being seen, the new privacy will be about "not [having] to look at or see anyone else" (Goldhaber, 1998b: n.p.).

Finally, gaining attention is relative to originality. It is difficult, says Goldhaber (1997: n.p.), to get new attention "by repeating exactly what you or someone else has done before." Consequently, the attention economy is based on "endless originality, or at least attempts at originality" (Goldhaber, 1997: n.p.). Attention is a function of "everything that makes you distinctly you and not somebody else" (Goldhaber, 1998b: n.p.).

Richard Lanham on the Attention Economy

In *The Economics of Attention* (1994), Lanham's focus is on the changing world of the library and, especially, on the changing role of librarians in the age of digitized information and communications technologies. According to Lanham, in order to address questions like, "How are libraries and librarians to negotiate the changing terrain of information?"; "What kind of changes are involved?"; and "Where should one look for clues to handling the changes?," it is important to understand the new economy of attention.

Lanham shares some common bases with Goldhaber. He begins by observing that we currently seem to be moving from "the goods economy" to "the information economy" (Lanham, 1994). Within the so-called information economy, however, information is not the scarce resource. On the contrary, "we are drowning in it" (Lanham, 1994: n.p.). At least, to put it more accurately, we are drowning in a particular order or kind of information—information as raw data.

Lanham argues that we use different terms for information, depending on how much attention—"the action that turns raw data into something humans can use"—has been given to it (Lanham, 1994: n.p.). No attention gives you "raw data." Some attention gives you "massaged data." Lots of attention gives you "useful information." Maximal attention gives you "wisdom." And so on (Lanham, 1994: n.p.). For simplicity's sake, Lanham reduces these to "data," "information" and "wisdom," and claims that of these wisdom and information are in shortest supply. In the face of the volumes of data coming at us "we do not have time [and] do not know how to construct the human-attention structures that would make data useful to us both for . . . private life and public life, domestic economy and political economy" (Lanham, 1994: n.p.). The scarce resource in the information economy, according to Lanham, is attention.

Like Goldhaber, Lanham claims that the key resource of the new economy is non-material (or what he calls "immaterial"). But when the most precious resource is non-material, "the economic doctrines, social structures, and political systems that evolved in a world devoted to the service of matter become rapidly ill-suited to cope with the new situation" (Wriston, 1997: 19–20). Similar to

Goldhaber, and along with growing numbers of others (e.g., John Perry Barlow, cited in Tunbridge, 1995), Lanham insists that we cannot continue to apply concepts, laws, practices, and the like that were developed to deal with the economic world of goods to the emerging economic world of information. Entertaining and exploring the notion of an emerging economy of attention looks like a step in the right direction.

From these similar starting points, however, Lanham's thought develops in a different direction from Goldhaber's. Rather than focusing on how to gain and maintain attention, Lanham is concerned with how to facilitate or enable attention to data by developing new human attention structures for attending to the flood of information-as-data that we face constantly. He notes that banks have been early starters here, out of necessity since the banks' traditional role of safeguarding clients' money and lending it out has largely been taken over by other institutions. "To survive, banks are now creating from the digital stuff of instantaneous global data new attention structures for savers and borrowers, new investment instruments [which banks call] 'securitization' " (Lanham, 1994: n.p.). These provide people with new frames for attending to the [financial part of their] world.

Lanham elucidates the concept of human-attention structures by reference to examples from contemporary conceptual art and pop art. In an environmental art exhibit which involved erecting many large umbrellas in two very different kinds of location—a rainy valley in Japan and a desert mountain pass in southern California—the artist, Christo, created "temporary attention structures to make us pause and ponder how we engage in large-scale collective human effort." The "product" was attention structures rather than objects. "The center of the project . . . became the contrast in how each culture went about its work, both social and geographic" (Lanham, 1994: n.p.).

Some decades earlier Roy Lichtenstein had taken popular attention structures like the comic strip as his subject matter. Andy Warhol, as much as conceptual artists in the mold of Christo, Rauschenberg, and Robert Irwin, along with today's leaders in the aesthetics of digital expression, recognized that organizing human attention was "the fundamental locus of art, not making physical objects" (Lanham, 1994: n.p.). Lanham notes that many of Warhol's best remembered observations indicate how far the Pop Art explosion comprised an "Arts of Attention Management": compare, "we weren't just *at* the exhibit, we *were* the art incarnate and the sixties were really about people, not about what they did," and "Fashion wasn't what you wore somewhere anymore, it was the whole reason for going" (Lanham, 1994: n.p.).

Within the information economy the scarce commodity is "how human attention sorts out an overpowering flow of information" (Lanham, 1994: n.p.). Examples from the worlds of conceptual and pop art reveal the macroeconomics of attention. From the perspective of the microeconomics of attention, Lanham asks how the overload of information carried by "the rich signal," which is the heart of the digital revolution, can be managed. This signal can be manifested as alphabetic text, as image and as sound, and "creates its own internal economy of attention" (Lanham, 1994: n.p.). Lanham illustrates this with his example of fighter-jet cockpit displays, where digital data arrives at quantum rates in alphabetical and numerical information, in iconic displays, and as audio signals. A design was needed to mix all this data-information into "a single functioning information structure" that, as in the rest of contemporary life, allows our minds to make sense of data coming at us "thick and fast" (Lanham, 1994: n.p.). This is a technical instance of the larger questions of how to develop structures—frames and organizers—that facilitate paying attention to data so that we can turn it into something useful, and who will develop these structures. To the extent that the world of information at large is becoming like the fighter cockpit displays as it falls increasingly under the logic of digital expression, the key questions for literacy and the answers to those questions will increasingly concern how to develop attention structures and to organize and manage attention.

As a new dominant metaphor for thinking about our world, as matter and energy were previously, the model of information directs us to attend to what lies behind or beneath "stuff"—the world of objects—and to see "hidden forces and forms . . . which those objects allegorize" (Lanham, 1994: n.p.). Similarly, a theory of communication based on stuff presupposes a model of simple exchange whereby a package of thought and feeling is transferred from one body and place to another or others. The same communication model, says Lanham, employs a "Clarity-Brevity-Sincerity" style of prose and expression (n.p.).

Lanham argues that this model no longer applies. In terms of style and expression, the transaction within an attention economy is no longer "simply the rational market . . . beloved by the economists of stuff" (n.p.). Rather, people bring with them to the free market of ideas "a complex calculus of pleasure" and "make all kinds of purchases" in the attention economy. Lanham suggests that our efforts to learn and understand how to handle the new conditions of "seeing" and thinking about the world, and of style and expression—in short, how to develop appropriate attention structures—may be usefully informed by earlier and long-standing arts and habits. These include the Western tradition of rhetoric and the medieval allegorical habit of life and thought that saw "the immanence of God as informing all things" (Lanham, 1994: n.p.).

In spatial terms, the information model is revolutionizing practices of literacy and thinking, which Lanham illustrates by reference to the library. No longer, says Lanham, can librarians see their role as one of "facilitating thinking done elsewhere," as was the case in the age of lending out books. Under that regime, the role of the librarian was a matter of librarians "maintaining the signifiers, and leaving the decryption of the signifieds to the readers" (Lanham, 1994: n.p.). The quotation is from a librarian named Atkinson, who corresponded with Lanham on this theme, arguing that the role of the librarian remained the same in the information age as it had been before. (Lanham repudiates this position.) Instead, in the world of superabundant information, thinking involves constructing attention structures and libraries and librarians are in the middle of this—as schools and teachers and academics should be (although Lanham holds out little short term hope for universities and does not mention schools).

The point, finally, is that gateways will need to be developed to facilitate attention to information, to turn it into something useful for users and to enable users to use it usefully in terms of their wants and goals. Lanham believes, however, that this involves much more than the current development of intelligent software agents like search engines, specialized bots, and the like. Rather, he says, there is a frame issue involved. Building attention structures is more than a software or "technical" issue alone. It calls for an architecture that incorporates frames, and for a "new kind of human architect" who will mediate the economics of attention. This will be far from a technical task and will comprise the highest order and most powerful, sought-after and rewarded literacy.

Advertising and the Attention Economy: The Aspen Institute and National Cash Register's Knowledge Lab

Not surprisingly, the notion of an economics of attention and the theme of how to gain attention as an increasingly scarce resource in proportion to the sources competing for it have entered advertising discourse during recent years. Advertising is a domain of human practice with a strong stake in the economics of attention: "The first challenge for every advertiser is to capture and hold the attention of the intended audience" (Adler, 1997: 5). Indeed, advertisers have to create attention to products in which the targets of advertising typically have no inherent interest. Despite the massive and increasing amount of time citizens in countries like the United States spend using or consuming media of one kind or another—calculated to be 60.5 per cent of the waking hours of the average U.S. person in 2000—advertising faces ever-increasing competition for attention. More is involved in this competition than the success of advertisers and products alone. The very fortunes of the media used for advertising—from TV (whether public broadcast or cable) to the world wide web, via newspapers, magazines, and radio—fluctuate with and depend upon levels and constancy of advertising revenue.

In 1996 the Aspen Institute hosted a seminar to assess the current state of and prospect for the field of advertising and to identify perspectives on how individuals choose to allocate their attention. The seminar made particular reference to the context of emerging new media, notably the internet and world wide web, which have the potential to challenge established media as advertising channels. As internet use has continued to grow rapidly in recent years, the Net has been transformed "from a non-profit medium for academic and personal communication into a dynamic commercial medium" in which most major corporations and many small companies have established an online presence (Adler, 1997: 20). Although internet advertising accounts for only a tiny proportion of total current advertising expenditure, it is growing rapidly and a hot search has been mounted by marketers and advertisers to create ever new and more effective means for gaining attention.

The internet, however, presents advertisers with differences in both degree and kind from other media. The Web, for example, produces a *massive* "fragmentation of channels" (Adler, 1997: 21). As the original situation of a very small core of television networks became dozens and then literally "hundreds of different cable- and satellite-delivered channels," advertisers had to switch from broadcast to narrowcast strategies. With the advent of the Net, however, "there are now potentially millions of channels available, with the conceivable end point being a separate, customized channel for each individual" (Adler, 1997: 21–22). The growth of new interactive media creates the possibility for one-to-one marketing. This involves a strategy which focuses less on building advertising market share than on "investigating a company's *best* customers and building a one-to-one relationship with them" in order to get more purchasing or consuming per customer by "treating them as individuals ... [to] build loyalty" (Adler, 1997: 24).

This is a context where there is much to play for and where old kinds of intermediaries and partnerships change and new ones are invented. For example, given that distribution expenses may account for 50 to 80 percent of the end cost of consumer products, if producers can bypass conventional marketing and distribution intermediaries and sell direct to the consumer via the internet, this has potentially huge advantages for the latter in terms of cost and ease. At the same time, however, internet users have greater potential than users of other media to actively control the information they receive. In Net advertising, the relative balance of power shifts from producers to consumers of advertising, since on the world wide web customers do not face the choice of sitting through intrusive ads (Adler, 1997: 37). The logic that has to operate in Net advertising is less one of how media users can opt out of advertisements to one of how advertisers can get users to opt in to marketing information.

This has seen the emergence of new kinds of intermediaries, like search engines, bots, the active creation of interest-based online communities with potential for commercial exploitation, collaborative filtering technology for sharing views and interests online, and so on. For example, marketers quickly saw and acted on the potential of creating and exploiting online communities concerned with specific topics that would attract key groups or niches of customers. Once these audiences are created and identified, marketers can interact with them to "sell and support products, provide customer service [and] conduct continuous market research" (Adler, 1997: 25). Ingenious devices and processes—as well as some utterly gross forms—have been developed to capture audience attention on the internet. Gross forms include "spam" and "push" strategies, as well as successive generations of eye-catching "gizmos" (animation, flashing signals, etc.). Subtler means include companies hiring marketers to create "ad bots" that inhabit chat rooms and similar spaces on the Net. These respond to trigger words and can engage potential customers in private conversation that has commercial relevance (Adler, 1997).

High-profile research work, backed by serious budgets, aimed at developing approaches to advertising grounded in the economics of attention are under way. At the time of writing a current leader in the field is the NCR Knowledge Lab. The lab's work in this area begins from the idea that consumers are saturated with potential information sources for practically any requirement and simply cannot use all the available options without eating heavily into time. For producers and vendors operating in the emerging network economy, this creates the challenge of how to get the attention of those consumers they want to attract and/or keep, and how to make their product or brand stand out amidst increasing competition for customer attention. According to the Knowledge Lab, as the network economy continues to grow, attention will become increasingly scarce. In an early statement, now superseded in the latest information contained on its site, the lab claimed that firms now "think of themselves as operating both in an Attention Market as well as their core market" (knowledgelab.ncr.com, 1999: n.p.; no longer available). The lab's introduction to its research focus claimed that:

> Attention will be hard to earn, but if it is viewed as a reciprocal flow, firms can use information about consumers and customers to stand out in a sea of content to increase profitability: pay attention to them and they pay attention to you. Relationships are likely to encompass attention transactions. As customers realize the value of their attention and their information needed to get it, we show that they may require payment of some kind for both.
>
> The Knowledge Lab is looking into how we can quantify, measure and track flows of attention in the Network Economy. (ibid.)

To this end, the Knowledge Lab has established consumer research as one of its five research foci. The program comprises research into a web of intersecting themes. These include (among others) the nature and role of online branding, the use of interfaces for interactions and relationships with customers, the adoption and diffusion of new technology, online communities and relationships, "connecting with kids" and "cashless kids," together with research on the attention economy.

The lab has also developed and trademarked the concept of "relationship technologies" and settled on a view of attention as "engagement with information." The key to successful business in the future, says the lab, will be the capacity to generate and maintain personal attention to new and existing customers. Advertising can create *opportunities* to gain attention, but it cannot actually secure, let alone maintain and build, ongoing attention (MacLeod, 2000). Early work by lab researchers suggests the importance of using personal information to gain initial attention and "harnessing [this] attention" to create successful "real relationships" with customers (MacLeod, 2000: 3) with the assistance of "relationship technologies." Successful relationships of all kinds "contain the elements of attraction, communication, 'being there' for the other party, and understanding." The lab's idea is that in business as well as in other areas of life, relationship technologies will serve to "enable, support and enhance these key elements of real relationships" (MacLeod, 2000: 3).

This will be achieved through attention transactions in which information flows back and forward between content providers (the business or commercial interest) and content users (potential and actual customers). Since attention is "engagement with information," both-way information flows grounded in reciprocal interest are, in effect, attention transactions that create and sustain relationships (MacLeod, 2000: 7). The lab puts its faith in the capacity of *paying* attention to gain and maintain attention. Its early research documents various mechanisms used to try and elicit customer attention (such as paying people to view content, providing free computers that come with content, offering free email via portals, which bombard users with advertising and other commercial information, and so on). Without dismissing these outright, the lab stresses the importance of attention transactions based on personal information. This requires customers to appreciate the advantages that can come from providing personal information that permits companies to pay personal attention to them in the course of creating and developing successful relationships (MacLeod, 2000). Reciprocally, it presupposes that companies will use this information fruitfully: "Acquiring personal information about a potential customer is useful only insofar as it can be translated into more personal attention" (MacLeod, 2000: 19).

MacLeod (2000: 19–20) identifies key strategic implications for businesses operating in the Network Economy. These include:

- Participation in the "network economy" presupposes participating in the Attention Market, since to develop relationships with customers it is necessary first to have their attention.
- To gain the attention of network users, companies must transform initial contacts (e.g., as obtained by advertising) into an attention transaction from which to develop relationships.

• At the beginning of a relationship it may be necessary to purchase information, and some consumers may be able to ask more for it than others. Hence, companies must be prepared to negotiate.

• Consumer information costs should be seen as investments that have value to be unlocked rather than as costs to be avoided.

• The "epitome of an attention-based relationship is to move from mass customization to engaging customers in the design of products for themselves."

• As pressure increases on people's time, companies best able to provide "intelligent agents or intermediaries" will get "first call on [a] consumer's attention."

• Companies will have to identify technologies best suited to capture consumers' attention, "and 'own' the newly emerging personal access points."

Overlaps and Differences: Multifaceted Attention

While there is much more to be said than is possible here, there seems to be significant overlap as well as significant differences among these positions. In terms of differences, the three perspectives pursue attention on behalf of quite different purposes and beneficiaries. Goldhaber's account focuses on individuals pursuing attention for their own purposes in terms of finding meaning for their lives under "post materialist" conditions. Lanham addresses the pursuit of attention structures that will enable other people to use information effectively in relation to what they are interested in. The work of the Aspen Institute and the NCR Knowledge Lab seeks in different ways to help companies mobilize attention in the interests of selling consumer items to customers who believe their purposes are served by purchasing them.

The main point of overlap seems to be the creation of effective attention structures, even though Lanham is the only one of the three to identify this construct explicitly. Managing attention is clearly where the action is for each perspective. The point of advertisers, producers, and vendors entering relationships with consumers and obtaining information on them directly or via research is to be able better to mobilize and organize their attention to what is available commercially as goods and services. Goldhaber's reference to the pursuit of endless originality seems also, albeit tacitly, to entail a search for frames that will draw or focus the attention of potential fans on would-be stars.

Digital Technologies and the Economics of Attention

Goldhaber (1997) highlights the distinctive significance of new information and communications technologies—especially, but by no means solely, the world wide web. He sees the capacity to send out multimedia or virtual reality signals via the web as a particularly effective and efficient means for attracting attention and paying illusory attention.

> Say you are primarily a writer of mere words, i.e., text; still, on the Web you [are] able to supplement your writings with your picture, with video images, with recordings of your voice, with interviews or pieces of autobiography. The advantages of doing that is that by offering potential readers a more vivid and rounded sense of who you are, you can both increase their sense of who it is who is offering them illusory attention, and have them have a clearer and more definite feeling than otherwise of what it is like to pay attention to you, rather than to some other writer of similar sounding words. Both these effects can help you hold their attention better. (Goldhaber, 1997: n.p.)

In this way the web is an ideal means for "transmitting and circulating attention" and is getting better for this all the time: a precondition, says Goldhaber, for a full-fledged attention economy to emerge. He contrasts the circumstances of Plato with those of any number of people today. Over the past two millennia, says Goldhaber, millions of people have read and studied (paid attention to) Plato. But apart from "contributing to his 'immortality,' the vast majority of that attention did him little personal good." It came after he was dead! Whilst very few of today's "attention getters" could aspire to be remembered for thousands of years, they are able to pursue the benefits of attention from many—maybe millions—of people via the web throughout their lives (Goldhaber, 1997: n.p.). This, says Goldhaber, is what will constitute living very well (on a sliding scale) in the new economy.

At the level of employing digital technologies, working the attention economy can take on very different forms. Two cases must suffice here. They will serve to make wider points as well.

Early in 2000, a number of online magazines (e.g., *Salon*, www.salon.com) described one young man's special mission and encouraged readers to help him meet his goal. Walter, a 16–year-old high school student, described his special mission on his web site, which was located within the Geocities community (now defunct). According to Walter's website, two girls from his school had told him that one would have sex with him if his web site received a designated (massive) number of hits within a given period. Pictures of Walter were published alongside the articles featuring his mission. They showed him to have an almost-shaved head, braces on his teeth, features that would conventionally be described as ungainly, and what would generally pass for an "unattractive air."

The articles urged readers to visit Walter's web site to help him complete his mission before his time ran out. The response was overwhelming. Some of these sources also stated that while Walter's special mission might be a hoax, people should visit his web site anyway, in case the endeavor was for real (cf., Suck, 2000). On the day we visited Walter's web site—well before the deadline set by his female peers—the only page that could be accessed told in huge letters that the mission had been accomplished. It also stated that due to still-heavy traffic to his web site, Walter had been forced to remove it from the internet.

In the second case, Steven Fitch, a graduate of MIT's Media Lab, developed a leather jacket containing in its back panel a complete Windows computer with a "233–MHz Pentium III processor, a 1 Gigabyte IBM micro hard drive, and a broadband wireless internet connection" (Kahney, 2000: 1). The jacket is being marketed as "wearable advertising" and even comes with "a built-in infrared motion detector that can tell how many people have seen it close up by sensing their body heat" (Kahney, 2000: 2). According to Fitch (cited in Kahney, 2000: 1), the jacket "allows people to use video as a form of self-expression."

The jacket could be used in diverse ways as a medium for initiating or mediating attention flows and transactions. Some uses might essentially serve the owner's own attention-seeking interests simply by attracting the gaze of passersby and engaging them in information (however briefly or superficially). Alternatively, the owner might use the display as an initial point of contact with potential customers for her or his own goods and services. Likewise companies, advertisers, and "stars" might hire "jacket space" as part of their contact-making and attention-attracting strategies. Many uses of the jacket display might serve multiple attention interests conjointly. For example, if the wearer were running a video for a popular band or an upcoming movie (that is, for "stars"), she or he would simultaneously be paying illusory attention to fans of the band or movie star, transferring attention to the star, giving the star an opportunity for paying illusory attention to the fans, and generating attention for herself or himself.

Fitch has formed a company called Hardwear International to market the video jacket. His main company tagline is "The revolution will be televised." Hollywood has already shown keen interest, planning to display trailers for upcoming movies on people's clothing. Fitch is currently also working on a range of video jackets for children, as well as lunchboxes, handbags, and hats, all of which incorporate his video technology. Fitch (cited in Kahney, 2000: 2) says, "I believe display technology will be incorporated into our lives as a form of personal expression."

New Literacies and the Economics of Attention

On the basis of the ideas sketched above, it is reasonably easy to identify a range of typical "new" literacies that might become increasingly significant within an emerging economy of attention. We will outline several such literacies here in embryonic ways that will provide a base for potential further exploration and development.

"Contact Displaying:" Jackets (and Similar Gadgets) that Work

This is the idea of using highly customizable, mobile, public media—like the video display jacket—to catch the eye and establish a basis for gaining attention. Not every jacket will "work" in an attention economy. Not every jacket owner/wearer will be able to use it successfully as a means to gain and sustain real attention. Moreover, the jacket itself (or any similar device) cannot be the medium for sustained attention unless its wearer can claim a "space" to which others "return" in order to see what she or he is up to today. More likely, a successful display will create an *opportunity* to gain attention in the manner described by MacLeod—by establishing initial contacts that may create the possibility to develop relationships via attention transactions. This could take diverse forms, ranging from broadcasting arresting or entertaining "display bytes" that achieve their task of establishing a sense of identity and presence instantaneously—in the moment of a passing by—to simply announcing a product or service that can be "taken down now" (e.g., a URL, phone contact, email address) or memorized for following up later. Part of displaying successfully is likely to be a matter of "immediate effects" (rhetorical, quirky, stunning), but much will likely be predicated on having something to say that is worth hearing, something to sell that is worth buying, and so on.

Meme-ing

As noted in the previous chapter, in his MEME email newsletter (see memex.org), David Bennahum defines "meme" thus:

> meme: (pron. "meem") A contagious idea that replicates like a virus, passed on from mind to mind. Memes function the same way genes and viruses do, propagating through communication networks and face-to-face contact between people. Root of the word "memetics," a field of study which postulates that the meme is the basic unit of cultural evolution. Examples of memes include melodies, icons, fashion statements and phrases.

We have extended this definition to suggest a kind of literacy that may prove very effective in gaining attention as well as in constructing attention structures along the lines described by Lanham (1994). Meme-ing may be seen as a meta level literacy whereby "writers" (e.g., displayers, conventional authors, advertisers, changemakers, publicizers, and so on) try to project into cultural evolution by imitating the behavioral logic (replication) of genes and viruses. This involves generating and transmitting a successful meme, or becoming a high-profile "carrier" of a successful meme. Meme-ing presupposes the existence or establishment of two necessary conditions: "susceptibility" (for contagion), and suitable conditions for replication to occur.

Susceptibility is tackled by way of "hooks" and "catches"—by conceiving something that is likely to catch on or that gets behind early warning systems and immunity (for example, even well-honed critical perspectives can be infiltrated by textual creations like Coca-Cola's white swirl on red, or by the Nike swoosh). Networks—for example, communities of scholars, electronic networks—provide ideal conditions for replication.

As noted in the previous chapter, examples of successful memes and their respective "cloners," "high profile carriers," or "taggers" include "cyberspace" (Gibson, 1984), "screenagers" (Rushkoff, 1996), "GenX" and "Microserfs" (Coupland, 1991, 1995), "the information superhighway" (Al Gore, as referred to in Gromov, 1995–2000), "global village" (McLuhan and Powers, 1989), "cyborgs" (Haraway, 1985), "clock of the long now" (Stewart Brand and colleagues, as referred to in Brand, 1999), "complexity" (the Santa Fe Group, as described by Waldrop, 1992), D/discourse (Gee, 1996), and so on. Obviously, for a meme to be a way both of gaining attention and of bringing attention to a particular individual or group, its cloners or key carriers must somehow or other lay claim to it or otherwise publicly establish their association with it.

"Scenariating"

Building or narrating scenarios is a good way of coming up with original or fresh ideas of the kind needed to attract and sustain attention. We think of it as a literacy because it is a way of reading and writing the world (of the future). Scenarios are catchy narratives that describe possible futures and alternative paths toward the future, based on plausible hypotheses and assumptions grounded in the present (see also Chapter 10, this volume). Scenarios are *not* predictions. Rather, building scenarios is a way of asking important "what if?" questions: a means of helping groups of people change the way they think about a problem (Rowan and Bigum, 1998: 73). Scenarios aim to perceive possible futures in the present, encourage us to question "conventional predictions of the future," help us to recognize "signs of change" when they occur, and establish standards for evaluating "continued use of different strategies under different conditions" (Rowan & Bigum, 1998: 73).

Scenarios must narrate particular and credible possible future worlds in the light of forces and influences that are apparent or inchoate in the present, and which are likely to steer the future in one direction or another if they get to play out. A typical approach to generating scenarios is to bring a group of participants together around a shared issue or concern and have them frame a focusing question or theme within this area. Once the question is framed, participants try to identify "driving forces" they see as operating and as being important in terms of their question or theme. When these have been thought through, participants identify those forces or influences that seem more or less "pre-determined," that will play out in more or less known ways. Participants then identify less predictable influences or uncertainties: key variables in shaping the future that could be influenced or influence others in quite different ways, but where we genuinely can't be confident about how they will play out. From this latter set, one or two are selected as "critical uncertainties" (Rowan and Bigum, 1998: 81). These are forces or influences that seem especially important in terms of the focusing question or theme but which are genuinely up for grabs and unpredictable. The "critical uncertainties" are then dimensionalized by plotting credible poles: between possibilities that, at one pole are not too unimaginative and, at the other, not too far-fetched as to be completely impossible. These become raw materials for building scenarios.

In relation to the economics of attention, "scenariating" is a potentially significant new literacy because it provides a basis not only for coming up with innovative, original, and interesting information, but also because it addresses a topic in which almost everybody has a keen interest: what the future might be like and how to prepare for it. Scenarios can work very well as attention structures, providing frames within which people can work on information in ways that make it useful. There are many reasons for engaging adolescent students in activities of building and narrating scenarios besides the potential value of such activities for helping prepare adolescents to participate effectively in an attention economy. The latter, however, would be sufficient reason on its own because of its fruitfulness as a way of balancing originality and freshness with sheer usefulness for human beings in most areas of everyday life.

"Attention Transacting"

This is based on MacLeod's (2000) idea of both-way information flows grounded in reciprocal interest that create and sustain relationships. "Attention transacting" need not be grounded in commercial or business motives. It is about knowing how to elicit information from others, encouraging them to provide it (with appropriate assurances), and knowing how to work with that information so that it becomes an instrument for meeting what the other party believes to be its needs or interests. These may be in terms of goods, services, or more interpersonal concerns. To a large extent there is nothing particularly *new* involved here. It is similar to the kind of thing talk-back radio hosts, psychoanalysts, therapists, market researchers, and diverse kinds of consultants have had to learn to do in the past using different media. What *is* new is largely the use of new information technologies to obtain, interpret, share, and act on information of a private nature, knowing how to build and honor trust in online settings, knowing how to divulge and interpret information obtained electronically in appropriate ways, and so on. So far as formal education is concerned, of course, this is an entirely new literacy because it projects into modes and domains of life with which schools have not typically been concerned—even in subject areas like business and commercial practice. Conventional curriculum and syllabus foci, however, have rarely encouraged serious movement into the kinds of "reading" and "writing" implicit in attention transacting. Many of these will have to be invented "on the fly" and by trial and error—as with so much that is important to know in any period of transition.

"Culture Jamming"

Culture jamming refers to counter-cultural practices that critique, spoof, and otherwise confront elements of mainstream or dominant culture. It relies on making incisive or telling "strikes" that manage to turn elements of mainstream culture against themselves in a manner reminiscent of Michel de Certeau's (1984) notion of "tactics." The logic of culture jamming tactics is of gaining maximum attention with minimum resources or inputs.

A good example of culture jamming is provided by Adbusters (see Adbusters Culture Jamming Headquarters at www.adbusters.org). A sequence of highly polished web pages comprising slick and clean designs present information about Adbusters and its purposes. They describe an array of culture jamming campaigns, subject-worthy media events and advertising, cultural practices, and overbloated corporate globalization to knife-sharp critiques in the form of parodies or exposés of corporate wheelings and dealings, and undertake online information tours that focus on social issues. The figure in the previous chapter (Figure 10.2), depicting Adbusters' rendition of the true colors of Benetton, is a typical example of culture jamming as literacy. It shows how the act of "tweaking" readily available resources in the form of images and texts can produce direct and bitingly honest social commentaries. These are commentaries that everyone everywhere is able to read—a form of global literacy, which has the potential to catch the attention of almost any population, whether or not they share the text's inherent values.

"Transferring" (or "Trickle Across")

The principle of transferring is apparent each time one uses a search engine to locate information on a well-recognized expert or authority and turns up a student assignment, or reads a journal article that takes the form of an interview with a well-known person conducted (and published) by a much less (or un)known person, or when one happens upon web pages and zines lovingly assembled by fans. Transferring is based on the principle that "you have to be in to win." If one has something to say or offer that might otherwise remain unrecognized and unknown, one has nothing to lose by hitching it to or bundling it up with a personality or theme that enjoys a good deal of attention. This literacy may involve nothing more than inserting references or hyperlinks into a text published on the

World Wide Web. At a more complex level, it may involve negotiating an interview, conducting, editing, and "thematizing" the interview, and then getting it placed for publication.

"Framing and "Encapsulating": Beyond Keywords

Lanham (1994) makes an important and interesting point in his comment on the "hot search for software intelligence agents that will create 'gateways' of one sort or another without further human intervention" (n.p.). This endeavor has developed in conjunction with attempts to define "information literacy" and identify the kinds of skills—e.g., locating maximally efficient keywords—integral to being informationally literate. The other side to this literacy, which relates more directly to attention, was evident in the discourse of "tricks" one can use when registering one's web site to try and ensure that it comes near the top of the list for keyword searches—or, at any rate, finds a place as often as possible within the kinds of searches people are likely to do about the things one has to offer.

This is useful so far as it goes. But Lanham (1994) is pitching for higher and richer stakes in his focus on attention structures: in short, ways of *framing* information that hook us into organizing our interests within an area in this way rather than that; or in ways of *encapsulating* information that stand out because they are especially attractive or interesting. This involves the kind of analytical and theoretical work that puts someone sufficiently "on top of" a subject or area to allow them to find "angles" that attract and compel. Notions like "a brief history of time," or of "Pythagoras' trousers," of "a language instinct" or of "things biting back" are reasonably familiar (if high-end) examples (see books by Hawking, 1998; Wertheim, 1997; Pinker, 2000; Tenner, 1998). The point here is that reading and writing the world (of information) is very different from the kind of approach evident in key words and the like. The same kind of difference is evident in titles for works: some (like the examples listed above) are frames and (en)capsules; others are more like keywords (accurate, functional, but short on inspiration). The best, of course, are both. Their production encapsulates the kind of literacy we have in mind here.

A Challenge for Schools

There are many other new literacies we have neither time nor space to think about and sketch here. For example, literacies that go to "smarts" in design, that get the mix right within "multimediated" productions, and so on. Hopefully, the examples sketched here will be sufficient to indicate and illustrate the nature and extent of the challenge facing formal education if we believe schools ought to be paying more attention to attention. It is worth noting that all the "new" literacies identified here are "higher order" and/or "meta literacies." Some are good for creating opportunities to gain attention, others for facilitating and structuring attention, and others for getting and maintaining attention. Some are good for a combination of these. Few are closely related to most of what passes for literacy in schools today.

Indeed, prior to even addressing the more specific issues of literacy in relation to preparing students for effective participation in an attention economy, it is important to note that attention is currently constituted mainly as a *problem* for schools. On the one hand, "attention seeking" is closely associated with—and often cited as a cause of—behavioral problems. On the other hand, learning difficulties are often attributed to "short attention spans" or "attention deficiency syndrome." Thus, schools are simultaneously caught between trying to reduce and increase attention.

Interestingly, the postmodern world of the web, channel surfing, and "playing the future" (Rushkoff, 1996) and the post-materialist world of the attention economy openly embrace tendencies that currently constitute problems for schools. Perhaps it is time for us in formal education to rethink the issue of attention, and quite possibly the interface between digital technologies and new literacies provides a good place to start.

For many of us this will almost certainly involve a challenge to existing mindsets. Cathie Walker, self-styled queen of the internet, creator of the Center for the Easily Amused and co-founder of Forkinthehead.com (now defunct; archived at: web.archive.org/web/*/http://www.amused.com), offers an early warning of what that challenge might involve. In "Short attention spans on the web" (reprinted at sitelaunch.net/attention.htm; now defunct), she confesses to having once read in a magazine that if you don't grab the average web surfer's attention within 10 seconds, they'll be out of your site. She immediately qualifies that claim by admitting that she doesn't remember whether the exact figure was 10 seconds because her attention span "isn't that great either" (sitelaunch.net/attention.htm; now defunct).

If we can "hack" that kind of entrée and accept her celebration of the short attention span as a basic assumption for effective web site design, the five short paragraphs that follow in Walker's statement provide an engaging perspective on literacy in relation to attention. It is a perspective that may well offer more to adolescent and young adult students than much that is to be found in our most venerated and most-cited tomes. If nothing else, it would provide a class with serious grist for problematizing "attention" and evaluating literacies that pitch for attention. Our own web site observes none of her recommendations, but it doubtless receives almost infinitely fewer visitors as well.

At the opposite extreme, we note that the sources we have cited on the economics of the attention economy point to the importance of having a good grasp of theory and analysis. This is not necessarily an explicit grasp of highbrow theory and analysis. The ideas surveyed owe as much to the tradition of "Geek" philosophers as to the tradition of Greek philosophers. They uniformly assume that an emphasis on content and lower-order skills is not enough. The kinds of competencies associated with successfully engaging the economics of attention are those that come with the capacity to *research* aspects of the world as opposed to merely looking at them or receiving them as content.

Once again, this does not imply a highbrow or academic approach to research, although these will be advantageous if other things are in place—such as an interest in "angles," an interest in originality, willingness to take risks, and so on. Rather, the generic sense of "research" that we have in mind is inherent in the very kinds of new literacies we have begun to identify here, and which we think it is now time for us to explore, develop, and encourage as the core of the high school literacy curriculum. Although imaginative and expansive use of new digital technologies is not a *necessary* facet of such literacies (c.f., scenariating, transferring, meme-ing, framing, encapsulating), they certainly enhance and enrich their scope and possibilities.

If we continue to believe that formal education has something to do with helping prepare (young) people for the world they will enter, it will be worth exploring further conceptions and implications of the economics of attention and relating them to our conceptions and practices of literacy education within formal settings.

References

Adler, R. (1997). *The Future of Advertising: New Approaches to the Attention Economy.* Washington, DC: The Aspen Institute.

Brand, S. (1999). *The Clock of the Long Now.* New York: Basic Books.

Coupland, D. (1991). *Generation X: Tales for an Accelerated Culture.* New York: St. Martin's Press.

Coupland, D. (1995). *Microserfs.* New York: Regan Books.

de Certeau, M. (1984). *The Practice of Everyday Life.* (Vol. 1). Berkeley: University of California Press.

Gee, J. P. (1996). *Social Linguistics and Literacies: Ideology in Discourses* (2nd ed.). Bristol, PA: Taylor & Francis.

Gibson, W. (1984). *Neuromancer.* New York: Ace Books.

Goldhaber, M. (1997). The attention economy and the net. *First Monday* 2(4). Retrieved August 3, 2000, from: firstmonday.org/htbin/cgiwrap/bin/ojs/index.php/fm/article/view/519/440

Goldhaber, M. (1998a). The attention economy will change everything. *Telepolis.* Retrieved August 3, 2000, from: heise.de/tp/english/inhalt/te/1419/1.html

Goldhaber, M. (1998b). M. H. Goldhaber's principles of the new economy. Retrieved August 3, 2000, from: well.com/user/mgoldh/principles.html

Gromov, G. (1995–2000). History of Internet and WWW: The roads and crossroads of Internet history. Retrieved July 13, 2001, from: netvalley.com/intval.html

Haraway, D. (1985). Manifesto for cyborgs: Science, technology, and socialist feminism in the 1980s. *Socialist Review,* 80: 65–108.

Hawking, S. (1998). *A Brief History of Time.* London: Bantam.

Kahney, L. (2000). Video clothes: 'Brand' new idea. *Wired Online.* June 7.

Lanham, R. (1994). The economics of attention. Proceedings of 124th annual meeting of the Association of Research Libraries. Retrieved July 25, 2000, from: http://www.arl.org/resources/pubs/mmproceedings/124mmlanham.shtml

MacLeod, R. (2000, September). Attention marketing in the network economy. Paper presented at The Impact of Networking: Marketing Relationships in the New Economy, Vienna, Austria.

Pinker, S. (2000). *The Language Instinct.* New York: Harper.

Rowan, L., & Bigum, C. (1998). The future of technology and literacy teaching in primary learning situations and contexts. In C. Lankshear, C. Bigum, C. Durrant, W. Green, E. Honan, J. Murray, W. Morgan, I. Snyder, & M. Wild, *Digital rhetorics: Literacies and technologies in education—Current practices and future directions.* (pp. 73–93). Canberra, Australia: Department of Employment, Education, Training and Youth Affairs.

Rushkoff, D. (1996). *Playing the Future: How Kids' Culture Can Teach Us to Survive in an Age of Chaos.* New York: HarperCollins.

Simon, H. (1971). Designing organizations for an information-rich world. In M. Greenberger (Ed.), *Computers, Communications and the Public Interest.* Baltimore: Johns Hopkins University Press. 37–72.

Suck (2000). Hit & Run CCXXXV. *Suck.* Retrieved July 6, 2000, from: suck.com/daily/2000/07/06/

Tenner, E. (1997). *Why Things Bite Back: Technology and the Revenge of Unintended Consequences.* New York: Vintage.

Tunbridge, N. (1995). The cyberspace cowboy. *Australian Personal Computer.* September, 2–4.

Wertheim, M. (1997). *Pythagoras' Trousers: God, Physics and the Gender War.* London: Fifth Estate.

Wriston, W. (1997). *The Twilight of Sovereignty: How the Information Revolution is Transforming our World.* Bridgewater, NJ: Replica Books.

TWELVE

Cut, Paste, Publish: The Production and Consumption of Zines (2002)

Michele Knobel and Colin Lankshear

Biography of the Text

Like the previous chapter, this one also began as an invited keynote address at the seminar "New Literacies and Digital Technologies: A Focus on Adolescent Learners," and was subsequently published in *Adolescents and Literacies in a Digital World* (Alvermann, 2002). We'd been interested in zines for quite some time as interesting texts to read in their own right, but it wasn't until we began working with the "cultural animateurs" Michael and Ludmila Doneman in the late 1990s that we really began to appreciate zine writing practices and to see them with fresh eyes (see also Chapter 10 in this volume). Ludmila and Michael worked with youth from a range of equity groups, along with professional artists, writers, performers and academics, to create more than a drop-in community youth centre, focusing instead on generating a series of projects that emphasized enterprise and self-sufficiency (see Chapter 8 in this volume). Many of these projects included creating individual and group zines, in hardcopy or digital copy (or both). These zines were deeply situated and "local literacies" (Barton and Hamilton, 1998)—including, for example, a zine of photographs and walking maps that included places and landmarks (e.g., a clock; the Salvation Army hostel; good, safe places to meet friends) important to local youth. This zine juxtaposed city council tourist maps that emphasized colonial history elements of the area. The location of the youth center in a then rather notorious and rundown inner-city section of Brisbane meant that these zines were a real vehicle for young people to use to have their say about a range of issues important to them. The Donemans' work showed very clearly how zines could be used effectively for authentic pedagogical purposes without compromising the nature and ethos of zines and zining.

Introduction

> Zine culture hit its stride in the mid-'80s with the mushrooming of thousands of tiny-edition photocopied publications distributed by mail, usually to other zine publishers. Many of these small, idiosyncratic hand-crafted publications no longer emphasized the idolized object of "fan action," but rather the zine creators themselves. They were proud amateurs—they loved what they did, even if few other readers (ranging from a couple of dozen to a couple of thousand) would ever appreciate their obsessive devotion to, for example, the respective subjects of *Eraser Carver's Quarterly* or *Thrift Shop News*. (Daly and Wice, 1995: 280)

Despite their direct relevance to studies of literacy practices, zines (pronounced "zeens") have scarcely featured in the literature of educational research. Zines *have* been taken seriously as a focus of inquiry mainly within studies of popular youth culture (cf. Chu, 1997; Duncombe, 1997; Williamson, 1994). This chapter is intended to provide a modest redress of the silence with respect to zines within literacy studies generally and the new literacy studies in particular. We believe that anyone interested in the nature, role, and significance of literacy practices under contemporary conditions has much of value to learn from zines and, especially, from thinking about them from a sociocultural perspective. Indeed, we think their significance extends beyond a focus on literacy *per se* to pedagogy at large. We begin from the premise that zines are an important but under-researched dimension of adolescent cultural practices and provide fertile ground for extending our understanding of new literacies and digital technologies.

We want to make one point as clear as possible from the outset. In what follows we do *not* want to be seen as advocating any attempt to "school" zines: to try and make the production and consumption of zines part of routine language and literacy education in the classroom in the kinds of ways that have befallen so many organic everyday literacy practices. The last thing we would want to see is a zines component within, say, a genre-based English syllabus, or a temporary "zines publication center" in the corner of the classroom. The best of zines are altogether too vital and interesting to be tamed and timetabled. After all, they are a do-it-yourself (DIY) countercultural form systematically opposed to conventional norms and values associated with publishing views of the "establishment" and "schooled" reading and writing. Rather, we think that many learners and teachers might benefit greatly simply from becoming more aware of zine culture. Beyond that, they can participate in zine culture in their own ways and to the extent of their interest (which may be zero), as they would engage with other learning resources and cultural practices in their lives outside school. Our aim here is simply to introduce zines to readers who may not be familiar with them and to advance a point of view about their significance as literate cultural practice. Our view is that zines exemplify some important dispositions and qualities that young and not-so-young people may find helpful as they negotiate jungle-like social conditions lying foreseeably ahead of us (cf., Friedman, 1999; Gee, 2002; Goldhaber, 1997).

Specifically, zines exemplify in varying degrees diverse forms of spiritedness (gutsiness); a DIY mindset; ability to seek, gain and build attention; alternative (often in-your-face anti-establishment, although not always nice) perspectives; street smarts; originality and being off-beat; acute appreciation of subjectivity; tactical sense; self-belief; enterprise; and a will to build and sustain communities of shared interest and solidarity. These are the kinds of themes that will arise in our account of zines as a characteristically contemporary literacy. In what follows, we will provide a general account of zines as a cultural phenomenon, using brief illustrations of their two main forms: hard copy and electronic zines. After that we will look at some zines we consider exemplary in relation to two main themes relevant to educational work: namely, a pedagogy of tactics and a pedagogy of subjectivity.

Zines and Zine-ing

As distinctive forms of publication, zines openly defy longstanding conventions. They often employ handwritten text. They very often subvert the cash nexus: zine purchasing currency is frequently a zine in trade or postage stamps. Among hard copy zines, smudgy photocopied products are common. Zines rarely break even financially on a print run, often running at a permanent loss (sometimes a mark of pride) borne by the self-publisher. Zines are usually accessed via networks of friends, reviews, or other zinesters without recourse to advertising budgets or distributors. It is typical for a zine to be written, illustrated, designed, published, and posted by one person.

Some writers date zines as an identifiable cultural form back to the 1940s (Duncombe, 1997, 1999). The kinds of zines we are concerned with here—perzines—are more recent, achieving "critical mass" from the mid-1980s. These zines grew out of the 1970s punk rock scene as fans put together "fanzines" about their favorite band—biographical details, appearance dates and venues, album reviews, and the like. These small-run magazines, "zines" for short, were originally typed texts that were cut and pasted by hand into booklet form and photocopied. They were distributed during concerts or via networks of friends and fans. Gradually, these zines evolved into more personalized locations of expression—and their topics and themes ranged far beyond the punk rock scene. Nowadays zines come in all shapes and sizes, forms and media:

> Some are just a page or two, others much longer. They can be photocopied or finely printed, done on the backs of discarded office papers or on pricey card stock, handwritten with collages or designed on a computer using different fonts. They can be purchased for anywhere from ten cents to ten dollars; some are free, or just the cost of a stamp. (Block and Carlip, 1998: 4)

Increasingly, zines are now being published on the internet, and conventional paper zine production also often involves computers. Mostly, zinesters retain the DIY ethos and the look and feel of original zines. So today, even when zine producers key and markup their texts using a computer, they will still cut and paste texts and images onto each page after it has been printed, and then scan or copy these pages as they are.

Young people, who are the majority of zine producers, become involved in zine-ing for all sorts of reasons, and their zines take diverse forms. For example, *Daddy's Girl*, by nine-year-old Veronica (a.k.a. Nikki) grew out of the death of her father when she was six, and was inspired by her older sister's zine making (Taryn Hipp, discussed later in this chapter). Veronica writes about herself, her family, and her friends. The first issue of her zine is 16 pages long and measures 4.5 inches by 5.5 inches (11 cm by 14 cm). She includes photos of her family and herself and lists her favorite things and what she would wish for if she had three wishes.

In his first issue of *archáologie francaise*, Caleb (19 at the time) wrote about the death of his grandfather. This issue is a series of photocopied and stapled pages of a size that reminds one of small religious tracts. Inside are copies of the death announcement of his grandfather and images of medicines and surgical tools. The zine is bound down one side with a supermarket "special" label. His second issue contains soul-searching poems apparently inspired by images found in a medical school resource catalogue and included in the zine ("Budget Hands-On Eyeball—give your students an in depth look into the organ of vision"). This issue is covered in thin, flesh-pink cardboard with a hand-printed three-color caduceus medical symbol. The cutout texts and pictures in this zine have been attached to pages by means of old photo corners and then photocopied. His third issue is a set of reflections on his relationships with girls, his friends, and himself. It comprises a burgundy cardboard, handsewn envelope containing two small booklets (approximately 2 inches by 2.5 inches, or 4 cm by 5 cm): part one and part two.

Fifteen-year-old Athena, a Filipino-Chinese living in Lungsod ng Makati, Manila, produced her online zine *Bombs for Breakfast* from early 2000 until mid-2001 (Athena, 2000). Her white text on a red background was stark and provocative, and her website included articles from her hardcopy zine, *Framing Historical Theft*, as well as journal entries, a well-used message board, a guest book for visitors to "sign," and a set of pages on the defunct sub-pop band *Hazel*. The website also included lists of books she's reading for pleasure (e.g., Hannah Arendt's *On*

Violence) and for English classes at the international school she attends (e.g., Joseph Conrad's *Heart of Darkness*), and her comments on these books, along with a collection of texts she has published in school magazines and so on. Her hardcopy zine is a vehicle for exploring and discussing "Flipino Chineseness," food, travel, and language. Her writing includes themes such as homophobia, racism, classism, imperialism, student-friendly teaching, the politics of golf, and the like.

Carla DeSantis's *ROCKRGRL* began as a disgusted response to the ways her fellow women musicians were portrayed in the rock media. *ROCKRGRL* is a zine about and for women in the rock industry (DeSantis, 1997).

Ciara (20 years old) published a queer/bisexual online zine for a number of years—now defunct at the time of writing—and continues to publish in a hardcopy format (Ciara, 2000). The main page of her website had an aqua-blue background with the text set onto a white inset column studded with pink stars. Her website was devoted to the personal and political: she critiqued rap music and racist lyrics, wrote about identity and ex-lovers, and posted "confessionals" about her enemies, likes, dislikes, and wrongdoings. The site also contained an archive of previous postings and an interactive message board where Ciara and readers of both the online and offline zines she produced could leave messages and comments. It also linked to a large number of other online zines.

Zines use a range of textual forms, including straight prose, poems (e.g., Paul's *Above Ground Testing*); literary and film narratives (e.g., *Deeply Shallow* edited by Jason Gurley); cartoons and comic strips (e.g., Jeff Kelly's *Temp Slave!*); clipart (e.g., Sean Tejaratchi's *Crap Hound*); collages; and so on. They are thematically diverse. A sample of zines we have surveyed deal with the following kinds of themes: personal tough times and lows (e.g., Steve Gevurtz in *Journal Song #1*); being bisexual or queer (e.g., Ciara, 2000); Abraham Katzman's *Flaming Jewboy* and his *I'm Over Being Dead*); dishwashing in restaurants and diners in the United States (e.g., Dishwasher Pete's *Dishwasher*); fine arts (e.g., *Cyberstudio*); thrift shop shopping (e.g., Al Hoff's *Thrift SCORE*); being fat (e.g., *FAT girl,* Marilyn Wann's *Fat!So?*); paganism (e.g., Madelaine Ray's *The Abyss*); the 1970s (e.g., Candi Strecker's *It's a Wonderful Lifestyle*); collecting things (e.g., Otto van Stroheim's *Tiki News,* Al Hoff's *Thrift SCORE*); being temp workers or work in general (e.g., Jeff Kelly's *Temp Slave,* Julie Peasley's *McJob,* various issues of *Cometbus*); true crimes and murder stories (e.g., John Marr's *Murder Can Be Fun*); feminism (e.g., *Riot Grrrl,* Mimi Nguyen's *Aim Your Dick* and *Slant,* Toad's *I'm Not Shy...I Just Hate People*); music, especially punk music (e.g., *Riot Grrrl,* Digitarts' *Losergurrl,* gutterbunny and others' *Bondage Girl*); popular media images (e.g., Betty Boob and Celina Hex's *Bust*); the "secret history" of wars, global companies, and so on (e.g., Iggy in *Scam*); movies and/or movie making (e.g., Russ Forster's *8–Track Mind*); death (e.g., Caleb, 2000, Kimberley in *the speak easy*); skateboarding/

snowboarding; visiting restricted- or no-access areas; UFOs; conspiracy theories; fetishes; and other zines (e.g., Angel, 1999; *Factsheet 5*).

A zine may specialize in a single theme across all its issues or cover diverse themes within single issues or across issues. In all instances, the writer-producers are passionate—at times to the point of obsession—about their subject matter and desire to share ideas, experiences, values, analyses, comments, and critiques with kindred spirits. Despite widespread claims that contemporary young people are apolitical or apathetically political (e.g., Craig and Bennett, 1997; Halstead, 1999), many zinesters write intensely and with a great deal of caring about the politics of alternative cultures and the politics of the everyday—race/ethnicity, class, sex, gender, work, identity, their bodies, eating, and so on. They voice their opinions loud and clear in their textual productions. According to Stephen Duncombe (1997), zinesters are busy *creating* culture more than consuming ready-made "culture," and many are interested in rewriting what counts as "success."

> They celebrate the everyperson in a world of celebrity, losers in a society that rewards the best and the brightest. Rejecting the corporate dream of an atomized population broken down into discrete and instrumental target markets, zine writers form networks and forge communities around diverse identities and interests. Employed within the grim new economy of service, temporary, and "flexible" work, they redefine work, setting out their creative labor done on zines as a protest against the drudgery of working for another's profit. And defining themselves against a society predicated on consumption, zinesters privilege the ethic of DIY, do-it-yourself: make your own culture and stop consuming that which is made for you. Refusing to believe the pundits and politicians who assure us that the laws of the market are synonymous with the laws of nature, the zine community is busy creating a culture whose value isn't calculated as profit and loss on ruled ledger pages, but is assembled in the margins, using criteria like control, connection, and authenticity. (Duncombe, 1997: 2)

To some extent businesses (corporate media) have muscled in on zines, as they have on "alternative cultures" more generally. Occasional television shows or books for young people feature a zinester as the main protagonist (CBC Television, 2000; Wittlinger, 1999). Other approaches include cajoling young people to produce their work as mainstream compilations or how-to-do-it books (e.g., Block and Carlip, 1998; Carlip, 1995), or by posting websites touted as "online zines" that are really for selling products (e.g., Abbey Records, 2000). Many "faux zines" now exist on the market. *Slant*, produced by the Urban Outfitters clothing chain, includes a "punk rock" issue, and the Body Shop's *Full Voice* praises those who are "rebelling against a system that just won't listen" (Duncombe, 1999).

Most zines and zine-related cultural practices remain steadfastly outside the publishing mainstream. They define themselves against conventional publishing culture and poach off it. As we have seen, corporate publishing culture itself has poached more or less successfully in its own terms off zine culture. So the defining and poaching goes two ways. There is, however, an important difference. Business corporate "faux zine-ing" tends to be highly *strategic*, in the sense developed by Michel de Certeau (1984), in relation to the everyday practices of consumers. By contrast, the operating logic of zines is often highly *tactical*—once more in the sense developed by de Certeau. One of our central concerns in this chapter is to explore zines in terms of a concept of tactics and to suggest how educators and learners might be able to draw insights from zine culture to develop *pedagogies of tactics*. We are interested in the extent to which pedagogies of tactics might be better adapted to preparing many young people—especially those from non-dominant social groups—for handling the "fast" world (Freidman, 1999) than more conventional pedagogical approaches, which buy more or less exclusively into a strategic logic of producers.

Zines and Pedagogies of Tactics

de Certeau (1997) is a wonderfully subversive and subtle writer. Perhaps it is on account of this that his work has remained relatively marginal within education. Whatever the reason, it is unfortunate because there is enormous potential in his approach to issues of power and subordination for critically informed educational practice. Two common postures within language and literacy education provide useful starting points for considering zines in relation to some of de Certeau's central ideas in ways that help point us toward potentially fruitful pedagogies of tactics.

The first posture might be summarized like this. We are moving into a postindustrial world in which large sections of the "middle" have disappeared and work and rewards have become increasingly polarized. For a few there will be high-skill, high value-added, well-rewarded work that draws on high-order symbolic-analytic knowledge and skills. Even getting lower-level work will require higher levels of literate and symbolic competence than in the past. As (literacy) educators, we must aim to teach higher-order skills to as many as can handle them, and make absolutely sure no learners fall through the basic literacy "net." Indeed, even basic literacy now needs to be seen in terms of problem-solving and trouble-shooting abilities that can be transferred to frontline work, as well as in terms of the traditional 3Rs.

The second posture concerns the study of media. According to this, media shape up individuals' understandings of the world as passive consumers of TV, newspapers, magazines, the internet, and advertising who absorb worldviews that at best dumb them down and that at worst undermine their own interests to the benefit of powerful groups. Hence, we need to teach (critical) media studies to help learners decode media messages so they can resist the way that these messages position us. Various techniques and procedures are adopted and adapted from fields like discourse analysis, critical language awareness, semiotics, critical literacy, and so forth, and taught as antidotes to being passive and/or duped. Without in any way wanting to denigrate such postures, not least because we (have) subscribe(d) to them ourselves, we also sense a need to come up with some new pedagogical crafts and orientations, including some that can be thought of as pedagogies of tactics.

In *The Practice of Everyday Life*, de Certeau (1984) develops a conceptual framework based on distinctions between producers and consumers, and strategies, uses, and tactics. Producers (the strong) are those who create, maintain, and impose disciplined spaces. They have the position and power to prescribe social orders and syntactical forms (discourses, timetables, procedures, the organization of space and things within it, etc.). Producers include governments, urban planners, corporations, professional associations, legislators, private utilities companies, scholarly and academic leaders, executives, and so on. Producers, in effect, shape dominant social structures. Consumers, on the other hand, are constrained to operate within these disciplined spaces or structures. (Of course, producers in one context are to some extent consumers in others, albeit typically consumers with greater power to negotiate these spaces than "everyday people"). Thus, for example, inhabitants of government housing consume what has been produced for them—as do users of public transportation and road networks, students, prisoners, and purchasers of diverse goods and services and media available on the market. Consumers are always and inevitably constrained by what producers serve up as disciplined discursive spaces and the commodities attaching to them.

The distinction between "strategies" and "uses and tactics" parallel that between producers and consumers. Strategy, according to de Certeau, is an art of the powerful—producers. These "subjects of will and power" operate from their own place (a "proper") that they have defined as their base for controlling and managing relations. This place (or "proper") is an enclosed institutional space within which producers regulate distributions and procedures, and which has "an exteriority comprised of targets or threats" (de Certeau, 1984: 36). For example, professional scientists define what counts as doing science, build science faculties within universities to police apprenticeships to science, and regulate who can receive qualifications and tickets to practice as scientists. The justice system defines the conditions under which convicted prisoners will live. Education departments regulate what students may and must acquire as formal education and how they must perform in order to be certified as successful, and so on. Strategy operates on a logic of closure and internal administration (Buchanan, 1993). "Strategy equals the institutional," says Ian Buchanan (1993: n.p.), and is the force "institutions must exact in order to remain institutions." Hence, the strategic "can never relax its vigilance; the surveillance of its parameters must be ceaseless. The strong must protect themselves and their institutions from the weak" (Buchanan, 1993: n.p.).

For de Certeau (1984), "uses" and "tactics" are arts of the weak, by means of which the weak make disciplined spaces "smooth" and "habitable" through forms of occupancy. Through uses and tactics consumers obtain "wins" within their practices of everyday life. de Certeau illustrates "uses" by reference to North African migrants obliged to live in a low-income housing estate in France and to use the French of, say, Paris or Roubaix. They may insinuate into the system imposed on them "the ways of 'dwelling' (in a house or in a language) peculiar to [their] native Kabylia" (de Certeau, 1984: 30). This introduces a degree of plurality into the system. Similarly, the indigenous peoples of Latin America often used

> the laws, practices, and representations imposed on them...to ends other than those of their conquerors...subverting them from within...by many different ways of using them in the service of rules, customs or convictions foreign to the colonization which they could not escape. (de Certeau, 1984: 32)

"Tactics" involve the art of "pulling tricks" through having a sense of opportunities presented by a particular occasion—possibly only a literal moment—within a repressive context created strategically by the powerful. Through uses and tactics "the place of the dominant is made available to the dominated" (Buchanan, 1993: n.p.). According to de Certeau (1984: 37), a tactic is

> a calculated action determined by the absence of a proper locus.... The space of a tactic is the space of the other. Thus it must play on and with a terrain imposed on it and organized by the law of a foreign power. It does not have the means to *keep to itself*, at a distance, in a position of withdrawal, foresight, and self-collection: it is a maneuver "within the enemy's field of vision,"...and within enemy territory. It does not, therefore, have the option of planning, general strategy.... It operates in isolated actions, blow by blow. It takes advantage of opportunities and depends on them, being without any base where it could stockpile its winnings, build up its own position, and plan raids.... This nowhere gives a tactic mobility, to be sure, but a mobility that must accept the chance offerings of the moment, and seize on the wing the possibilities that offer themselves at any given moment. It must vigilantly make use of the cracks that particular conjunctions open in the surveillance of proprietary powers. It poaches them. It creates surprises in them.... It is a guileful ruse. (emphasis in original)

Buchanan (1993) helps clarify what is at stake here by distinguishing between "place" and "space." Buchanan construes "place" as the "proper" of the strategy of the powerful. Place is "dominated space" (Lefevbre) or "disciplined space" (Foucault). Space, on the other hand, is used by Buchanan to refer to *appropriated* space. Tactics, says Buchanan, are means by which consumers convert places into spaces. In this, consumers employ tactics like "bricolage" and "perruque" to "make do" by "constantly manipulating events in order to turn them into 'opportunities'" (de Certeau, 1984: xviii). Very ordinary examples of tactics include stretching one's pay packet to allow for a few "luxuries" every now and then, producing a dinner party out of a few simple and available ingredients, inventing words on the spur of the moment, and so on.

de Certeau (1984) thinks of consumers' everyday creativity in terms of trajectories that can be mapped as a dynamic tracing of temporal events and acts (the precise obverse of passive receiving and absorbing). "In the technocratically constructed, written, and functionalized space in which consumers move about [i.e., the *place* of producers and their productions], their trajectories form unforeseeable sentences, partly unreadable paths across a space" (ibid.: xviii). These trajectories, or transcriptions of everyday ways of operating, "trace out the ruses of other interests and desires that are neither determined nor captured by the systems in which they develop" (ibid.).

Thinking about zines in terms of trajectories adds a dynamic that can move our analyses beyond zines as merely exotic and static artifacts. We look at them, instead, as vibrant, volatile, thriving social practices that describe deep currents and concerns within youth culture. We can explore zines as enactments of tactics on enemy terrain, and on a number of levels. We may begin this kind of exploration by considering how zines often employ tactical maneuvers of *bricolage* and *la perruque* (de Certeau, 1984). Bricolage refers to the "artisan-like inventiveness" of consumers' everyday practices whereby they use whatever comes to hand in carrying out these practices. de Certeau refers to bricolage as "poetic ways of 'making do' " (1984: 66), and as "mixtures of rituals and makeshifts" (1984: xvi). He celebrates the bricolage-like practices of consumers as they go about their everyday lives. Such bricolages are often extraordinarily ordinary, yet underwrite effective modes of living and being on unfriendly terrain. The life of a community, for example, is made from the harvest of miniscule observations, a sum of microinformation being compared, verified, and exchanged in daily conversations among the inhabitants who refer both to the past and to the future of this space. As an old lady who lives in the center of Paris leads her life:

> Every afternoon she goes out for a walk that ends at sunset and that never goes beyond the boundaries of her universe: the Seine in the south, the stock market to the west, the Place de la République to the east.... She knows everything about the cafés on the boulevard, the comparative prices, the age of the clients and the time that they spend there, the lives of the waiters, the rhythm and style of people circulating and meeting each other. She knows the price and the quality of the restaurants in which she will never lift a fork. (de Certeau, 1997: 96)

The "mixtures of rituals and makeshifts" that are bricolages—like those orchestrated in the old lady's walks—are integral to the practice of zines as creative appropriations rather than strategic productions. To use de Certeau's concepts, zines are mostly "miniscule observations" and conglomerations of "microinformation." A good example is provided by Dishwasher Pete and his zine *Dishwasher*. This zine literally traces (documents) a trajectory of poetic ways of making do on a daily basis.

Pete's life goal is to work as a dishwasher in every U.S. state. His zine *Dishwasher* provides accounts of his work in various restaurants and his reflections on life. Pete does not own a car or have a fixed address. He stays with people he meets via his zine—crashing on their lounge room floors until he quits his job and moves on. Much of the detailed commentary in *Dishwasher* focuses on inequities in the food service industry, behind-the-scenes critiques of restaurant owners, work anecdotes from other dishwashers, and so forth. His bricolage is a "critique of class and privilege from a unique viewpoint which preserves [his] personal freedom, self esteem, and well-being" (Vale, 1997: 11).

Interestingly, much of de Certeau's (1984: 25) work traces the collapse of revolution—overthrowing oppressive regimes by force—as a viable means for transforming "the laws of history" and suggests, instead, that the art of "putting one over" on the established order on its own home ground is a means for undermining these orders from within. One way of doing this is through a tactic identified by de Certeau as *la perruque*—French for "the wig." This is a "[a] worker's own work disguised as work for his [or her] employer" (1984: 25)

La perruque differs from stealing or pilfering because nothing of significant material value is actually stolen (the worker uses scraps or leftovers that would ordinarily be thrown out). Likewise, it is not absenteeism because the worker is "officially on the job" (de Certeau, 1984: 25). Instead, the worker diverts time to his or her own needs and engages in work that is free and creative and "precisely not directed toward profit" (de Certeau, 1984: 25). It may be something as simple as a secretary writing a love letter on company time (and using a company computer and their paper and mailing system) to something much more complex, such as a cabinet maker using a work lathe to create a piece of furniture for his home (using timber offcuts from the for-profit-work, which he picks up from the scrap heap to build his chair). Thus, "[i]n the very place where the machine he [or she] must serve reigns supreme, he [or she] cunningly takes pleasure in finding a way to create gratuitous products whose sole purpose is to signify his own capabilities through his [or her] *work* and to confirm his [or her] solidarity with other workers or his family through *spending* his [or her] time this way" (emphases in original; de Certeau, 1984: 25–26).

La perruque captures the deviousness of tactics and captures ways in which "[e]veryday life invents itself by *poaching* in countless ways on the property of others" (de Certeau, 1984: xii). Many hardcopy zines are, in fact, perruques, and would not exist without the possibility of poaching on others' property. To some extent this involves poaching on material resources. A not-for-profit ethos can be sustained, subversively, by means of *la perruque*, as shown in the following example:

> I had a temp job working in the mail room of an insurance company that was promising me full-time employment. I thought, "Hey—this will be good. I can deal with this work; it's easy, I get benefits, I get a regular paycheck ..." then they reneged and said they were bringing in someone from another department to take over my job. Anger and access to paper and copiers motivated me to produce the first issue [of the now-famous *Temp Slave*]—everything coalesced at once. (Jeff Kelly in conversation with Vale, 1996: 22–23)

La perruque even can help us understand young people's job choices: "I was showing a zine to a friend and coincidentally its producer was employed in her office mailroom. She'd always thought he was too talented for the job but suddenly realized why he stayed there..." (Bail, 1997: 44).

Sometimes the material resources that are poached actually become the substance of the zine. R. Collision, for example, worked in a photocopying shop and was amazed at the kinds of images people brought in for copying—everything from mugshots to photos of operation scars to pictures of body parts and pornography. Collision was so fascinated by these windows into the human condition that he made double copies of interesting images and kept one copy for himself. Then, as he describes it, from "the graphics I had accumulated at work, I decided to publish an image compilation book that would say 'Recycle this' on its cover, and began copying as many pages as I could at work [without paying for them]. Eventually I had enough sheets to publish 200 copies of a 300–page book" (R. Collision in conversation with Vale, 1996: 43).

In other cases, zinesters' practices of *la perruque* involve poaching on abstract or intellectual "property" in order to appropriate space. Vale speaks of zines as a grassroots response to a crisis in the media landscape: "What was formerly communication has become a fully implemented control process. Corporate-produced advertising, television programming and the PR campaigns dictate the 21st century 'anything goes' consumer lifestyle" (Vale, 1996: 6). Numerous zine and zine-like productions poach upon and subvert corporate media productions as exercises in "culture jamming," parody, and exposé.

At one level this is evident in practices as direct and straightforward as literally turning media images in on themselves, or by combining images and tweaking texts to produce bitingly honest social commentaries that everyone everywhere can read and understand—a kind of global literacy. This kind of tactic, wonderfully employed in the Adbusters critique of Benetton's attempt to evoke an "equality" and "global village" ethos in the fashion world (see Lankshear & Knobel, 2002; see also Chapter 10 in this volume) is widely practiced within zine culture.

At another level, strategic productions—or enacted strategies on the part of producers—in the form of 'official' versions of how we should be and do are poached, preyed upon, and otherwise made into opportunities to turn place into space by tactical means. For example, Taryn Hipp writes in the first edition of her zine *girl swirl fanzine*: "Being an 'overweight' girl is not easy. When I look around all I see are these pictures of skinny women in revealing clothes standing next to a handsome man" (Hipp, 1999: 1). Taryn uses her zine as a personal space: she critiques images of women in the media; candidly discusses her relationship with her boyfriend, Josh; openly describes being a member of a rather unconventional family; and so on. While not a direct "attack" on or resistance to popular media, *girl swirl fanzine* is the product of Taryn's "making space" in the niches and crevices of institutions such as mainstream magazines and television by thumbing her nose at the formal structures and strategies of these institutions. Her hand-crafted paper zine sits nicely alongside her website (Hipp, 2001), which in addition to showcasing issues of and excerpts from her zine, also includes a web log (similar in concept to a diary, which can be added to at will) and is often asynchronously interactive, thanks to email and other responses from readers. Her online zine and social commentaries are further supported by an email discussion list. Taryn is not so much out to change the world as to declare her position within it:

> I am happy with the way I am. I am happy with the way I look. I am happy being "overweight." I used to worry about what other people thought of me. I have pretty much gotten over that. It wasn't easy. It never is. (Hipp, 1999: 1)

In his inimitable way, Dishwasher Pete also deftly creates his own "space" within the formal world of work and communicates this for a wider audience in *Dishwasher*. Using texts, images, and his own experiences in creating his zine, Pete critiques mainstream mindsets about what young people "should" do and be. For example, he recounts critiquing social assumptions and institutions from a very young age—which in large part he attributes to growing up desperately poor. While he was still in primary school, Pete recalls analyzing and "busting" the myth of upward social mobility through education by means of his observations of the microinformation of everyday life. He recalls:

> No matter how poor you are, you're expected to pretend that someday you'll be a doctor. Every year the nuns at our school would ask, "What are you going to be when you grow up?" Destitute kids would get up and crow about how they were going to be some great lawyer—this is what you were *supposed* to say. I would always say I wanted to be a house painter, because I remembered watching one with a paintbrush in one hand, a sandwich in the other, his transistor radio

> playing while he sat on a plank brushing away in the sun. I thought, "That's the job for me—I could do that!" The nuns were never happy when they heard this: "A house painter?! Are you sure you don't want to be a doctor?" "No, ma'am." (emphasis in original; Dishwasher Pete in conversation with Vale, 1997, p. 8)

In addition to critiquing social institutions and myths, and as we've mentioned already, Pete's zine is not just about dishwashing in countless restaurants across the United States, but is also a deeply thoughtful and thought-provoking critique of work and economic inequality. Indeed, Dishwasher Pete himself actively sidesteps "baby-boomer" work ethics and turns the proliferation of "McJobs" to his own ends (cf. Howe and Strauss, 1993). As he puts it:

> I'm addicted to that feeling of quitting; walking out the door, yelling "Hurrah!" and running through the streets. Maybe I need to have jobs in order to appreciate my free leisure time or just life in general.... Nowadays, I can't believe how *personally* employers take it when I quit. I think, "What did you expect? Did you expect me to grow old and die here in your restaurant?" There seems to be a growing obsession with job security, a feeling that if you have a job you'd better stick with it and "count your blessings." (Dishwasher Pete in conversation with Vale, 1997: 6; see also Duncombe, 1997)

By no means do all zines employ tactics in the kinds of ways we have illustrated here. Many zines reflect sophisticated expertise in the use of tactics in the sense that their author-producers "[pinch] the meanings they need from the cultural commodities...offered to them" (Underwood, 2000: n.p.). Zinesters are often highly adept at appropriating spaces of dominant culture for their own uses, or of otherwise making these spaces "habitable."

Some important points for educational practice generally and literacy education specifically flow from our attempt to explore zines in the light of de Certeau's (1984) conceptual frame. One fairly obvious implication is that for all the value there is in addressing critical analyses of media texts and other cultural artifacts within curricular learning, it is also important to understand how consumers *take up* these commodities. Doubtless the world will and should be transformed. Meanwhile we need to make it "habitable." There is much to be learned from those whom we classify as learners and/or in need of learning in terms of how they make places habitable, how they borrow meanings to make do, and how having *enough* people making do successfully might *act back* on dominant culture.

Buchanan (1993) makes an important series of points here. He notes that theorists often see strategy and tactics as oppositional terms, and thereby assume that de Certeau's approach belongs to a weaker category of resistance. In other words, it is often thought that tactics are merely "reactive forces, a practice of response" (n.p.). Buchanan notes that, on the contrary, tactics "define the limits of strategy" and force "the strategic to respond to the tactical" (Buchanan, 1993: n.p.). Hence, tactics contain an active as well as a reactive dimension. So, for example, prisoners determine the level of security required in a given prison. Users of non-standard Englishes determine the degree of policing needed on behalf of Standard English. Zinesters help to determine the degree of diversity required in establishment publisher lists. In a context where tactics are strong, healthy, many, and pervasive, the fact that the strategic machines are always one step behind when they need to be one step ahead becomes apparent (Buchanan, 1993: n.p.). The situation could become stressful for producers. Could "armies" of tacticians up the ante to the point where strategies pop? Our hunch is that it is worth testing this out.

Perhaps in schools we spend too much time trying to set kids up to perform within *strategically* defined parameters of success. This, paradoxically, often leads to engaging in practices that actually dumb kids down—such as enlisting them in moribund basic literacy remediation programs or engaging them in painting by numbers activities to familiarize them with dominant genres. This kind of approach can subvert many genuine "*smarts*" that extraordinarily ordinary practitioners of tactics have—including practitioners who are the so-called literacy disabled—and which could productively be built on in classrooms.

One of our favorite examples here concerns a Year 7 student, Jacques, who told us "I'm not keen on language and that. I hate reading. I'm like my Dad, I'm not a pencil man" (Knobel, 1999, 2001; see also Chapter 9 in this volume). His teacher concurred, describing Jacques as "having serious difficulties with literacy." Jacques did all he could to avoid reading and writing in class, although he collaborated with family members to engage successfully in a range of challenging literate practices outside school. These included producing fliers to attract customers to his lucrative holiday lawn mowing round, and, as a Jehovah's Witness, participating in Theocratic School each week, where Jacques regularly had to read, explain, and give commentaries on texts from the Bible to groups of up to 100 people.

Jacques's literacy-avoidance behavior in class yielded a classic use of tactics with respect to the Writers' Center his teacher had established in one corner of the classroom, where students could work on the narratives they had to produce for their teacher. During a two-week period we observed him spending several hours at this Writers' Center making a tiny book (6 cm by 4 cm, or 2 inches by 1 inch) containing several stapled pages. On each page he wrote two or three words which made up a "narrative" of 15 to 20 words (for example: "This is J.P.'s truck. J.P. is going on holiday in his truck. J.P. likes holidays in his truck. The End"). Other students found these hilarious when he read them out loud to them, and he eventually produced a series of six "J. P. Stories."

His teacher's response was negative and highly critical. She was not impressed and saw his activities as "very childish" and as a means of avoiding writing and of not taking too seriously something he could not do. Yet Jacques's tactical approach to making this literacy learning context "habitable" showed precisely the kind of "spark" that could serve him well in all kinds of real-world contexts. It also inchoately contains a critique of much classroom activity (what's the point of it? How is it relevant?) that is consistent with formal research-based critiques of non-efficacious learning (cf. Gee, Hull, and Lankshear, 1996). A teacher who could appreciate and celebrate tactics might have been able to reward the potentially fruitful and genuinely subversive element of Jacques's "trick" and extend it pedagogically.

We want to argue that zines provide the kind of tactical orientation that would help teachers and learners develop pedagogies of tactics to supplement pedagogies that render unto producers. Such pedagogies would identify, reinforce, and celebrate tactics when they occur, and invite other participants to consider alternative possible tactical responses to the same situation. This might take some time out of being on task within formal learning activities, but with the chance of stimulating and enhancing native wit, survival potential, critical thinking, and creative subversion. It may be worth contrasting here the capacity of a Dishwasher Pete to handle the impact of a new work order where many middle-level workers and managers experience their lives and worlds collapsing when their jobs no longer exist. A good tactician always has somewhere to move. Under current and foreseeable conditions of work, teachers as much as their students (will) need well-honed tactical proficiency in order to obtain the meanings they need. Many of us in education might benefit by refining our capacity to pinch and poach on the property of education producers. In so doing we might contribute something to the tactical prowess of all who are *compelled* to be education consumers. Our argument is that zine culture is a likely place to include in our efforts to understand and develop pedagogies of tactics. This work is greatly assisted by the close study of subjectivity in relation to zining. An individual's sense and enactment of self is tied intimately to his or her ability to celebrate the "everyperson" and the microinformation of everyday life, and to practice poetic ways of "making do."

Zines, Subjectivity, and Pedagogy[1]

The role of education in relation to personal development has been massively complicated during the past two decades. Such phenomena as intensified migration and intercultural exchange, the demise of former longstanding "models" and "pillars" of identity (e.g., well-defined gender norms) and the linear life course, displacement of modernist/structuralist ways of thinking about persons and the world by postmodern/poststructuralist/postcolonialist perspectives, the rise of radically new forms and processes of media, and an emerging new globalization have been prime movers of this complication. They have intersected in ways that generate profound challenges to knowing how and what to be in the world at the level of subjecthood. They have also helped to complicate aspiring, emerging, and established educational reform agendas in areas of equity, gender reform, and the like.

The individual's sense of self, now commonly referred to as subjectivity rather than identity, is shaped at the confluence of diverse sociocultural practices and discourses (Rowan et al.: 2001). "Subjectivity" refers to "our ways of knowing (emotionally and intellectually) about ourselves in the world. It describes who we are and how we understand ourselves, consciously and unconsciously" (MacNaughton, cited in Rowan et al., 2001: 67). Individuals negotiate cultural understandings about acceptable, proper, or otherwise valued modes of gendered or ethnic *being* in the course of shaping and reshaping their own senses of themselves. Cultures circulate meanings about what it is to be a *valued* kind of girl or boy, or member of a particular ethnic grouping, and so on. Poststructuralist perspectives in particular have re-emphasized the point that while powerful and regulatory social fictions about gender and ethnicity are circulated and endorsed by diverse institutions and discourses, it is also possible for alternative and less restrictive representations to be constructed, circulated, and validated. For example, strands of feminist research have focused on the personal and political significance of alternative representations and images of being a girl or a woman. Donna Haraway (1985) speaks here of new "figurations" (such as her notion of *cyborg*). Such "figurations" are not merely "pretty metaphors [but] politically informed maps [that] aim at redesigning female subjectivity" (Braidotti, 1994, p. 181; Rowan et al., 2001).

From this kind of standpoint, reform agendas within education in areas like gender and ethnicity involve identifying dominant narratives of gender and ethnicity and then working to develop, promote, and validate counternarrratives that recognize there are multiple ways of being, say, a girl or a boy. Moreover, such counternarratives work from the premise that individuals may align themselves with more than one version of being a girl/woman or boy/man in the course of their life or, even, in the course of a day (Rowan et al., 2001: 72.). This is to see "the self" or one's personhood as "continually constituted through multiple and contradictory discourses that one takes up as one's own" (Davies, 1993: 57).

The educational implications of this are clear enough. Teachers and learners concerned with moving beyond limitations of dominant cultural fictions of *valued* modes of gendered and ethnic being are necessarily involved in entertaining, discussing, acting out, and producing counternarrative representations. Many zines, especially the burgeoning array of electronic zines, offer fruitful and diverse insights into how different people try to work out or create their subjectivities. Many online zines make available spaces for discussing, critiquing, and reporting different people's experiences of negotiating subjectivity. Two examples are indicative here.

Slant/Slander

Mimi Nguyen is a self-labelled Asian American bi-queer feminist anarchist who has created a range of hardcopy zines (e.g., *Slant, Slander*) and cyberzines (e.g., *Slander, Worse Than Queer*). Nguyen, refugeed from Vietnam when she was one year old, identifies punk rock as the original driving force behind her zines. More recently, however, she has focused on issues and injustices occurring at the interstices of race/ethnicity and sexism. Her zines grew out of her desire to network with people of color in the punk music scene who—like her—were struggling with identity issues. She uses her online zine, *Worse than Queer*, to deconstruct "Asian-ness" as an anarchist, feminism as a bi-queer, and race in general as a young graduate student at University of California, Berkeley. Her goal is to turn long-standing assumptions about Asian women on their head by refusing to submit to the "Oriental sex secrets" and "Suzy Wong" Asian personae that people foist upon her (cf. Nguyen in conversation with Vale, 1997: 54). Nguyen draws herself as a punk rocker complete with piercings, as shaven-headed and toting a gun, and in martial arts poses that are definitely "in your face." Her zine *Slander* is definitely "in your face" as well—no holds are barred, and Nguyen refuses to throw dummy punches:

> In a phone interview over three years ago I was asked, "What do you think of Asian women who bleach or dye their hair; do you think they're trying to be white?".... That day my hair was chin-length, a faded green. I said, "No.".... It is already suggested by dominant "common sense" that anything we do is hopelessly derivative: we only *mimic* whiteness. This is the smug arrogance underlying the issue—the accusation, the assumption—of assimilation: we would do anything to be a *poor* copy of the white wo/man. Do *you* buy this? Are you, too, suspicious of "unnatural" Asian hair: permed, dyed, bleached? But if I assert the position that *all* hair-styles are physically *and* socially constructed, even "plain" Asian hair, how do we then imagine hair as politics? Who defines what's "natural"? Does our hair have history? What does my hair say about my power? How does the way you "read" my hair articulate yours?.... Asian/American women's hair already functions as a fetish object in the colonial Western imaginary, a racial signifier for the "silky" "seductive" "Orient." Our hair, when "natural," is semiotically commodified, a signal that screams "this is exotic/erotic." As figments of the European imperial imagination, Suzie Wong, Madame Butterfly, and Miss Saigon are uniformly racially sexualized *and* sexually racialized by flowing cascades of long, black shiny hair. Is this "natural" hair? Or is hair always already socially-constructed to be "read" a certain way in relation to historical colonial discourse? Is this "natural" hair politically preferable? "Purer," as my interviewer implicitly suggests? (Angel, 1999: 91; Nguyen, 1998: n.p.).

Slander is a bricolage of Mimi's views about race and gender, articles written by friends and colleagues, bold and evocative sketches she has done herself, and in the hardcopy version of the zine, collages and other artwork done by her or by friends, and so on, making *Slander* more than an "amateurish" cut-and-paste production. It qualifies in more than one sense as a "poetic way of making do." Nguyen's writing and artwork are loud voices of protest, as are her other projects such as "exoticize this!" (members.aol.com/critchicks; no longer available), a virtual Asian American feminist community she founded in the late 1990s, and a 1997 compilation zine entitled *Evolution of a Race Riot*. This zine was and is "for and about people of color in various stages of p[unk]-rock writing about race, 'identity,' and community" (Nguyen, 2000: n.p.).

Nguyen is producing a new literacy in her zine *Slander* (and elsewhere) that is rewriting traditional conceptions of and roles for Asian American women. This literacy concerns finding ways to draw attention to assumptions and stereotypes of Asian and Asian American women that are currently at work in popular media. This includes critiquing texts in "underground" magazines that profess to be anti-establishment and pro-young people (e.g., *Maximumrocknroll*, 1998, issue 198), but which often simply perpetuate images of Asian women as sex toys or as exotica. She also carries her message in the strong, line-drawn images she creates herself for her zine. In these ways, Nguyen is creating a space for herself that grows directly out of the microinformation of her everyday life as a punk, bi-queer, Asian American woman who grew up in Minnesota speaking Vietnamese and who recently has given over her shaved head and combat fatigues for red lip gloss and spiky heels. Mimi does not claim that she is speaking for, or even to, everyone and refuses to make concessions to non-Asian readers of her zine:

> [Mimi] wrote about how someone didn't enjoy her zine because they claimed they "couldn't relate" (being some hip white riot grrl type), but Mimi says "duh, of course you can't relate" (Squeaky, n.d.).

Digitarts: Grrrowling

Although numerous reports (e.g., National Science Foundation, 1997; Roper Starch Worldwide, 1998) indicate that boys and young men spend more time on the internet than girls and young women, the number of online zines created by young women appears to greatly outnumber those created and maintained by young men. Internet searches using advanced search engines and techniques, along with consulting a series of popular online zine web rings and indices,[2] suggest that young women dominate the online zine world, unlike in the offline, meatspace world where young men seem to publish more zines than women.

Digitarts—as described in chapter 10 in this volume—is an online multimedia project space constructed originally by young women for young women, but now also encompasses disadvantaged youth and people with disabilities (Digitarts, 2000a). The *Digitarts* website explores different conceptions and constructions of female identity through poems, narratives, journal pages, "how-to-do" texts, and digital images, and presents alternative perspectives on style, food, everyday life and commodities. This Australian-based project is dedicated to providing young women who are emerging artists and/or cultural workers with access to knowledge, expertise, and hardware necessary for the development of their arts and cultural practices in the area of new technologies. It aims to "provide young women and artists with the knowledge and resources to create a world wide website for the creation, distribution and promotion of their own cultural work and that of their peers" (Digitarts, 2000b: n.p.). Digitarts provides a venue for emerging multimedia artists to showcase their work and seeks to attract young women to the field by providing web development courses and beginner and advanced levels, and by publishing a cyberzine called *grrrowl.*

grrrowl (Digitarts, 2000c)—as mentioned in a previous chapter—is an ongoing, collaborative publishing endeavor, remarkable for its long life (many zines on the internet only ever reach the "first issue" stage). Like all authentic (not-for-profit, DIY) zines, *grrrowl's* production is not regular. It follows the beat of projects conducted by Digitarts. Its first issue focused on grrrls and machines. Each contributor constructed a page that is either a personal introduction—in the style of a self-introduction at a party—or contains poems or anecdotes about women and technology. Hyperlinks to websites engaging with a similar theme also define each writer's online self, and her self as connected with other selves.

grrrowl #4 (Digitarts, 2000d) investigates the theme "Simply Lifeless" and documents online "the everyday lives of young women in Darwin and Brisbane." Its thesis is: "Our culture informs our everyday activity. Our everyday activity informs our culture" (n.p.). The issue celebrates the "everyperson" and everydayness of their lives (cf. de Certeau, 1984; Duncombe, 1997), with eight young women—ranging in age from 12 years to 25 years—broadcasting web page-based "snapshots" of their lives. These snapshots include digital videos of personally important events such as composing music on a much-loved guitar, a daughter feeding a pet chicken and so on, or hypertext journals that span a day or a week and that also include photographic images such as digitized family album snaps, scanned hand-drawn graphics, 3D digital artwork, and so on. For example, 12–year-old Gabrielle writes about a typical few days in her life that involve waking early, dressing and going to school, who she plays with at school during lunch and snack breaks and what they do, and what she does after school. She talks a little about what she usually has for dinner and about going to stay with her father every Saturday night. He lives near her mother and her partner, Stephen. In documenting the "banal" and "everyday," this issue of *grrrowl* aims at "increasing the range of criteria by which our cultures are measured and defined" (Digitarts 2000d: n.p.).

grrrowl #5 is subtitled *Circle/Cycle* and focuses on "things that are round and things that go round" (Digitarts, 2000e: n.p.). The main menu is a spoof of a woman's diet menu that uses images from a 1960s *Australian Women's Weekly* magazine. The food items listed for various times of day (breakfast, beauty break—morning, lunch, beauty break—afternoon, dinner) are hyperlinked to interviews with interesting women such as comic-strip artists (dubbed "ladies of the black ink"), circus performers, bookstore owners, and so on. Other entries in the zine include a range of summer recipes, a detailed account of how to get rid of cockroaches in the house, recounts of food explorations and adventures, and a zine-within-a-zine link to the *Losergurrl* zine (no longer available): one young woman's personal offshoot of Digitarts projects.

This fifth issue employs a diverse range of text and image genres. The front page for *Losergurrl*, for example, is a collage of images cut from 1960s and 1970s women's magazines. Each image is hyperlinked to reviews of grrrl punk rock music; interviews with women in the music industry; rants about personal demons, safety issues, and women's comic books; book reviews; online games; treats such as recipes for natural beauty products; DIY files that deal with everything from DIY-Cryonics, to gardening and getting rid of pests in ecologically sound ways.

Items in the *grrrowl* issues are steeped in cultural analyses of everyday life and subjectivity. The zine presents online magazine-type commentaries and is used to establish and nurture interactive networks of relations between like-minded people. It is used to explore and present cultural membership and self-identity through digital and textual bricolages of writing, images, and hyperlinks. The digitarts' work is also a keen-edged critique of "mainstream" discourses in Australia and elsewhere. For instance, the editorial in the third issue of *grrrowl* explains how to subvert the default settings on readers' internet browser software, and encourages young women to override or side-step other socially constructed "default settings" that may be operating in their lives. Digitart projects challenge social scripts that allocate various speaking and acting roles for young women that cast them as passive social objects or as victims (e.g., "This is not about framing women as victims—mass media vehicles already do a pretty good job of that" (National Library of Australia, 2010: n.p.) and that write certain types of girls (or grrrls) out of the picture altogether (cf., Cross, 1996; Green and Taormino, 1997).

Indeed, "bricolage" is a key concept in this tactical work: the girls and young women involved in producing the various issues of *grrrowl* experiment with new technological literacy skills that have recently emerged (e.g., Virtual Reality Markup Language, Perl script, and shockwave applications). They use whatever technological equipment they can access at the time, or they poach, scan, and insert images from found texts—often placing mainstream images of women or objects often associated with women beside non-mainstream commentaries or narratives in order to underscore the different worldviews from which the digitarts themselves are operating. In this way, *grrrowl*—along with the other Digitarts projects—offers a coherent alternative to the commodification of youth culture, and the concept of "youth" as a market category is made too complex for corporations to use. *grrrowl* is a cyberspace in which young women can become *producers*, and not merely consumers, of texts and culture (cf., Duncombe, 1997, 1999; Knobel, 1998, 1999).

Just as many zines can provide graphic and hard-hitting insights into everyday uses of tactics in the practice of social critique and commentary and in the enactment of alternative politics, so they provide equally valuable insights into the nature and politics of subjectivity. As will be obvious by now, however, vexed issues converge around the place and roles zines might assume within classrooms in publicly funded schools. We will turn to this and other issues briefly in our concluding section. Meanwhile, it seems clear that teachers and learners who happen one way or another to become familiar with zines and zine culture will be helped in their efforts to negotiate subjectivity and subject positions within classroom pedagogy, as well as to bring a range of perspectives and familiarity with diverse and hybrid text forms to themes and tasks arising within the formal curriculum.

Issues and Possibilities

Zines provide firm ground from which to interrogate literacy education as currently practiced in schools and offer hard evidence that young people are not held necessarily in a "consumer trance" or are without sophisticated critical capacities. Even large corporations recognize that many young people are media smart to a degree that their parents were not and never will be. For example, the Nike faux zine *U Don't Stop* avoids including the globally famous Nike swoosh logo on any of its pages. It seems that the absence of the logo is an intentional nod to young people's "media savviness." Duncombe (1999) explains:

> When I called Wieden & Kennedy's Jimmy Smith and asked him why the Nike logo was conspicuously absent from *U Don't Stop* he explained that, "The reason [the zine] is done without a swoosh is that kids are very sophisticated. It ain't like back in the day when you could do a commercial that showed a hammer hitting a brain: Pounding Headache. You know, it's gotta be something cool that they can get into." (n.p.)

It may well be that no matter what teachers try to do in bringing young people's literacy practices into the world, it will never be sophisticated enough for their students. Or, as happens all too often with "new" literacies, zine literacy will become domesticated within the classroom so that the zines are produced according to the teacher's vision and purposes, rather than according to the grassroots, personal motivations of authentic zines.

For our own part, we remain unclear about exactly what *direct* implications zine literacy has for *schools*. In optimistic moments we think that the proper literacy business of schools should be to take proper account of any new literacy that is demonstrably efficacious. From this perspective, the role of people involved in studying and interpreting new literacies is to continue politicizing literacy education and research. Protesting claims that all young people today are politically apathetic and unmotivated would be another way of approaching zines in education. This would entail reading and discussing meatspace and cyberspace zines in classrooms.

On the other hand, for all their potential for fruitful educational appropriation, zines are often controversial, visually and mentally confronting, and regularly deal with topics taboo to classrooms. If some parents get up in arms about witches in storybooks, imagine how they would react to articles and zines entitled *Murder Can be Fun, Sex and Sexuality*, and *Why I Jack Off So Much Instead of Talking to Girls, Real Skinheads Take a Stand...A Feature on Red, Anarchist, Anti-Fascist and Activist Skinheads*, and so forth (cf., Williamson, 1994: 2). One way out of this dilemma might be to focus on the ethos of zines—the potent do-it-yourself writing and reading ethic for young people—and acknowledge the manner in which and extent to which new literacy practices evinced in hardcopy and cyberzines engage young people as active and often critically sophisticated participants in and creators of culture.

Alternatively, perhaps a revamped critical literacy that is enacted as "tactics," "clever tricks," a knowledge of how to get away with things, a suspicion of grand narratives, and not simply as critical analysis of media texts as commonly practiced, offers a way of maximizing students' media smarts in literacy education. Projects could include a public radio segment conducted by students that critiques some element of media culture each day over a four-week period; a commercially published booklet of interviews with local zinesters about their zines and what zines enable them to do on a day-to-day basis, and organized into themes that speak to young people; and a Mavis McKenzie-type letter-writing campaign (see Bail, 1997) that subtly spoofs large corporations or institutions. For example, students could write to a munitions company asking for their magazine catalogue, to the department of education or large hospital asking for a copy of their recycling policy, to local town councils asking for their youth policy, etc. These letters could then become the basis for a multimedia "position paper" or commentary.

The trick, we believe, is to approach the place and role of zines within school-based (literacy) education *tactically*. Here as well, the medium is the message. Whatever other capacities and dispositions they display, smart teachers and smart learners are tactically adept. Zines present us with a tactical challenge; an ideal learning and implementation problematic for new times. How can we get the kinds of orientations, ethos, perspectives, worldviews, and insights encapsulated in zines into classroom education when to do so necessarily involves maneuvering on enemy terrain? If we cannot work out how to do this and get away with it—with the assistance of endless models of tactics available within the practices of everyday life, of which zines are but one—we probably should not try to incorporate zines and core zine culture values into formal learning. By the same token, if we cannot engage in tactics of this kind it might be time to question our credentials for being educators under current and foreseeable conditions. For it seems likely that in the "fast" world that is now upon us, those who survive well will increasingly be "tactically competent."

Endnotes

1. We are grateful to Dr. Leonie Rowan, then of Central Queensland University and now of Griffith University (Gold Coast), Australia, for her generous insights and input into this section.
2. Indices used included:
   http://www.zinebook.com/index.html
   http://www.zinos.com
   http://www.altzines.tripod.com/index_t.html (now defunct)
   http://www.sleazefest.com/sleaze (now defunct)
   http://www.meer.net/~john1/e-zine-list now defunct)
   http://www.ilovepisces.bigstep.com/businesspartners.html (now defunct)
   http://www.geocities.com/SoHo/Café/7423/zineog2.html (now defunct)

Bibliography

Abbey Records (2000). *MusicEzine.* Retrieved December 26, 2000, from: musicezine.com (no longer available).

Alvermann, D. (ed.) (2002). *Adolescents and Literacies in a Digital World.* New York: Peter Lang.

Athena (2000). *Bombs for Breakfast.* Retrieved November 19, 2000, from: bombsforbreakfast.com (no longer available).

Angel, J. (1999). *The Zine Yearbook: Vol. 3.* Mentor, OH: Become the Media.

Bail, K. (1997). Deskbottom publishing. *The Australian Magazine,* May 3–4: 44.

Barton, D., and Hamilton, M. (1998). *Local Literacies: Reading and Writing in One Community.* London: Routledge.

Block, F. and Carlip, H. (1998). *Zine Scene: The Do It Yourself Guide to Zines.* Los Angeles: Girl Press.

Braidotti, R. (1994). *Nomadic Subjects: Embodiment and Sexual Difference in Contemporary Feminist Theory.* New York: Columbia University Press.

Buchanan, I. (1993). Extraordinary spaces in ordinary places: de Certeau and the space of postcolonialism. *SPAN,* 36. Retrieved December 4, 2000, from: wwwmcc.murdoch.edu.au/ReadingRoom/litserv/SPAN/36/Jabba.html.

Caleb (2000). *archáologie francaise.* (Issue 1). (zine; no publication source given).

Carlip, H. (1995). *Girl Power: Young Women Speak Out.* New York: Warner Books.

CBC Television (2000). *Our Hero.* Retrieved 22 November, 2000, from: ourherotv.com/home.shtml (no longer available).

Chu, J. (1997). Navigating the media environment: How youth claim a place through zines. *Social Justice*, 24(3): 71–85.
Ciara. (2000). *A Renegade's Handbook to Love and Sabotage.* (Issue 1). Portland, OR: Ciara.
Craig, S. and Bennett, S. (eds) (1997). *After the Boom: The Politics of Generation X.* Lanham, MD: Rowman and Littlefield.
Cross, R. (1996). Geekgirl: Why grrrls need modems. In K. Bail (Ed.), *DIY Feminism.* Sydney: Allen & Unwin. 77–86.
Daly, S. and Wice, N. (1995). *alt.culture: An A-to-Z Guide to the '90s—Underground, Online, and Over-the-Counter.* New York: HarperCollins.
Davies, B. (1993). *Shards of Glass: Children Reading and Writing Beyond Gendered Identities.* Sydney: Allen & Unwin.
de Certeau, M. (1984). *The Practice of Everyday Life. Vol. 1.* Berkeley: University of California Press.
de Certeau, M. (1997). *The Capture of Speech and Other Political Writings.* (Trans. Tom Conley). Minneapolis: University of Minnesota Press.
DeSantis, C. (1997). Foreword. In V. Kalmar, *Start Your Own Zine.* New York: Hyperion.
Digitarts (2000a). Front page. Retrieved January 1, 2000, from: digitarts.va.com.au (no longer available; see instead: pandora.nla.gov.au/nph-arch/Z1998–Dec-15/http://digitarts.va.com.au/why.html)
Digitarts (2000b). *grrrowl #1: Women and Technology.* Retrieved January 1, 2000, from: digitarts.va.com.au/grrrowl1/front.htm (no longer available; see instead: pandora.nla.gov.au/nph-arch/Z1998–Dec-15/http://digitarts.va.com.au/grrrowl1/front.htm)
Digitarts (2000c). *grrrowl.* Retrieved January 1, 2000, from: digitarts.va.com.au/frames2.htm (no longer available; see instead: pandora.nla.gov.au/tep/10258)
Digitarts (2000d). *grrrowl #4: Simply Lifeless.* Retrieved January 1, 2000, from: digitarts.va.com.au/grrrowl4 (no longer available; see instead: pandora.nla.gov.au/nph-arch/Z1998–Dec-15/http://digitarts.va.com.au/grrrowl4/index.htm)
Digitarts (2000e). *grrrowl #5: Cycle/Circle.* Retrieved January 1, 2000, from: digitarts.va.com.au/grrrowl5 (no longer available; see instead: pandora.nla.gov.au/nph-arch/2000/Z2000–Feb-21/http://digitarts.va.com.au/grrrowl5/index.html)
Duncombe, S. (1997). *Notes from the Underground: Zines and the Politics of Alternative Culture.* New York: Verso.
Duncombe, S. (1999, December). DIY Nike style: Zines and the corporate world. *Z Magazine.* Retrieved January 4, 2001, from: zcommunications.org/diy-nike-style-by-stephen-duncombe
Friedman, T. (1999). *The Lexus and the Olive Tree: Understanding Globalization.* New York: Anchor Books.
Gee, J. (2002). Millenials and Bobos, *Blue's Clues* and *Sesame Street*: A story for out times. In D. Alvermann (ed.), *Adolescents and Literacies in a Digital World.* New York: Peter Lang. 51–67.
Gee, J. P., Hull, G., and Lankshear, C. (1996). *The New Work Order: Behind the Language of the New Capitalism.* Boulder, CO: Westview Press.
Goldhaber, M. (1997). The attention economy and the net. *First Monday*, 2(4). Retrieved August 3, 2000, from: firstmonday.org/htbin/cgiwrap/bin/ojs/index.php/fm/article/view/519/440
Green, K. and Taormino, T. (eds) (1997). *A Girl's Guide to Taking Over the World: Writings from the Girl Zine Revolution.* New York: St. Martin's Press.
Halstead, T. (1999). A politics for Generation X. *The Atlantic Monthly.* August. Retrieved November 23, 2000, from: theatlantic.com/issues/99aug/9908genx.htm
Haraway, D. (1985). Manifesto for cyborgs: Science, technology, and socialist feminism in the 1980s. *Socialist Review*, 80: 65–108.
Hipp, T. (1999). Editorial. *girl swirl fanzine.* (Issue 1). Retrieved March 21, 1999, from: www.girlswirl.net (no longer available; see instead: zinewiki.com/Girl_Swirl)
Hipp, T. (2001). *girl swirl.* Retrieved July, 27, 2001, from: www.girlswirl.net (no longer available; see instead: zinewiki.com/Girl_Swirl)
Howe, N. and Strauss, R. (1993). *13th Gen: Abort, Retry, Ignore, Fail?* New York: Vintage Books.

Knobel, M. (1998). Paulo Freire e a juventude digital em espacos marginais. In M. Gadotti (Ed.), *Poder e Desejo: Paulo Freire a as Memorias Perigosas de Libertacao.* Porto Allegre, Brazil: Artes Medicos.

Knobel, M. (1999). *Everyday Literacies: Students, Discourse and Social Practice.* New York: Peter Lang.

Knobel, M. (2001). "I'm not a pencil man": How one student challenges our notions of literacy "failure" in school. *Journal of Adolescent & Adult Literacy,* 44(5): 404–419.

Lankshear, C. and Knobel, M. (2002). Do we have your attention? New literacies, digital technologies and the education of adolescents. In Alvermann, D. (ed), *Adolescents and Literacies in a Digital World.* New York: Peter Lang. 19–39.

National Library of Australia (2010). *Digitarts: Girls in Space.* Retrieved August 19, 2010, from: pandora.nla.gov.au/nph-arch/Z1998–Dec-15/http://digitarts.va.com.au/gis/gis.html

National Science Foundation (1997). *U.S. Teens and Technology: Gallup Poll Executive Summary.* Retrieved July 13, 2000, from: nsf.gov/od/lpa/nstw/teenov.htm (no longer available).

Nguyen, M. (1998). Me and my hair trauma. *Slander.* Retrieved January 4, 2001, from: worsethanqueer.com/slander/hair.html (no longer available; see instead: zinewiki.com/Slander).

Nguyen, M. (2000). *Slander.* Retrieved January 4, 2001, from: worsethanqueer.com/slander/zine.html (no longer available; see instead: zinewiki.com/Slander).

Roper Starch Worldwide (1998). *Roper Youth Report.* Princeton, NJ: Roper Starch Worldwide.

Rowan, L., Knobel, M., Bigum, C., & Lankshear, C. (2001). *Boys, literacies and schooling: The dangerous territories of gender based literacy reform.* Buckingham, England: Open University Press.

Squeaky, J. (n.d.). Non-music Reviews. Retrieved January 5, 2001, from: misterridiculous.com/reviews (no longer available)

Underwood, M. (2000). Semiotic guerrilla tactics—Michel de Certeau. Retrieved December 4, 2000, from: cultsock.org/index.php?page=media/decviews.html

Vale, V. (ed.) (1996). *Zines. Vol. 1.* San Francisco: V/Search.

Vale, V. (ed.) (1997). *Zines. Vol. 2.* San Francisco: V/Search.

Williamson, J. (1994). Engaging resistant writers through zines in the classroom. *Rhetnet: A Cyberjournal for Rhetoric and Writing.* October. Retrieved January 1, 2001, from: wac.colostate.edu/rhetnet/judyw_zines.html

Wittlinger, E. (1999). *Hard Love.* New York: Simon and Schuster.

THIRTEEN

The Ratings Game: From eBay to Plastic (2003)

Michele Knobel and Colin Lankshear

Biography of the Text

This chapter was originally published by Open University Press in the first edition of *New Literacies*, although we had previously published a piece on reading, writing and ratings on eBay.com in Ilana Snyder's (2002) collection, *Silicon Literacies*. Like many of the themes we began exploring from the standpoint of what might be significantly "new" in contemporary conceptions and practices of literacy, the focus on ratings grew directly out of our everyday lives living in Mexico as researchers and writers associated with the National Autonomous University of Mexico (UNAM) during 1999–2001.

In part, we became strongly dependent on online retailers and sales mediators like Amazon and eBay—often making online purchases and having them shipped to a U.S. address (e.g., a conference hotel, friends' homes) and picking them up on the next visit "north." We read closely the customer reviews and ratings of items in the case of Amazon and paid careful attention to sellers' ratings on eBay. At the same time, and more to the point so far as our writing was concerned, we were interested in *trendspotting*. From the inception of the mass internet in the early nineties we had actively followed people and initiatives identified as "shapers" (Friedman, 2000) by leading edge sources like *Wired* magazine and stories about influential people on the web. This had got us into reading online magazines like *Suck* (suck.com) and, later, trendspotting blogs like BoingBoing.net. When *Suck* and *Feed* (feedmag.com) merged and produced *Plastic* (plastic.com) we were struck by the way the site's use of ratings and karma points resonated with our interest in attention economics and with how participants acquired status within identifiable online communities, along with how seriously they took their ratings and karma. In addition, we had been interested for some years in trends we identified with the idea of "digital epistemologies"—ways in which, and the extent to which, phenomena like hits and ratings come to constitute indices of "truth" and "value" alongside, or in some cases in place of, more conventional epistemological criteria.

These intersecting interests led to us taking "ratings" seriously as a theme to be explored in relation to new literacies.

Introduction

Current educational literature is awash with talk of "new literacies," "technoliteracies," "multiliteracies" and the like in response to the deep incursion of new information and communications technologies into everyday routines within modern societies. Much of this talk is general and impressionistic, however. Considerably less documentation of new literacy practices engendered and mediated by the internet has been forthcoming, let alone of what social issues and responsibilities such new practices may evoke.

This chapter focuses on the emergence of the community ratings feedback systems on eBay and Plastic as cases of a new literacy—a new way of reading and writing aspects of the world that are important to participants in these online activities. It explores the rise of ratings systems as regulating devices within online communities and how these are taken up (or not taken up) by community members.

In particular, the ratings system used on eBay will be examined from two standpoints that are evolving to a considerable extent in tension with one another. One standpoint is that of its creators—the owners and operators of eBay—and their communitarian "visionary" purposes for developing it. The other standpoint is that of its users, among whom many seem to be appropriating the community ratings feedback system for personal purposes—some of which appear mean-spirited and intentionally self-serving to say the least.

The ratings system on Plastic—whose tagline reads, "Recycling the Web in real time"—is an online forum devoted to posting and discussing the "best content from all over the Web for discussion" (Plastic, 2002a: n.p.). Its rating and filtering system is one of the striking things about Plastic. Participants can use this to screen out comments with low ratings and read only those rated highly. This saves them time they would otherwise have to spend sifting through postings to sort out those that are worth reading from those that aren't.

The use of rating systems as public evaluations of "worth" (moral, commercial, intellectual, etc.) is not confined to these two sites alone. Rather, it is spreading rapidly across a range of different web-based communities. For example, Amazon—the "Earth's Biggest Selection™ of products" (Amazon.com, 2002: n.p.)—has set in place a 5–star rating system that users can use to evaluate products and Amazon Marketplace sellers (where purchases are made directly from companies, rather than Amazon.com per se), and a numerical system that alerts buyers to the sales ranking of a product within the Amazon.com system.

Our aim in this chapter is to capture something of the dialectic between *strategies* of producers and *uses* of consumers (de Certeau, 1984) at play in the emergence of a distinctively contemporary practice of everyday life online.

What *Is* eBay?

eBay was among the first person-to-person auction venues to go online. It is currently the most popular and successful internet trading community in the world (Friedman, 2000; Multex.com, 2001: 1). As of early 2002, eBay has over 42.4 million registered users and each day users list millions of items for auction in more than 8,000 categories—and it is still growing (eBay, 2002a: 1; MSN Money, 2002: 1). Categories range from premium artworks, through real estate and cars, to clothing, jewelry, toys, comics and trading cards (with one person recently auctioning off his soul . . .). eBay describes itself as

> the world's personal trading community, [which] has changed the way people buy and sell collectibles and unique items. By providing a safe trading place on the internet, the eBay community has flourished. Not only does eBay provide an efficient medium for people to buy or sell items directly from or to a large number of people, it's a forum where buyers and sellers develop reputations and, in some cases, it can change people's lives (eBay, 2002b: 1).

eBay (ebay.com) comprises sets of internet pages that are basically long lists of new and used items that people have posted to the eBay internet site for sale by auction. Sellers are responsible for writing item descriptions and for generating pictures of the items that are then inserted into an eBay page template and posted on the eBay website under a self-selected category heading (and where it is automatically allocated an item number). Potential buyers—who must be registered with eBay—browse these lists or use the eBay search function to locate items of interest. They can then bid on or "watch" these items. Watching involves clicking on the "watch this item" hyperlink, and the item is then hotlinked to one's personal "eBay" space (i.e., "my eBay").

Bidding works in two ways, similarly to conventional auctions. Bidders may make the lowest viable bid possible at that particular point in time and wait to see what happens (or place a new minimum bid after being outbid by someone else). Alternatively, they can place a "proxy" bid—which is the maximum amount they are willing to pay for the item, and eBay acts for them as a proxy bidder: bidding in their place until the item is "won" or their specified maximum amount has been exceeded by another bidder. Sellers—who are also always registered members of eBay—pay to list their items with eBay. The fee depends on the starting price or reserve set for an item (e.g., a $0.01 to $9.99 starting price costs $0.30 to list, and a reserve price of $0.01 to $24.99 costs $0.50 to list), or on the type of item being listed (real estate comes with a $50.00 listing set fee, as does a vehicle). Items can also be sold for a fixed price (i.e., not auctioned off) under the eBay stores service. These items are subject to a different listing fee scale. Commission on auctioned and other sold items is charged at 5.25 per cent of the first $25.00 and an additional 2.75 per cent after that up to $1,000. From $1,000 onwards, 1.5 per cent commission is paid on the remaining amount (eBay, 2002c: 1). eBay membership is free, as is bidding and browsing.

eBay currently operates in 22 countries and 11 languages. Although it is certainly advantageous to access an eBay site in your home country (language, currency, dates, time and shipping wise), it is possible to bid from anywhere in the world whenever payment options and shipping agreements permit.

What's *New* About eBay

While some people might claim that eBay is just an old physical space (auctions) in virtual get-up, we think it is spawning some genuinely new social practices and new literacies associated with them.

We will make our case for the newness of some of the social practices and literacies associated with them from two angles. The first simply identifies some new features of reading relevant aspects of the world occasioned by moving the familiar social practice of auctioning into an unfamiliar space, namely the virtual space of the internet. One or two brief examples must suffice here.

eBay calls for interesting new constellations or "batteries" of ways of reading and writing in order to achieve one's purposes as an online buyer or seller. For example, the eBay venue operates as a "transaction medium." Nobody at eBay sees or handles what is being bought and sold. And there is nobody to tell one where to go to find what one is looking for (or might want to look for if one knew it could be available). Hence, it is not a matter simply of knowing how to read or write the text of item descriptions. Participants need also to know how to navigate through or add to the website. For example, they need to know how to access and read the battery of "how-to" texts provided on the eBay website (e.g., how to bid, how to post an item for sale, how to leave feedback, how to lodge formal complaints about a buyer or seller, how to access user-to-user help and advice discussion lists). Users also need to know how to read and write "taxonomically" in the sense of knowing what is likely to be in or should be in each category—of which there are thousands. They need to be able to read between the lines in item descriptions (e.g., a Clarice Cliff style crocus jug is *not* a Clarice Cliff crocus jug). In many cases it is necessary to be able to read digitized images accurately (e.g., know that color is often not true-to-life in digital images of objects, understand depth of field and the effects it has on objects, be wary of out-of-focus or soft focus images or lighting effects). Knowing (how) to convert from imperial to metric measures, or even one currency to another, is often required for international dealings and so on. Fakes and forgeries are much easier to disguise on eBay than in meatspace. Collectors appear to have developed a whole new set of criteria for judging the authenticity of an item. These include evaluating the source of the product (e.g., if the seller is the daughter or son of a famous sportsperson, then it's likely the sports memorabilia he or she is selling is genuine), judging the seller by location (e.g., someone selling art deco ceramics and living in England or ex-English colonies is most likely to be selling authentic pieces in good condition), judging the seller by the other products he or she has listed for auction (cf. Smith and Smith, c.2002). They also include a wariness of what some call, "overdocumentation." This is the presence of too many documents "verifying" the authenticity of an item (Sherman, 2001: 63). Perhaps most importantly, however, the reader of item descriptions and images has to pay careful attention to what is *not* said or shown. For example, sellers who list high-end designer handbags (e.g., Gucci, Hermés, Coach) without mentioning that the bags have serial numbers, or who does not list the silver content stamp (.925 or higher) for a purported solid silver item, may be less likely to be selling an authentic item.

Moreover, physical or meatspace literacy practices often *mean* different things within eBay. For example, one regular eBay user we interviewed said she loves coming across item descriptions that include misspelled words. To her it means she is more likely to "win" a bargain from this person than from someone who spells correctly. Non-standard spelling indicates to her someone who is less likely to be in a professional job or to own a shop and, hence in her eyes, to be less likely to know the real value of the ceramics or other objects they are offering for sale.

The second way of considering what is new about eBay is by reference to Bezos' distinction between first and second phase automation introduced in Chapter 3. Indeed, eBay provided Bezos with an exemplar of the kind of thing he wanted to do. We may recall Bezos' distinction here by way of Robert Spector's (2000: 16) account of Bezo's passion for "second phase automation."

> Bezos has described second-phase automation as "the common theme that has run through my life. The first phase of automation is where you use technology to do the same old business processes, but just faster and more efficiently." A

> typical first phase of automation in the e-commerce field would be barcode scanners and point-of-sale systems. With the internet "you're doing the same process you've always done, but just more efficiently." He described the second phase of automation as "when you can fundamentally change the underlying business process and do things in a completely new way. So it's more of a revolution instead of an evolution."

Bezos' distinction enables us to distinguish further between processes and practices that have simply become "digitized" and those practices and processes that exist only because digital technologies do. As we can see by reference to its community feedback ratings system, eBay is a case of the latter.

In his analysis of globalization, Thomas Freidman (2000) distinguishes between two roles available for companies, governments and institutions in the "Evernet world" of globalized networks of communication, service and power. He calls these roles "shapers" and "adapters" respectively. Shapers are agents that shape up activities within a globalized world of networked coalitions and practices—whether that activity is "making a profit, making war or making a government or corporation respect human rights"(Freidman, 2000: 202). Shapers design rules, create interaction frameworks and set new standards for global practices. Adapters, on the other hand, follow shapers' leads and adapt to the "scene" being created.

Friedman identifies eBay as a foremost and highly original shaper. He sees it as having been a leader in creating a whole new marketplace and instigating an entirely new set of interaction protocols for buyers and sellers. eBay, says Friedman, "came out of nowhere and within three years created a new set of rules and forms of interaction by which consumers would buy and sell things on the World Wide Web" (Friedman, 2000: 202). At the core of eBay's business process is a simple rating scale and 80–character feedback system by which buyers can rate and respond to the effectiveness of sellers over the course of a transaction, and *vice versa*. This ratings system has been absolutely integral to eBay's success in its enterprise. It has simultaneously transformed relations between buyers and sellers on the internet, and been elevated to prominence in the identity-shaping and reputation making behavior of many individuals in the practice of pursuing a positive ratings profile.

eBay's rating system involves a three point rating scale—positive, neutral and negative—that serves as a public judgment of a person's reputation, trustworthiness and reliability. Once an auction transaction has been completed (the winning bidder has paid for and received the item) the buyer can leave feedback about the seller and *vice versa* by means of the item number. Only the buyer and seller are authorized to comment formally on a particular transaction. Feedback consists of the actual rating (positive, neutral, negative) and a written recommendation.

eBay's website reminds eBayers that "[h]onest feedback shapes the community" (eBay 2002d: 1). The higher the positive ratings a person has, the more "reputable," "trustworthy" and "reliable" they are in eBay terms. On the other hand, accumulating a net feedback rating of -4 (minus four), means an individual can be excluded from the eBay community. Exclusion is not automatic, however, since it is up to users to notify eBay that someone has received four or more negative ratings.

As an aside, eBay's success has spawned a diverse range of complementary products and services, many of which entail literacies of one kind or another. For example, *eBay a-go-go*™ has been purpose-designed to be an eBay wireless auction alerting service that operates via one's mobile phone or pager. It alerts users when they have been outbid, or won or sold an item. There are also various auction "tracking" and bidding software programs and online services (e.g., Amherst Robots, 2001; eSnipe, 2001), online mediation services for auction transactions that go wrong, escrow and e-cash services (e.g., BidPay and Billpoint), a range of how-to-bid-successfully books (e.g., Collier and Woerner, 2000; Reno, Reno, and Butler, 2000), and online beginner's introductions to eBay (e.g., SoYouWanna, 2000). Finally, for those who are truly serious about learning how to read and write the world according to eBay, there is eBay University.

Why *Ratings*?

eBay's overt intention in devising and implementing the feedback ratings system is to build a self-monitoring ethical community of eBay users—or "eBayers" for short. We would argue that the feedback ratings system might actually be read as an embodied ideological induction into a certain community-based "cyber space." That is, eBay is not only a shaper within the new technologies arena, but it is also an "educator" in that it "teaches" people how they *should* act within this new cyber space, and therefore, how they should act in relation to each other. It is, therefore, a space of induction. It plays a role in shoring up new discursive norms. It socializes people about what counts as an exemplary global space, and helps generate good global citizens by encouraging the "right" kind of cyber practices that lead to a well-organized and civil World Wide Web.

Indeed, even this side of the eBay experience is held up as an exemplary model for other companies to emulate on the internet.

> Besides being creative about using the internet, Kanter [a Harvard Business School professor] says, there are other ingredients of the eBay model that companies should study as they expand their Web initiatives. She emphasizes principles rooted as much in social interaction as in the tenets of business: a sense of community where people can talk to one another as well as with the company they

> patronize and a corporate culture that reinforces those connections and serves all members.
>
> As an example, Kanter points to eBay's Giving Board, where users can post a problem and receive advice from other eBay members. "You get a lot of loyalty by treating people as members of a community," she says. "Analysts can't quite quantify that, but it sure shows up in eBay's profitability." (Heun, 2001: 1)

eBayers and Their Ratings

eBayers are very clear about the importance of their ratings. Many go to extraordinary lengths to obtain positive ratings. Some item postings contain a "customer assurance statement" that resembles an airline "thanks for flying with us" patter to stand in as a "bid confidently" statement. For example,

> A Word of Thanks . . .. We at Lorelei's Jewelry would like to Thank all of our Customers for their Patronage over the last 4 years. Our number 1 priority is to give you the best Customer Service in the Business. We know that you have choices and appreciate your business. Our Goal is to provide an Exceptional Line of Jewelry at the Absolute Lowest Prices. We are here to answer any Questions that you may have in a Timely Manner Via Telephone or Email. All Winning Bidders are Notified Promptly and Items are Normally Shipped the Day Payment is received. We hope that you will join our long list of Satisfied Customers . . .Over 10,000 Feedbacks and Growing Daily. (Lorelei's, 2000: 1)

The reference to 10,000 feedbacks is the clincher here. It is worn like a badge of honor (although canny eBayers will note that the company does not advertise "10,000 *positive* feedbacks" and will immediately go to the company's ratings page to verify the ratings are positive as implied). Some sellers email successful bidders at the end of a transaction to let them know the seller has left them positive feedback. The email can even contain a hyperlink to automated feedback forms. Customers need only fill in the actual rating and the written feedback line.

Many individual eBayers have constructed elaborate processes that aim at ensuring as many positive feedback statements and ratings as possible:

> I have a spreadsheet that i use to keep track of my items, buying and selling and there is a space for me to check off that i have left feedback for a buyer/seller. When the buyer/seller leaves feedback for me in return, i circle the check mark, letting me know the transaction has come full circle. when i sell something, i include a thank you card with the item number listed, the item name listed, my ebay name and a note stating that i have left positive feedback for them and would appreciate the same in kind and i still have problems getting them to leave me feedback! So every month, i go down the spreadsheet and e-mail those who have failed to leave feedback asking them why they have not done so and if there were problems i was not aware of. this is very time consuming but it has

> worked on most of the delinquents. it more or less embarrasses them into leaving feedback (eBay message board, 2001).

Having even one negative feedback is perceived as bad for business:

> [Ratings and feedback] are very important as it's the only real way of knowing how good sellers are. I have never bought off someone with a bad rating and there are quite a few of them out there. . . . I have had to give out a few bad ratings to people who have won auctions and have never paid me or contacted me for that matter (*arkanoid2020* email interview 12/02/2001).

> [Ratings] are extremely important. I don't want to buy from vendors with negative feedback, and I don't expect people to want to buy from me if I have any. Those comments are listed in red, and they show up like a neon sign!! (*bea1997* email interview 25/09/2000).

Ratings have actually become a "currency" for the eBay community, assuming the kind of role local community networks and character references have in physical space. One of our interviewees, *susygirl*, says:

> I really take pride in [my ratings]. And for me it is the alter ego—it is susygirl's not mine. And so I get pissed [off] if someone doesn't send me a positive feedback. But I never write and ask them to. Some sellers do that and I usually don't respond to that (email interview 1/02/2001).

Others, like *bea1997*, a long-term and very experienced eBayer, have preferred to be "duped" by buyers than risk negative feedback. *bea1997* explained to us,

> Sometimes I lose money from customers who break an item and ask for money back. I just don't want to risk having my good reputation ruined for a few lousy bucks so I just take the blame and send their money back (email interview 25/09/2000).

bea1997's experience tallies with others reported elsewhere. For example, Erick Sherman recounts,

> [b]oth buyers and sellers get burned from time to time, but usually not badly. Shamus remembers someone who bought a $25 trading card from him on eBay then returned it, but with a corner newly bent. "He said, 'That's what you sent me'," says Shamus, who didn't argue because the amount was too small and negative feedback would hurt his future sales (Sherman, 2001: 63).

Others have vigorously fought with eBay to have what the eBayer regards as unjust negative feedback removed (which is almost impossible to have done), and various eBayers have established entire websites devoted to explaining the events behind any negative feedback they have received.

Ratings are considered by most eBayers to be so important that the dedicated discussion board attached to the eBay website (located on a server in the U.S.) for discussing feedback is a popular and much-used service. This "board" is a web-based service that allows people to post messages (or responses to messages) about their ratings and feedback problems, warnings about "deadbeat" sellers or buyers, "sniping"(bidders waiting until the very last moment to place a winning bid), how to go about lodging a complaint about an unfair negative rating, and so on. Despite eBay's emphasis on *community*, however, the rating and feedback system has not made for close-knit and harmonious fellowship.

Reciprocity is a key value enacted on eBay in relation to ratings. As *susygirl* observed, she is "pissed (off)"if she completes a transaction and the seller doesn't leave feedback for her. Reciprocity in ratings is likewise important to *arkanoid2020*:

> I have also had the problem of people not giving me a rating after a successful transaction, which is a shame because I always make the effort (email interview 12/02/2001).

Other eBayers express their feelings about a lack of reciprocity very strongly. Much of the eBay-based discussion about ratings is taken up with who should leave feedback and a rating first, and why. For example,

> I figured out feedback right away. When I receive an item I immediately leave feedback. That's my way of keeping track of things. I then immediately email the seller and thank them for good service (I've been very fortunate in this regard.) and ask them to leave feedback. It seems to me that sellers will only leave feedback if requested to do so and _if_ I leave positive feedback. Sellers should leave feedback when they get my prompt payment in my opinion. Why do I have to gently nudge them and leave my feedback first? They get my money first (eBay feedback discussion board, 2001).

In bad-case scenarios, the power of leaving feedback is held over the buyer or seller. For example, when a buyer has received an item and for some reason wants to return it to the seller, but the seller does not agree to receive it back, the buyer may threaten to leave negative feedback if the seller does not comply with his or her wishes. eBay members refer to this as "feedback hostage taking"—where the seller (or buyer) is held hostage to receiving feedback (e.g., "I'll leave you feedback only when you've left me feedback"). In worst-case scenarios, this is what eBayers refer to as "feedback extortion," and it is taken very seriously by eBay itself. The eBay feedback system has actually generated a metalanguage for talking about participant practices. This vocabulary comprises mostly new, mostly pejorative terms that eBayers invent and use freely and fluently (see Table 12.1).

Term	**Definition**
feedback bombing	Two senses: (1) The process whereby two or more people gang up on someone, purchase products, then leave negative feedback (2) The process whereby two or more people gang up on someone, purchase products, then leave positive feedback (and is usually reciprocal—those in on the scam positively bomb each other's auction)
feedback padding	One person creates two eBay bidding accounts and uses one account to pad out the feedback on his or her other account.
feedback extortion	eBay defines feedback extortion as "demanding any action from a fellow user that he or she is not required to do, at the threat of leaving negative feedback."
retaliatory negative feedback	When a negative feedback rating is given to one person in an eBay transaction by the other, the first responds with a negative rating—regardless of the quality of service received. This often goes hand-in-hand with feedback hostage taking (e.g., "I was really unhappy with this transaction but can't leave feedback until the other one does because I want to leave a negative feedback but am worried that if I leave it first then the person I'm dealing with will give me a negative feedback in response!").
(to be) neutraled	To receive a "neutral" rating (it is also possible to "neutral" someone, too; that is, to give them a neutral rating).
(to be) NEGed (also "neg")	To receive a "negative" rating (it is also possible to 'NEG' someone, too; that is, to give them a negative rating).
Deadbeat bidder or seller (also referred to as "deadbeats")	This is the term eBayers use to describe people who do not deliver on their half of the transaction (i.e., do not pay or do not send the item or send the item in poor condition etc.). eBay's official term of bidders who skip out on a deal is "Non-Paying Bidders," or NPBs for short.

Table 12.1: Shared metalanguage on eBay's Feedback Discussion Board.

Not surprisingly, exchanges on the feedback discussion board can become heated, with little evidence of the kind of tolerance one would expect in a community of the kind eBay aims to foster:

> and i agree if you knew the answer why bother asking? i get lots of people asking stupid ? [trans: questions] like what does it measure? when it is already posted on my auctions . . .i tell them to go back and read the description. i don't find that to be rude. (eBay feedback discussion board, 2001)

And responses from two different people:

> not rude? must be why you have so many successful transactions . . ..
>
> Why not just answer the question and accept that stupid people make up a big percentage of customers?
>
> The guy's sarcastic, not rude. Read his very limited, posted feedback for a good laugh.

The reference to the first poster's "successful transactions" and "very limited, posted feedback" are snide comments on the poster's beginner status: one positive rating. Such reactions indicate a tendency for eBayers to read ratings both as statements of their public reputations and as indicators of "wisdom" and knowledge where all things eBay are concerned.

eBay's response to the soap opera-like dimensions of the community feedback and ratings system is to continue holding out for a self-regulating, "trustworthy" and intelligent community:

> Hello folks,
> Thanks for the discussion. Let me offer you eBay's perspective on Feedback for consideration:
>
> The real value in Feedback is in the trends that it reveals. While it is an admirable goal to work towards a perfect rating, it is IMPOSSIBLE to always please everyone all the time anywhere in life, right? An occasional isolated negative will not impact the VAST majority of users when they are deciding whether or not to bid or accept a bid. (I would say "ANY" users, but then someone would post to prove me wrong, hehehe).
>
> We hope you will use the Feedback forum faithfully, despite the risk of receiving a negative that you feel you don't deserve, because in this way our whole community is served best. The purpose of Feedback is to help keep the site safe. If we use it appropriately, the good guys are always going to have FAR more positive comments than the less-scrupulous users who will quickly earn track records that show their true colors for all to see, as well.
>
> Daphne will step down from her soapbox now. :)
>
> Daphne

eBay Community Support
(eBay feedback discussion board, 10/03/2000)

Interestingly, eBay has recently instituted a feedback service that alerts participants to items they have yet to leave feedback on. It is also possible to access a list of feedback each user *leaves* others. This adds a second dimension to a user's feedback and rating profile and often makes for interesting exchanges in the discussion spaces of eBay.

Plastic

eBay users are not the only ones to take the ratings game seriously. eBay's rating system has impacted powerfully on internet-based social interactions, with numerous other interactive internet sites using ratings systems as public reputation markers. Plastic is a good example of this. It is somewhat unusual on the internet, however, since it evaluates quality of thinking and expression rather than business conduct.

Plastic began in January 2001, with the aim of being a "new model" of news delivery: "anarchy vs. hierarchy, and so on and so forth" (Joey, 2001: 1), and promising "the best content from all over the Web for discussion" (Schroedinger's Cat, 2002: 1). This new model of news delivery puts "the audience in charge of the news cycle as much as possible without devolving into the kind of ear-splitting echo chamber that's turned 'community' into such a dirty word" (Joey, 2001: 1). Historically speaking, Plastic was the offspring of a merger between Suck and Feed—two popular but culturally edgy content provider web services established in 1995 (Greenstein, 2000: 1), combining the quirkiness of Suck newsletters with the insightful and wide-ranging discussions of Feed (cf. Anuff and Cox, 1997; Johnson, 1997). Plastic was launched in partnership with the editors and services of ten news and content providers—Spin, The New Republic, Inside, Movieline, Gamers.com, Modern Humorist, TeeVee, Netslaves, Nerve and Wired News—who helped to choose from among member submissions which news items would be posted on the Plastic website for discussion (Greenstein, 2000). In December 2001, Plastic decided to go it alone with the help of "user-editors" (Plastic members who act as volunteer submissions editors) after the web service was bought by Carl Steadman—known generically as "Carl" by Plastic users—who remains Chief-Editor and self-appointed site janitor.

The Plastic community is highly heterogeneous. Judging by the comments posted and the historical and cultural reference points used, however, the majority of users appear to be male, North American, and mostly 20– and 30–somethings. Or, as one anonymous poster described the typical Plastic user:

1. Age: 18 - 35

> 2. Education: Liberal Arts College, 3.5 years, no diploma.
> 3. Career: 2 - 10 years pulling lattes at Starbucks, 18 months at a failed dot-com, now an "independent Web-design consultant."
> 4. Likes: Digerati, biscotti, anime
> 5. Dislikes: Math, finance, work
> 6. Gender: One of six available choices
> 7. Politics: 60% lefty gonzo, 40% libertarian nutcake
> 8. Pastimes: S&M, B&D, T&A, LS/MFT (no weirdos, please!)
> (Anonymous Idiot, 2001: 1).

In other words, Plasticians tend to be self-styled members of an erudite, ironic and humorous "plugged in" crowd, interested in quirky takes on anything newsworthy—particularly anything connected with popular culture—as well as in serious and informed discussion of current events. Estimates place the number of regular Plastic users at around 15,000 (McKinnon 2001: 1). And anything that will provoke discussion is regarded as post-worthy (Carl, in an interview with Honan, 2001:1).

Although Plastic emulates a long-existing technology news and discussion website service devoted to a technogeek audience—Slashdot.com—it is "new" in the sense that it turns "push media" like email-posted newspaper headlines and news websites on their head by having members propose content and comment publicly upon it. "Plastic's original contribution is a forum to discuss the diverse news pieces it promotes. At Plastic, readers' comments are what it's all about" (Barrett, 2001: 1).

Items are written up by users and can be posted to 8 topic categories: Etcetera, Film&TV, Games, Media, Music, Politics, Tech and Work. Those whose news items are accepted for posting and/or who post comments on the website are awarded ratings on two dimensions. One of these is "karma," which is used to rate a participant as an active member of the community relative to the number of newsworthy postings—both in terms of submitting stories and posting comments on stories—she or he has made to the site overall. A karma rating of 50 or over generally elevates the poster to (volunteer) submissions editor status.

The other rating system—which is linked directly to karma—is peer moderation that operates on a scale of –1 to +5 for a posting overall. Non-registered posters are allocated a default initial rating of "0" when they first post a comment, while the rating baseline for registered users is +1. Moderation points are awarded by Plastic's editors and by a changing group of registered Plastic members who have been randomly assigned the role by Plastic's editors; or, as the message alerting members to their new moderator status puts it: "Congratulations! You've wasted so much time on Plastic that for the next 4 days we're making you a moderator" (Plastic, 2002a: 1). Each moderator gets 10 moderating points to award to posted comments on a Plastic news item, and the possible ratings each moderator can allocate are:

Whatever 0
Irrelevant –1
Incoherent –1
Obnoxious –1
Astute+1
Clever +1
Informative +1
Funny +1
Genius +1
Over-rated +1
Under-rated +1

The moderation points awarded to each post are tallied and the final score is automatically updated and posted in the subject line of the message for readers to see. In other words, "if four or five moderators think a comment is brilliant, it may end up with a +5; useless comments are moderated down to a –1" (Plastic, 2002b: 2).

This ranking practice is based on formal recognition by the site that users cannot read everything that is posted on a topic. With a peer ranking system in place, users can set filters to screen out postings that fall outside a ranking range of their choice. For example, setting the filter threshold at +3 means only those comments that have been moderated and score at or above +3 will be displayed. Conversely, setting the filter threshold at –1 means every comment posted will be displayed. Plastic offers this ranking and filtering function as a means for helping users practice selective reading and to help enhance the quality of postings to the site.

Like the ratings system used on eBay, the moderation and karma system on Plastic does not necessarily ensure a harmonious community of users. "Meta-discussions" involving litanies of complaints about being "modded down" unfairly—or being a victim of "downmod" attacks, where a moderator flushes out all your postings and moderates them down regardless of content—are common. One disgruntled user even went so far as to equate downmod attacks with terrorism:

> Well, to get back to the point I was making, there is a type of attack for which Plastic.com is almost uniquely susceptible. In this type of attack, the terrorist chooses a victim. Then, when he [is] given the opportunity to moderate, he strikes. He goes through the victim's list of comments and then, ignoring Plastic's moderation guidelines, he moderates each of the comments downward. He does not care whether his moderation votes make sense, only that it drives down the rating of the comment. The terrorist's goal is to drive the victim's presence on Plastic "under the Radar," below the filtering level of most of the Plastic audience. He also wants to put the victim's Karma into a nose-dive. (Gravityzone, 2001: 1)

Other users are accused of being "karma whores" if they appear to be "sucking up to" or worming their way into the favor of Plastic editors in the hopes of getting more stories accepted than other users or having their comments modded up. As one poster put it bluntly to another, "Tyler, kissing Bart's ass won't get you more stories posted" (jbou, 2002: 1) in response to Tyler's comment in a heated discussion of the U.S.'s threats to invade Iraq again. Tyler made reference to Bart, one of Plastic's editors and had written in part: "My, we're feeling self-important today, aren't we jbou? Could you please point to Bart's 'warmongering' posts? And why should he have to answer to you, or to anyone else, on command?" (tylerh, 2002: 1).

And, despite some posters loudly and repeatedly protesting that they don't care about their overall karma, karma ratings—and the moderation system—are indeed "a new arithmetic of self-esteem" on Plastic (Shroedinger's Cat, 2002: 1). At stake is public recognition of a poster's incisive mind, keen-edged humor, "innate hipness," and of being "plugged in" (Plastic, 2002a: 1).

Complexities

The point at which we began, with the idea of website communities that are organized around easy-to-use and read ratings systems as being new socializing spaces that shape people into becoming appropriate users of new cyber spaces, now appears much more complex—indeed, contradictory.

On one hand, eBay's community feedback ratings system has been an important factor in its stunning success to date. In part this is because it has helped establish eBay's mission and identity as a helpful broker, with its clients' best interests at heart, and as a responsible cyber force with whom people who want to be part of the project of building a successful tradition of e-commerce seek to be associated. Moreover, as emulators like Plastic have found, engaging participants in active roles of evaluators—and content producers—encourages further participation, "hooking" people in by publicly valuing their contributions. In addition, however, it appears that part of the success of ratings systems in web spaces such as eBay and Plastic has to do with the fact that it helps meet a range of personal needs, including identity and esteem needs. Both "services" actively recruit membership to an affinity group with which one can identify (Gee, 2001) and offer individuals and groups a way of attaining a visible and enviable presence. *susygirl* sums up this aspect of eBay nicely:

> For me it is a kind of therapy. I like it too because i become susygirl and not some English professor. I like to hide behind my new identity. (email interview, 12/02/2001)

MayorBob, a well respected member of the Plastic community, explains it this way: "The nice thing about the karma [rating system] is that, when you're not getting downmod assaulted, you do get a little feedback on whether you are making sense or getting your point across" (MayorBob, 2001: 1). In other words, besides providing a means for mediating responsible and satisfying commercial or intellectual exchange, the ratings system also offers a service to personal identity formation and to what is fast becoming a highly valued "currency"—an exemplary personal ratings profile.

On the other hand, however, the practice of promoting written feedback and ratings in response to eBay transactions has become a space in which many participants engage in purposes that do not merely contradict the "cyber civic" goal of eBay, but actually involve a range of malicious, preying, nasty, hurt-causing acts toward others (some of whom doubtless contribute to their own pain by investing more than is wise in the discourse and otherwise taking their "profile" or "identity" more seriously than the context merits). Some of the data concerning eBay we have presented smells of interpersonal power-tripping, petty acts of malice, and the desire to belittle others (which is endemic in internet spaces). Similarly, Plastic discussions readily collapse into searing vitriolic exchanges of hate-laced postings and taunting challenges.

It goes without saying, then, that dynamic relationships exist between technologies and the practices in which they are employed. On one hand, the development of new technologies creates conditions in which people can change existing social practices and develop new ones, as well as change and develop new literacies that are integral parts of these new or changing practices. On the other hand, these practices simultaneously "constitute" the technologies involved as cultural tools and shape what they *mean* and, indeed, what they *are* within the various contexts in which people use them. In its complexity and contradictoriness, the "ratings game" is par for the course so far as literacy and technology are concerned. The point here is simple and well rehearsed, but bears reinforcing in the present context, since there are still many people who think the internet unleashes all sorts of undesirable forces that are not equally present in the social practices of physical space.

Literacy and technology are never "singular," never the "same thing." They are always "so many things" when in so many hands. The same alphabetic code can be used for writing notes to one's children or for publishing sophisticated experimental findings in learned journals. It can be used for writing good wishes to friends and for writing extortion notes to intended victims. The same kind of ambiguity and range is open to practically any tool or body of knowledge and information we care to name. The same is true of more specific literacies, including different forms of feedback and rating genres. We need only to think of the uses to which various kinds of referees' reports can be put for the point to be perfectly clear.

The particular "new literacy" of producing (or withholding) ratings and feedback shares the formal character of all literacies (different people put it to different uses, understand it differently, etc.). It is susceptible, then, to the same "play" of moral, civic, and emotional forces—the way that people are and how they live out their (in)securities, pleasures and pains, values and aspirations, and so on.

Strategies and Uses: Reading the Social Practice of Rating Others and Feeding Back

Among the multiple ways we might try to describe and understand some of this complexity, we find the option offered by Michel de Certeau's (1984) concept of consumer "uses" particularly fruitful (see also Chapter 12 in this volume).

de Certeau develops a distinction between "producers" and "consumers" that is much wider than—but incorporates—our usual distinction drawn from the domain of commodity production. In this larger sense, producers are those with the power—and, hence, a "place" (a "proper")—from which to shape discourses and discursive formation in all spheres of human life. Institutions like universities mediate the producer role to the extent that they are sites where social actors with the acknowledged power to do so can maintain and police what counts as "science." Consumers are those (scientists themselves) who "consume" the discourses by participating in them. The same distinction works all the way down to the level at which consumers consume specific artifacts of commodified popular culture, such as television fare as packaged entertainment. The strength of de Certeau's formulation is that it directs attention away from a narrow focus on particular acts of consuming artifacts and toward a wider and deeper understanding of social *practices*—in which individual acts *participate* but by no means *constitute*. Just as the "operations" of producers are deeper and larger than we often think—they *produce* the discourse that *constitutes* a TV program as *entertainment* in the first place, not merely the program—so too are the "operations" (*practices*) of consumers.

de Certeau develops a set of concepts including "strategies," "uses," and "tactics" as part of a framework for investigating the nature and politics of cultural production within the practice of everyday life. He is keen to redress perceptions of consumers as passive effects or reflexes of the practices of producers, without denying the relations of differential power that play out across social and cultural groupings everywhere. He nonetheless wants to identify, understand, and explain the power by which the "weak" maneuver within the spaces constituted strategically by producers to make them habitable and to meet their own purposes as best they can.

Producers, who have established and defined their own place from which to manage relations with an *exteriority* composed of targets or threats, can develop *strategies* to this end. Strategy is an art of the strong (cf. scientific institutions that define and regulate "knowledge" through the power to provide themselves with their own place). Through strategic practices producers define the spaces to be lived in by all. Consumers, or the "weak," cannot strategize. Instead, they can maneuver within the constraining order of regulatory fields within which they are obliged to operate by "making use of" the constraining order and by employing "tactics." We will focus here only on "uses."

de Certeau illustrates "use" by examples like that of North African migrants being obliged to live in a low-income housing estate in France and to use the French of Paris or Roubaix (see de Certeau, 1984: 30–32). These people might insinuate into the system imposed on them "the ways of 'dwelling' (in a house or in a language) peculiar to [their] native Kabylia" (de Certeau ,1984: 32). This introduces a degree of plurality into the system. It also confirms consumers as *active* to that extent—albeit still *subordinate*—in working to make such spaces "habitable."

We want to argue that this dialectic is present in every case of literacy and technology. In the present context, eBay's "feedback and ratings" practice and the specific practices of literacy it engenders is a case in point. Where Friedman (2000) talks of eBay as a "shaper" we may equally speak of eBay as a producer. As a constitutive element of shaping the field of commercial exchange in cyberspace—even if it is trying to do so in "good" and "civil" ways according to recognized discursive constructions of these; after all, there is nothing inherently "wicked" about producers and their productions, since we are talking contingencies of power here as distinct from ethics *per se*—the community feedback and ratings system forms part of a constraining order. One can choose whether or not to *be* a "consumer" within this space, but if one chooses to participate in this space, then its order applies.

What we think we see in the snippets of data presented above are varying "ways" of consumers "making use of" the ratings game. They are "insinuating" into the system produced for them ways of "dwelling" with which they are familiar, adept, or which they otherwise find satisfying or reinforcing—no matter how unpleasant we may find some of these. The "new literacy" of ratings, then, can best be understood as endlessly complex and multiple. "It" is flexed into myriad uses. "It" is susceptible to policing and "moralizing" on the part of producers and other consumers alike, just as much as and in parallel manners to the literacies of physical spaces like schools—where the "players" involved are also inclined to invoke notions of fairness, propriety, and "getting it right."

In the end, online community feedback and ratings systems are often an illuminating microcosm of literacy and social practice at large. We may, if we choose, use it as a reference point from which to consider the dialectics of production and consumption of the official literacies of school. We might consider where we, personally, are positioned in these and with what consequences for learners whose "right" to consume is, precisely, an obligation in the way that the participation rights of eBay members or Plastic users are not.

Bibliography

Amazon.com (2002). About Amazon.com. Retrieved July 25, 2002, from: amazon.com/exec/obidos/subst/misc/company-info.html/ref=gw_bt_aa/102–1153992–2446513 (no longer available).

Amherst Robots (2001). Auction data analysis and delivery. Retrieved March 21, 2001, from: www.vrane.com.

Anonymous Idiot (2001). Typical Plastic member profile. *Plastic*. Retrieved February 10, 2002, from: www.plastic.com/media/01/09/04/147231.shtml (no longer available).

Anuff, J. and Cox, A. (1997). *Suck: Worst-Case Scenarios in Media, Culture, Advertising, and the Internet*. San Francisco: Hardwired.

Barrett, A. (2001). Plastic.com Invites Visitors to Shape the Site. *PCWorld.com*. Monday, January 22. Retrieved February 10, 2001, from: pcworld.com/news/article/0,aid,39013,00.asp.

Collier, M. and Woerner, R. (2000). *eBay for Dummies* (2nd edn). New York: Hungry Minds.

de Certeau, M. (1984). *The Practice of Everyday Life*. Berkeley, CA: University of California Press.

eBay (2002a). About eBay: Company Overview. Retrieved February 10, 2002, from: pages.ebay.com/community/aboutebay/overview/index.html (no longer available; similar information now at: http://www.ebayinc.com/who).

eBay (2002b). About eBay: The eBay Community. Retrieved February 10, 2002, from: pages.ebay.com/community/aboutebay/community/profiles.html (no longer available; similar pages now at: http://hub.ebay.com/community).

eBay (2002c). Seller Guide: Fees. Retrieved February 10, 2002, from: pages.ebay.com/help/sellerguide/selling-fees.html (no longer available; similar pages now at: http://pages.ebay.com/sellerinformation/starting/startselling.html).

eBay (2002d). Frequently Asked Questions about Feedback. Retrieved February 10, 2002, from: pages.ebay.com/help/basics/f-feedback.html (no longer available; similar pages now at: http://pages.ebay.com/services/forum/changes.html).

eBay Feedback Discussion Board (2001). Retrieved July 25, 2002, from: forums.ebay.com/dwc?14@1017295028187@.ee7b9c6 (no longer available).

eSnipe (2001). Retrieved March 21, 2001, from: www.esnipe.com.

Friedman, T. (2000). *The Lexus and the Olive Tree*. New York: Anchor Books.

Gee, J. (2001). Reading as a situated practice: A sociocognitive perspective. *Journal of Adolescent & Adult Literacy*, 44(8): 714–25.

Gravityzone (2001). " 'Terrorist' Strikes Plastic.com—Editors Helpless." *Plastic*. Retrieved February 13, 2002, from: www.plastic.com/media/01/09/06/2124202.shtml (no longer available).

Greenstein, J. (2000). *Feed* and *Suck* link up. *The Industry Standard*. Retrieved February 10, 2002, from: thestandard.com/article/display/0,1151,16708,00.html (no longer available).

Heun, C. (2001). What the rest of us can learn from eBay. *InformationWeek*. March 12. Retrieved March 15, 2001, from: www.informationweek.com/828/rbebusiness_side.htm.

Honan, M. (2001). "Plastic is all I do." Part 2: Plastic's Return. Online Journalism Review. Retrieved February 10, 2002, from: http://www.ojr.org/ojr/workplace/1017862577.php.

Joey (2001). Anatomy of a Plastic Story. *Plastic*. Retrieved February 13, 2002, from: www.plastic.com/01/01/15/0223203.shtml (no longer available).

Johnson, S. (1997). *Interface Culture: How New Technology Transforms the Way We Create and Communicate*. San Francisco: HarperEdge.

Lorlei's (2000). Customer Care Assurance. Retrieved May 11, 2000, from: cgi.ebay.com/aw-cgi/eBayISAPI.dll?ViewItem&item=412824070 (no longer available).

MayorBob (2002). "This kind of karma attack has happened…" *Plastic*. Retrieved February 13, 2002, from: plastic.com/media/01/10/03/2149248.shtml (no longer available).

McKinnon, M. (2001). One Plastic Day. *Shift Magazine*. 9(4): 1–3. Retrieved February 10, 2002, from: http://www.shift.com/toc/9.4 (no longer available).
MSN Money (2002). *eBay Inc.: Company Report*. Retrieved February 10, 2002, from: moneycentral.msn.com/investor/research/profile.asp?symbol=EBAY.
Multex.com (2001). Market guide: Business Description. eBay Inc. Retrieved February 27, 2001, from: 2yahoo.marketguide.com/mgi/busidesc.asp?rt=busidesc&rn=A1C7E (no longer available).
Plastic (2002a). Retrieved March 21, 2002, from: www.plastic.com.
Plastic (2002b). Guide to Moderating Comments at Plastic. Plastic. Retrieved February 13, 2002, from: plastic.com/moderation.shtml (no longer available).
Reno, D., Reno, B., and Butler, M. (eds) (2000). *The Unofficial Guide to eBay and Online Auctions*. New York: Hungry Minds.
Schroedinger's Cat (2002). "First Plastiversary—Happy Birthday, Plastic!" *Plastic*. Retrieved January 15, 2002, from: plastic.com/article.pl?sid=02/01/09/0235209&mode=thread&threshold=0 (no longer available).
Sherman, E. (2001). The world's largest yard sale: Online auctions—the sale of collectibles is migrating to the Web. It's the place where niche buyers meet niche sellers. *Newsweek*. March 19. 62–64.
Smith, C. and Smith, N. (c.2002). Buyer Beware! Retrieved February 10, 2002, from: home.earthlink.net/~cardking/buyer.htm (no longer available).
SoYouWanna (2000). SoYouWanna use eBay (and not get ripped off)? Retrieved March 12, 2001, from: www.soyouwanna.com/site/syws/ebay/ebay.html.
Spector, R. (2000). *Amazon.com: Get Big Fast*. San Francisco, CA: Harper Business.
tylerh (2002) "My, we're feeling self-important today…." *Plastic*. Retrieved February 13, 2002, from: www.plastic.com/article.pl?sid=02/02/13/1731211&mode=nested (no longer available).

Part Four

FOURTEEN

Researching New Literacies: Web 2.0 Practices and Insider Perspectives (2006/2007)

Colin Lankshear and Michele Knobel

Biography of the text

This paper grew out of our tenure as adjunct professors and visiting scholar (Colin) at McGill University, where we got involved with numerous Canadian colleagues in a range of roles, including journal reviewing, collaborative publishing, and participating in various proposals for funded projects. This process generated several successful and enjoyable collaborations, including involvement in a two-day workshop on the theme of "Researching New Literacies: Consolidating Knowledge and Defining New Directions," at the Memorial University of Newfoundland. This workshop, which brought together researchers from several Canadian provinces, was supported by grants from the Social Sciences and Humanities Research Council of Canada and the Canadian Society for the Study of Education. The workshop organizers had a session accepted for the 2007 annual conference of the Canadian Society for the Study of Education in Saskatoon, and we presented our account of researching new literacies. The papers from the conference were subsequently reworked for publication as a special issue of the online journal *e-Learning* (2007, volume 4, issue 3). The present chapter is the version of the paper published in *e-Learning.*

"Literacies"

In *New Literacies: Everyday Literacies and Classroom Learning* (Lankshear and Knobel 2006: 64) we define literacies as "socially recognized ways of generating, communicating and negotiating meaningful content through the medium of encoded texts within contexts of participation in Discourses." This definition is intended to emphasize three main points.

First, by "socially recognized ways" we mean something close to the concept of "practice" as developed by Scribner and Cole (1981: 236) in relation to literacy. They define practices as "socially developed and patterned ways of using technology and knowledge to accomplish tasks." When people participate in tasks that direct them "to socially recognized goals and make use of a shared technology and knowledge system, they are engaged in a social *practice*" (Scribner and Cole, 1981: 236, our emphasis). Practices comprise technology, knowledge, and skills organized in *ways* that participants recognize, follow, and modify as changes emerge in tasks and purposes as well as technology and knowledge.

A distinctive feature of today's literacy scene is the extent to which, and the pace with which, *new* socially recognized ways of pursuing familiar and novel tasks by means of exchanging and negotiating meanings via encoded artifacts are emerging and being refined. Moreover, these new ways are being developed and matured as socially recognized patterns of engagement with a good deal of creative consciousness on the part of those who are developing and refining them. Much of this conscious creation and refinement is being done by "tech savvy" people, many of whom are young. In light of this we find Scribner and Cole's account of practice more useful than a number of more recent alternatives that have been advanced within literacy studies. Scribner and Cole put technology up front in their account of "practice." Subsequently, this visibility often slipped into the background as concepts of literacy practices increasingly centered on *texts*, and their linguistic-semiotic dimensions. We want to put the technology dimension of practices squarely back in the frame.

Second, encoding involves much more than "letteracy." Encoding means rendering texts in forms that allow them to be retrieved, worked with, and made available independently of the physical presence of an enunciator. The particular kinds of codes employed in literacy practices are varied and contingent. In our view, someone who "freezes" language as a digitally encoded passage of speech and uploads it to the internet as a podcast is engaging in literacy. So, equally, is someone who photoshops an image, whether or not it includes a written text component.

Third, social practices of literacy are *discursive*. Discourse can be seen as the underlying principle of meaning and meaningfulness. We "do life" as individuals and as members of social and cultural groups—always as what Gee (1997) calls "situated selves"—in and through Discourses, which can be understood as meaningful co-ordinations of human and non-human elements. Meaning-making draws on knowledge of Discourses and insider perspectives, which often goes beyond what is "literally" in the sign. In other words, "outsiders" to a Discourse will make different sense of something important within that Discourse when compared with the meaning that discourse "insiders" can make (e.g., different groups' interpretations of religious texts). Part of the importance of defining literacies explicitly in relation to Discourses, then, is that it speaks to the meanings that insiders and outsiders to particular practices can and cannot make respectively. It reminds us that texts evoke interpretation on all kinds of levels that may only *partially* be "tappable" or "accessible" *linguistically*.

"*New* Literacies"

"New literacies" is a useful construct when understood from a *historical* rather than a *temporal* perspective. On the one hand, there is little to be gained from speaking of new literacies in merely temporal terms, such that as soon as Instant Messaging appears, email seems like an "old" literacy. Certainly, it would not be viable to try and build a research agenda on anything as fleeting as that. On the other hand, we are clearly at an important historical conjuncture. We are witnessing a "surpassing"—although *not* a displacement—of the mechanical age by digital electronics and other micro-technologies (e.g., in biology, in manufacturing, in communications). Far from disappearing, many mechanical devices are being accompanied and augmented by diverse electronic devices and, in many cases, "spliced" with them, yielding transcendent technologies and processes.

The same is occurring at social, economic and cultural levels as well. The various "posts"—like "post-industrialism," "postmodernism" and "post-capitalism"—reflect attempts to theorize certain changes in material circumstances, in ways of doing things, and in ways of understanding socio-historical and cultural phenomena. Integral to the kinds of shifts being mapped in such ways is what we think of as changes in sensibilities and ethos. These simultaneously *respond* to and *help to shape* processes and outcomes of change (Castells, 2000)—including social practices and conceptions of literacies.

Accordingly, we think of new literacies having new "*technical* stuff" and new "*ethos* stuff" that are dynamically inter-related. The significance of the new technical stuff largely has to do with how it enables people to build and participate in literacy practices that involve different kinds of values, sensibilities, norms and procedures, and so on, from those that characterize conventional literacies. These values, sensibilities, etc., comprise the "new ethos stuff" of new literacies.

New "Technical Stuff"

Much of what is germane to "new technical stuff" is summarized in Mary Kalantzis' idea that "You click for 'A' and you click for 'red'" (Cope et al., 2005: 200). Basically, programmers write source code that is stored as binary code (combinations of 0s and 1s) and that drives different kinds of applications (for text, sound, image, animation, communications functions, etc.) on digital-electronic apparatuses (computers, games hardware, CD and mp3 players, etc.). Anyone with access to a fairly standard computer and internet connection, and who has fairly elementary knowledge of basic software applications and functions, can create a diverse range of meaningful artifacts using a strictly finite set of physical operations or techniques (keying, clicking, cropping, dragging), in a tiny space, with just one or two (albeit complex) "tools." They can, for example, create a multimodal text and send it to a person, a group, or an entire internet community of global reach in next to no time and at next to no cost.

Machinima animations are a good example of what we mean here. Until recently such productions required expensive, high-end 3D graphics and animation engines that were usually the preserve of professional animators. Today, however, a laptop computer, $30.00 game (e.g., The *Neverwinter Nights* Diamond Pack), video and audio editing software (often part of the software bundle that comes with a new computer), and some free or low cost video recording software (e.g., Fraps) comprise ample resources for creating polished animated movies.

Red vs. Blue, created by Rooster Teeth Productions, is widely acknowledged as one of the first amateur machinima to capture sustained "mainstream" attention (Kellend, Morris and Lloyd, 2005). This serialized, science fiction soap opera, which ended in June 2008 with the completion of 100 episodes spread over five seasons, was filmed entirely within the massively-multiplayer online game, *Halo*. Each episode was created first as an audio track, using a range of friends' voices for the characters appearing in each scene (with some additional spoken lines that developed during the actual filming sometimes spliced in later). Taking the resulting audiotrack as their guide and working on multiple networked computers, the producers "puppeteered" their Halo characters, using multi-player options within the same game in synch with the audiotrack. While the puppeteering was taking place, the series creators recorded or "filmed" the action using either separate software (e.g., Fraps), or onscreen recording software built into the Halo game itself. This recorded action and audio soundtrack were then spliced together and edited using video editing software. Each 5–minute movie took hours to complete and polish, but did not require expensive outlays in terms of buying computers, digital recording devices, or software.

Music remix practices are another good example of hobbyists being able to produce high-quality artifacts, this time in the form of audio files. Software that comes bundled with most computers allows users to convert music files from a CD into an editable format (e.g., wav), edit and splice sections of different songs together, to convert the final music files back into a highly portable format (e.g., mp3) and upload them to the internet for others to access or, alternatively, use them as background soundtracks in larger do-it-yourself multimedia projects. A good example is provided by the anime music videos (AMV) created by a 17–year-old high school student, DynamiteBeakdown. His work is very well regarded and "Konoha Memory Book," was registering over one million views on YouTube in mid-September 2007 (see youtube.com/user/maguma for Dynamite-Breakdown's account, with drummerjdm's posting of this same AMV attracting the most views). This same AMV won the "popular vote" at the 2006 Anime Expo in Los Angeles. DynamiteBreakdown admits he doesn't have a sophisticated computer, doesn't own a large library of anime DVDs, and that he uses free video editing software in the form of Windows Movie Maker. His results reflect what he puts into his work in terms of time and conception, not hardware and software. His current AMV projects absorb months of time and hundreds of anime clips.

The kinds of technological trends and developments we think of as comprising new technical stuff represent a quantum shift beyond typographic means of text production as well as beyond analogue forms of sound and image production. New technical stuff can, of course, be employed to do in new ways "the same kinds of things we have previously known and done," and often is (Bigum, 2003; Hodas, 1996). Equally, however, this new technical stuff can be integrated into literacy practices (and other kinds of social practices) that in some significant sense represent *new* phenomena. The extent to which they are integrated into literacy practices that can be seen as being "new" in a significant sense will reflect the extent to which these literacy practices involve different kinds of values, emphases, priorities, perspectives, orientations and sensibilities from those typifying conventional literacy practices that became established during the era of print and analogue forms of representation and, in some cases, even earlier.

New "Ethos Stuff"

The idea that many contemporary social practices involve new "*ethos* stuff" from that which often characterized earlier ways of doing things refers to the intensely "participatory," "collaborative," and "distributed" nature of many current and emerging practices within formal and non-formal spheres of everyday engagements. We understand this difference in "ethos" between conventional and new literacies in terms of large scale historical and social change phenomena (Lankshear and Bigum, 1999; Lankshear and Knobel, 2006: Ch. 2). In economic terms these changes correspond broadly to the transition from industrialism to post-industrialism, informationalism (Castells, 2000), and the emergence of "knowledge economies." Technologically, they correspond with the emergence of digital-electronics, biotechnologies, and nanotechnologies that transcend, but do not displace, technologies of the earlier mechanical era. Culturally and epistemologically, the phenomena in question correspond to trends and patterns encapsulated in distinctions between the modern (or "modernity") and the postmodern (or "postmodernity"), as well as with that between "structuralism" and "post-structuralism" (cf. Lankshear and Knobel, 2006: Ch.s 2–3).

For present purposes, much of what we regard as new "ethos stuff" in contemporary practices is crystallized in current talk of "Web 1.0" and "Web 2.0" as different sets of design patterns and business models in software development, and in concrete examples of how the distinction plays out in real life cases and practices mediated by the internet (O'Reilly 2005; see Figure 13.1).

Web 1.0		Web 2.0
Ofoto	▷▷	Flickr
Britannica Online	▷▷	Wikipedia
Personal websites	▷▷	Blogging
Publishing	▷▷	Participation
Content management systems	▷▷	Wikis
Directories (taxonomy)	▷▷	Tagging ("folksonomy")
Netscape	▷▷	Google

Figure 13.1: Web 1.0 and Web 2.0 (adapted from O'Reilly 2005: n.p.)

The first generation of the Web has much in common with an "industrial" approach to material productive activity. Companies and developers worked to produce artifacts for consumption. There was a strong divide between producer and consumer. Products were developed by finite experts whose reputed credibility and expertise underpinned the take up of their products. *Britannica Online* stacked up the same authority and expertise—individuals reputed to be experts on their topic and recruited by the company on that basis—as the paper version of yore. Netscape browser development proceeded along similar lines to those of Microsoft, even though the browser constituted free software. Production drew on company infrastructure and labor, albeit highly dispersed rather than bound to a single physical site.

The picture is very different with Web 2.0. Part of the difference concerns *the kind of products* characteristic of Web 2.0. Unlike the "industrial" artifactual nature of Web 1.0 products, Web 2.0 is defined by a "post-industrial" worldview focused much more on "services" and "enabling" than on production and sale of material artifacts for private consumption. Production is based on "leverage," "collective participation," "collaboration" and distributed expertise and intelligence, much more than on manufacture of finished commodities by designated individuals and work teams operating in official production zones and/or drawing on concentrated expertise and intelligence within a shared physical setting.

The free, collaboratively produced online encyclopedia, Wikipedia.org, provides a good example of collaborative writing that leverages collective intelligence for knowledge production in the public domain. Whereas an "official" encyclopedia is produced on the principle of recognized experts being contracted to write entries on designated topics and the collected entries being formally *published* by a company, Wikipedia entries are written by anyone who wants to contribute their knowledge and understanding and are edited by anyone else who thinks they can improve on what is already there. Wikipedia provides a short policy statement and a minimal set of guidelines to guide participants in their writing and editing. It is, then, an encyclopedia created by *participation* rather than via publishing; it "embraces the power of the web to harness collective intelligence" (O'Reilly 2005: no page). This is collaborative writing supported by the "technical stuff" of a "wiki" platform or some other kind of collaborative writing software like Writely.com (or similar). It builds on distributed expertise and decenters authorship. In terms of ethos it celebrates inclusion (everyone in), mass participation, distributed expertise, valid and reward-able roles for all who pitch in. It reaches out to "all-of-the-Web," regardless of distinction (Anderson 2006).

Many popular literacy practices—like fanfiction, fan manga and anime works, and multiplayer online gaming—reflect Wikipedia's commitment to inclusion, collaboration, and participation, while going somewhat further in explicating what counts as successful performance and providing guidelines for participants. Gee (2004) and others (e.g., Black, 2005, 2006, 2007, 2008; Lankshear and Knobel, 2006: Ch. 3) describe how participants in various online affinity spaces share their expertise, make as explicit as possible the norms and criteria for success in the enterprise, and actively provide online real time support for novices and, indeed, participants at all levels of proficiency. These range from statements about how to develop plausible characters and plots in fanfiction, to elaborate walkthroughs for games produced for the sheer love of the practice and shared with all online. Practices and relationships are widely marked by generosity, reciprocity, and a sense that the more who participate the richer the experience will be (cf. Ito, 2005). In terms of "ethos," the ontology of practices like blogging, writing fanfiction and collaborating in Wikipedia celebrate free support and advice, building the practice, collective benefit, co-operation before competition, everyone a winner rather than a zero-sum game, and transparent rules and procedures.

"New" Literacies: A Summary

To summarize, we believe that the more a literacy practice integrates new technical stuff with the kinds of qualities and values currently associated with the concept of Web 2.0—but which are, of course, aspects of a much larger historical "moment" that has been playing out, and will continue to play out, over decades—the more appropriate it is to regard it as a *new* literacy. The more a literacy practice that is mediated by digital encoding privileges participation over publishing, distributed expertise over centralized expertise, collective intelligence over individual possessive intelligence, collaboration over individuated authorship, dispersion over scarcity, sharing over ownership, experimentation over "normalization," innovation and evolution over stability and fixity, creative-innovative rule breaking over generic purity and policing, relationship over information broadcast, DIY creative production over professional service delivery, and so on, the more sense we think it makes to regard it as a *new* literacy.

This means that being an "insider" to a new literacy practice presupposes sharing the ethos values in question—identifying with them personally. Consequently, what may look on the surface like engagement in a new literacy may well turn out upon closer examination not to be. For example, simply downloading video clips from a popular participatory site like youtube.com to accompany lectures, without otherwise engaging in any of the forms of participation that characterize engagement in a fan practice site does not, for us, rank as a new literacy practice. It is the cultural equivalent of cutting a picture out of a magazine to use as an illustration in a handwritten story or project. As we have noted elsewhere (Lankshear and Knobel 2003, 2006), in contexts of using new technologies of a lot of old wine comes in new bottles at the interfaces of literacy and new technologies.

Elements of a New Literacies Research Agenda from a Sociocultural Perspective

When taken in conjunction with "theory" and "research methodology" as additional variables, these ideas provide some co-ordinates for mapping elements of a sociocultural studies research agenda for new literacies.

Our definition of "literacies" yields four substantial constructs toward framing up a field:

- "recognized ways," construed in a social practice sense
- "meaningful content"
- "encodification"
- "discourse membership"

These are all well established research constructs within sociocultural research and beyond. But they can take on distinctive nuances and parameters when directed to the study of *new* literacies—as distinct from familiar literacies that have colonized digital media. There is all the difference in the world between, on the one hand, the kinds of "recognized ways" involved in colonizing the internet for "doing webquests" or presenting narratives as a series of web pages and, on the other hand, "mashing up" multiple web resources to make customized smart tools for specific affinity purposes, or "hacking" game engines or sampling software to invent novel fan practices that negotiate and transcend linguistic and cultural differences in intricate ways. Taking these same examples, we find similar degrees of qualitative difference ranging over what constitutes "meaningful content" and the manner of its articulation, interpretation and negotiation, as well as over the nature and processes of codification and discourse membership.

Scribner and Cole's frame of "technology," "knowledge," "tasks" and "skills" provides an overlapping (and augmenting) variation on the four previously mentioned constructs, and how inquiry might be opened up productively into identifiable literacies that integrate new technical and new ethos stuff. A viable study could be as seemingly finite and focused as investigating how particular "new technical stuff" is recruited within the development of some particular new "way" of creating meanings "collaboratively," such as within a photoshopping affinity space, or among a group of anime music video remixers, or between participants sharing photos within a community like Flickr.com. Equally, studies might focus on particular configurations of "ethos" within boundable literacy practices.

Alternatively, a study could look at a particular "realization" of a tool or a resource within a practice, from learning to work with it, to refining and elaborating it, to mashing it up. In this vein, the "tool" presented at Pandoralicious (http://sites.google.com/site/pandoralicious; a mash up of Pandora.com, Delicious.com and the now defunct Grazr.com feed aggregator), for instance, could be researched as an in depth case study of "leverage" or of small scale "innovation" or purposeful "creativity" (Sawyer 2006).

Adding the dimensions of theory and research methodology to the framing mix generates diverse possibilities and raises some interesting current issues. In terms of the diversity of the potential research field, it is obvious that studies focusing on, say, the development and/or appropriation of new technical stuff within the context of a particular practice might take very different turns from one another if undertaken from, say, the standpoints of activity theory, actor network theory, or a classical Vygotskian approach to cultural tools, respectively.

Of greater immediate interest, however, is the question of the extent to which investigating new literacies—notably, perhaps, those with substantial online components—might call for developing innovative theoretical and/or methodological approaches and mixes. Alternatively, a second question arises of the extent to which the emergence of new literacies creates contexts and opportunities to identify and address issues of theory and method that may have been incipient or even evident, for some time.

Regarding the first of these questions, from the standpoint of *methodological* innovation, Kevin Leander (2008) refers to the work undertaken during recent years in the "adaptation paradigm" that problematizes "the transfer of familiar methods to the internet" and works toward developing new methods (p. 36). A specific example here involves work being developed around problems concerning participant observation. Leander discusses initiatives in methodology that attempt to reckon with the fact that practices travel across spaces typically treated as binaries—online/offline, virtual world/real world, cyberspace/physical space—and so, therefore, must ethnography. Furthermore, he asks what form ethnography can take under conditions where it is less a matter of ethnographers physically displacing themselves than it is of displacing themselves experientially within a process of following connections. What, he asks, might be involved if ethnographers take seriously concepts like "it takes a village to study a village" (2008: 61).

By contrast, Sonia Livingstone and her colleagues (2008: 119) observe that "[b]road trends in media and communications research ... lean toward the elaboration of existing methods rather than their replacement with wholly new methods." They note, however, that emerging research challenges in the field "may require some new approaches to method" (ibid.). Examples include using link analysis to map the blogosphere and combining random telephone surveying, observation and experiments to explore public understanding of phenomena pertaining to search engines.

At the level of *theoretical* innovation, it is evident that current research is encouraging a push into domains of theory substantially new to literacy research. These include an interest in games studies (see, for example, Gamestudies.org), Actor Network Theory (e.g., Andrews 2006), recent developments within theories of space and time hitherto most commonly associated with fields like geography and architecture (e.g., Appadurai, 1996), "flow theory" (e.g., Csikszentmihalyi, 1990, 1996), developments in socio-technical studies (e.g., Perkel, 2006), social network theory (e.g., Wellman, 2001), social informatics, and so on.

With regard to the second question, Andrew Burn (2008) reports recent research he has conducted on the meaning of multimodal texts in the form of video games being developed by young people. He notes that while cultural studies radically shifted the emphasis within media research from textual structures to lived cultures, it failed to develop a new way to think about signification and text. Moreover, the idea of merging cultural studies and social semiotics has to date failed to generate research that connects a semiotic analysis of media texts with research into the cultures of those who produced them and those who received them. The need for this connection derives from the fact that textual analysis alone can at most demonstrate *potential* meanings. Burn's work seeks to connect social semiotic analysis of multimedia texts with analysis of interviews with cultural producers and receivers of multimodal texts that afford access to "insider" meanings and perspectives associated with the social practices in question—in Burn's case, designing and producing videogames. This approach resonates strongly with our definitional link between literacies and participation in discourses and of the meanings cultural "insiders" and "outsiders" can and cannot make respectively from textual artifacts.

Options for Research Orientations in the Sociocultural Study of New Literacies

Various intellectual, purposive, and procedural or organizational options are available for investigating new literacies. As educationists interested in new literacies we are aware that researchers in this area often sense an expectation that their research should try to make some active and more or less direct contribution toward enhancing teaching and learning within formal education settings. While this is a valuable research outcome we think it is important to acknowledge that the very "newness" of the phenomena under investigation, plus the fact that to a considerable extent the field of literacy studies needs to re-invent itself in order to address the changes going on around us, caution against adopting unduly goal-directed and functional/applied orientations at the outset. We envisage a range of legitimate orientations toward the study of new literacies and will briefly describe some of them here.

(a) "Let's see" Research

Viable research includes studies conducted for their own sake that aim primarily at understanding in depth a "new" social practice and the literacies associated with or mobilized within this practice. It adopts a "let's see" attitude that encourages the researcher to get as close as possible to viewing a new practice from the perspectives and sensibilities—or "mindsets" (Lankshear and Knobel, 2003, 2006)—of "insiders." The existence of striking differences in mindsets with respect to social practices involving new technologies should alert researchers to the importance of attending to how "insiders" engage with new literacies on *their* (i.e., insider) terms (cf. Jenkins, 2006). This involves attending to the ways that meaningful content and socially recognized ways of interacting, using expressive resources, and conveying meanings are engaged, monitored, "realized" and thought about by those who are "inside" the practice in question.

The growing field of game studies presents many good examples of the "let's see" orientation towards new literacies research (e.g., Shaffer, 2005, 2006; Squire, 2006, 2008; Steinkuehler, 2006, 2008). New forms of social expression like anime music videos and machinima (short films made with the help of digital game content and game play engines) similarly provide rich terrain for a "let's see" approach, which could also focus on examining how norms of participation and interaction are established, challenged, and evolve within a meaning-making practice and community (e.g., how the mores of effective participation in a given virtual world are established, transmitted and adhered to; how players learn to participate effectively within a dedicated gaming discussion forum or chat channel). This would be done in order to better understand what might be entailed, for example, in participating in Discourses that include "new" forms of social interaction, identity presentation, and meaning making resources.

Another potentially helpful research dimension of the "let's see" type would focus on teachers' awareness of political ramifications of their practices when they try to take account of young people's literacy interests in ways that put them at odds with the politics and policies of education systems. Indeed, the kind of ethos we associate with new literacies will often—if not typically—run counter to systemic thinking and norms. For this reason, teachers who seek to adapt their practices to take account of new literacies "insider" perspectives may well find themselves stepping into "minefields of local education-system politics" (Jill McClay, personal communication). They may do this consciously or unconsciously, strategically or not strategically, in smart or not so smart ways, and successfully or unsuccessfully. In terms of a new literacies research agenda, investigations of teacher thinking and action in such cases could be very helpful.

(b) "Try on" Research

Research possibilities also include studies that "try on" different theories, or, better yet, develop new theories for explaining what's "new." This is not to say that "old" theories have passed their use-by date. Rather, the point here is that the convergences we see between Web 2.0, new technical and ethos stuff, and the second mindset are to some extent proportionately related to new convergences in discipline-based theories and methods.

For example, second language acquisition theory can be brought together with post-colonial and postmodern identity theories in an analysis of a young man's online fansite dedicated to a popular Japanese band (Lam, 2000). Researching and theorizing multimodality is being extended by researchers like Andrew Burn (2008, 2009) beyond the social semiotics of texts *per se* to include insights into the accomplishment of meaning making gleaned from interview data provided by text producers and consumers. Narrative theory and game theory are being applied to collaborative, real time narrative construction in live role playing contexts in order to better understand layers of narrative construction and agency (cf., Hammer, 2007; Ito, 2007).

Concepts from human geography and space theory are being applied to studies of young people's literacy practices to better understand the *dimensionality* of these practices (e.g., Leander and Sheehy, 2004). Sociolinguistic analysis, discourse analysis, and conversation analysis techniques are brought to bear variously on transcripts of instant messaging conversations or other interactive online texts to explore distributed project collaboration or collaborative writing processes (cf., Black, 2005; Thomas, 2007).

Further possibilities include using Actor Network Theory (Latour, 2005) to analyze participation in virtual worlds; literary analysis to examine new forms of narrative emerging in and across fictional blogs, wikis, and video diaries accessed via video hosting services like Youtube.com etc. (e.g., real blogs written by fictional authors; fiction narratives told using the medium of blogs; wikis dedicated to documenting fictional worlds); or social network analysis theory or network systems theory to examine collaborative online spaces (e.g., Myspace.com, Flickr.com, the blogosphere), among others. Guy Merchant (2007), for example, investigates the usefulness of social network theory in his examination of everyday digital literacy practices. He brings this theoretical position to bear on Bourdieu's concept of "cultural capital" to develop and explore the concept of "digital capital" and its possible role in new forms of social and civic participation.

(c) Educationally Applicable Research

A third research orientation focuses more directly and self-consciously on pursuing findings that can potentially be applied to better understanding or enabling learning in school and other formal learning spaces or, perhaps, to applying ideas and findings from extant studies to formal learning settings (e.g., Alvermann 2002, Alvermann et al. 2007). Studies like those of Chandler-Olcott and Maher (2003) and Black (2005), which address the nature, role and efficacy of reviewer feedback in honing young people's artistic craft and Standard English written narrative expression, respectively, might be trawled for clues about how to mobilize effective features of reviewer feedback for school learning purposes. Other examples might include case studies of participants who are working collaboratively with others on projects requiring them to learn through participation. Foci might include examining and documenting self-directed or do-it-yourself learning in participants' everyday lives, such as learning a range of highly valued, sophisticated digital processes and/or language-related practices, such as specific programming languages, photoshopping techniques, or learning how to use sophisticated software (e.g., Leander and Mills, 2007; Thomas, 2007).

Alternatively, work done at the interstices of games, learning and society by researchers at the University of Wisconsin has obvious potential application for enhancing learning within formal contexts. Like the examples mentioned above, this is not applied research in any pragmatic, functional, or *direct* sense. The point is not to research games and related phenomena with a view to seeing how games can be imported into schools. Rather, the point is to examine games and gaming with a view to better understanding the kinds of principles underlying effective games design, to explore patterns of engagement that seem to be associated with good practice within virtual environments. Findings provide concepts and principles that can be "interpreted and translated" into possible approaches to creating good learning environments. The research does not produce any "off the shelf" options or even remote approximations to these—although they may eventually provide some resources that well-informed educators could put to good use within well-designed approaches to formal learning. Rather, findings from this research provide evidence-based starting points for developing innovative approaches to formal learning and to augment existing approaches that begin from similar principles, goals and assumptions. Hence, the concept of an "affinity space" has been developed out of empirical investigations of gaming (Gee, 2004, 2007). This poses questions about what affinity spaces for learning science might look like, or what might be involved in trying to develop affinity spaces for learning science. At the point where one species of hard work ends, a space emerges for undertaking a different but related species of hard work (see, for example, Shaffer et al., 2005).

(d) A Research Program Orientation

Before presenting some brief indicative cases of new literacies research that can be seen as developments out of the matrix we have described we think it is important to endorse the idea of adopting a *research program* orientation to developing the field wherever and whenever appropriate opportunities exist. Without in any way wanting to under estimate the value of individual researchers conducting single studies independently, or of virtual networks of individual researchers sharing and interacting, we nonetheless see considerable potential benefits deriving from substantial programs emanating out of Centers. The GLS (Games + Learning + Society) program (formerly known as GAPPs) developed by James Paul Gee and colleagues at the University of Wisconsin, Madison; centers developed by Mitch Resnick and Henry Jenkins at MIT, Boston; the graduate research program of the Annenberg Center for Communication at the University of Southern California; research being undertaken at the London Knowledge Lab (http://www.lkl.ac.uk/cms/index.php); and the ITU Center at the University of Oslo, among numerous others, exemplify the kind of advances that can be made in building components of a field when critical masses and economies of scale are achieved.

Some Brief Indicative Cases of New Literacies Research

(i) Researching Collaborative Online Game Development as a New Literacy

Within the context of a larger research project (SYNchrony) conducted by a team of researchers, Kevin Leander (Leander, 2005; Leander and Lovvorn, 2006; Leander and Mills, 2007) investigated retrospectively aspects of a collaborative endeavor to design and develop an online game. Leander's account indicates very clearly how real time study of the kind of phenomenon he captured retrospectively would constitute a paradigm case of new literacies research.

The Phenomenon

Leander's informant, Steven, recounted his experiences over an 18 month period—which began when Steven was 13 years old—of collaborating online to design and build a massively multiplayer game. He teamed with Jake (then aged 9), a British friend he'd met online, and they recruited others from the U.S., England and Australia to form a core of 4 game builders and a peripheral crew of three additional designers and builders, along with free access to an experienced programming consultant. Their game, "Perathnia," was modeled on successful online roleplaying games like Runescape. The project was based on members' enjoyment of games like Runescape—with their rich imaginaries or game universes of characters and foes—as well as on particular limitations within these games (e.g., characters couldn't jump or fly in Runescape). The group hoped to transcend such limitations by building their own game and, at the same time, were spurred on by the possibility of making good money in the event of hatching a successful subscription-based game. Unfortunately for the group, however, the project ended prematurely, at which time they had created a number of parts for the game, including 3D models for most of the player character types, different clothing models and skin textures, "designs for 50 different weapons ..., designs for a few game structures, parts of the game landscape, a number of animations, and some preliminary testing of the game program or 'engine' " itself (Leander and Mills, 2007: 180).

Discourse and Discourse Membership

The participants could have been studied in "real time" as members of both games playing communities and as members of a games design/development/production discourse. This dimension could have been opened out into a focus on identity, for example, which could well constitute a study in its own right. Equally, an approach that looked at how the discourse coordinates its members and how its members organize various elements of the discourse in sync, could provide another orientation. Leander obtained some interview-based clues on this, especially with respect to the development of Steven's identity as a game developer, but a full-blown real time ethnography could capture insider perspectives and understandings about what was going on at different "levels" of engagement within the discourse and at the interfaces between membership of one discourse and membership of others, and so on.

Tools, Techniques—New "Technical Stuff"

Building the game components involved accessing and becoming proficient with a range of technical tools and processes. This included, for example, obtaining copies of useful software via social networks or online stores, by having their game consultant create a small program to solve a tricky file-sharing problem. It also included learning how to use a range of software to render objects in three dimensions (e.g., *3D Studio Max*) and which involved referring to manuals and other guide texts in the process, how to animate 3D objects, how to create and add textures to objects (e.g., skin textures, sword surface textures), and how to divide tasks up in ways that best matched people's areas of digital expertise.

The technical dimension of game development also involved the group in working out ways to work collaboratively across time and space, to troubleshoot coding snags, in object development, to deal with bandwidth and data transfer issues, to work effectively in a context where not everyone had the same suite of tools and software, and so on. For example, Steven was responsible for all the 3D object development, and Sid, a 21–year-old graphic artist in England, was responsible for creating textures for different objects. Steven gives a sense of how they collaborated across software applications (*Adobe Photoshop* and *3D Studio Max*), geographical distance, and game developer roles in solving an object development and file-sharing issue.

> STEVEN: See, I just send him this little thing [referring to a "face" file that can be created inside 3D Studio Max], cause that's easy to send, and he uses Photoshop on it and sends it back to me. And I take this little bitmap [sent by Sid], and I apply it in 3–D Studio Max, and it shows up on the [character] model. And then I see where it looks a little bit weird, and then I say, "It looks weird on the nose," and he didn't know what I was talking about, so I took screen shots, and I drew little arrows and showed him (Leander and Mills, 2007: 183).

Encoding Meanings

In fact, Steven's explanation of the group's game development process also gestures toward discursive aspects of what is involved in encoding meanings successfully. The audience would not entertain a character's skin looking "weird on the nose." Moreover, the meanings to be encoded were such as to call for a specialist on "texture." The conceptual and material division of labor involved in encoding meaningful content in this example is interesting and may differ in significant ways from everyday literacies in earlier times. The division of encoding labor under conditions of "new technical stuff" may be worth investigating as a theme in its own right.

Theory Choices

Leander uses elements of space theory (cf. Leander and Sheehy, 2004), and Appadurai's theory of *flows* in making sense of his data. He argues that hitherto literacy research has been overly *situated* in terms of the scope and contextualization of the practices being studied (see also Leander and McKim, 2003): "[w]e have ... held literacy too far apart from the flows of materials, bodies and embodied practices [and] privileged a reading of their world as being organized by literacy" (in Leander and Mills, 2007: 184). Leander argues for a conception of literacy practices that includes distributed systems and the movement of ideas, resources, media, money, knowledge, and people around the world. The case of Steven is all the richer for this conceptual framing around "flows." Not content to focus simply on the "textual" design of the game itself, Leander identifies three digital flows that played significant roles in shaping, enabling and constraining Steven's project:

Realizing the project required at least three forms of digital flow: digital knowledge (skills and programming code), digital resources (programs, servers, networks) and design for data flow (economizing on digital file size so that the game will be kept mobile). (Leander and Mills, 2007: 185)

Focusing on the team members' identities as game developers, on the goal of their project, and the knowledges on which they draw affords a powerful and overdue critique of text-centric concepts of "design."

We want to push the notion that in a distributed, digital project such as this, the challenge for team members was not simply to acquire skills, tools, and resources for design, as has been imagined in multi-modal design (e.g., New London Group, 1996); rather, their challenge was to make knowledge, resources, and data move across national borders. (Leander and Mills, 2007: 197)

New "Ethos Stuff"

Developing the game also involved attention to new "ethos" stuff. This included the importance of choice within a game, which effectively backgrounded narrative plotlines. Foregrounded were opportunities for players to develop their own in-game goals (cf. Gee, 2003) and to take multiple paths through the game.

Design for data flow offers another angle on "new ethos stuff." Design involved paying attention to file sizes, internet bandwidth, baseline hardware requirements for users, ease of use, finding a compromise between detail and speed of online action, and the like.

> [T]he look of the game was a compromise between the artistic abilities of the team, their desire to improve on the graphics of Runescape, the limited resources of the kind of server and bandwidth that could be run by their start-up company, and the computational load that the game would place on prospective subscriber's computers. A leaner, simpler game would be easier to serve and easier for subscribers to run. The beauty of Perathnia had to be achieved in a compromise with its mobility. (Leander and Mills, 2007: 178)

Leander's work shows clearly and powerfully how fruitful new ways of conceptualizing literacy can be. By paying attention to the ways in which a group of young people in various countries used and shared ideas, resources, and expertise, the study demonstrates how understanding new literacies may well call for "new" theorizing and conceptualizing. Pure speculation on our part suggests other theoretical and conceptual framings that might also have proved fruitful for understanding new literacies include—among many others—Actor Network Theory (e.g., Latour, 2005), activity theory (e.g., Engeström, Miettinen, and Punamäki,1999; Kaptelinin and Nardi, 2006), Gee's principles of effective learning and his concept of affinity spaces (Gee, 2003, 2004), ludology or the study of gaming and play activities (cf. Gonzalo, 2003; Squire, 2008), approaches to multimodality that are more "user" and less "text" focused (e.g., Burn, 2006, 2008, 2009), along with theories from business studies, economics, the sociology of work, etc., that discuss leverage, collaboration, networks, as well as attention economics (Goldhaber, 1997), theories underlying software, hardware and network development, and the like, and coherent combinations and hybridizations of these and other options. This is not to say that any of these theoretical orientations would have been "better" or "more effective" than the position developed by Leander; instead, our point here is that each one of these positions not traditionally hooked up to studies of literacy would have afforded potentially fruitful insights into the same data.

Working from our position here, a range of research questions arise out of work that "tries on" new theories and new ways of thinking about literacy practices. Another set of questions also arises out of a focus on young people's new technology production. These include:

- What can we learn about literacy from the ways in which young people take up and use digital tools and skills to work on collaborative projects? In what ways do social networks assist with the technical dimensions of achieving one's design and product development aims and goals?
- What can we learn from the strategies young people employ to troubleshoot design and programming problems encountered in building a digital game?
- What new forms of collaboration are being enacted by young people involved in distributed game design? How are distributed groups formed and sustained over time? How are new collaborators recruited to the group and non-contributing collaborators ejected from the project? What effects does this seem to have on the project itself?
- What design practices are being developed in collaborative project spaces online and what might this mean for education?
- In what ways do literacy-learning pathways developed by young people in non-school settings challenge established assumptions about effective classroom learning?
- What are some of the powerful literacies to be found in Web 2.0 practices?

(ii) Analyzing Writing and Identity Online

Fan fiction ("fanfic") involves devotees of some media or literary phenomenon, like a television show, movie, video game, anime series, or book, writing "alternative" stories based on its characters or plotlines (Black, 2005; Jenkins, 1992, 2006). Stories relate alternative adventures, mishaps, histories/futures, and locations for main characters, create "prequels" for shows or movies, or realize previously non-existent relationships between characters. Fanfic predates the internet and considerable fanfic activity continues outside online environments. Nonetheless, the internet has enabled almost infinitely more people to actively participate in contributing and reviewing fanfic than was previously possible.

Fanfic research is gaining visibility within literacy studies (cf. Black, 2005; Chandler-Olcott and Mahar, 2003; Thomas, 2007; Trainor, 2003). Rebecca Black's research into a popular online fanfiction archive and review forum, Fanfiction.net, provides a perspective on how studies of affinity spaces might be framed and implemented as a substantive focus within new literacies research.

Fanfic, Meanings and Discourse Membership

Black examines the practices of posting and reviewing fanfictions on Fanfiction.net, emphasizing the discursive nature of being fanfic writers and reviewers and how this is integral to doing meaning work in that space. Fan narratives must establish their authors as people with close knowledge of the original sources sparking their narratives and a strong sense of what can be done within parameters set by borrowed characters, plotlines and settings (cf., Black, 2008; Lankshear and Knobel, 2006). Reviewers must likewise demonstrate knowledge of the original sources for fics they are reviewing by commenting on, say, how "well" (or otherwise) the author has changed or enhanced a familiar set of characters or added to an established storyline. In short, fan *affiliations* shape how things are written *and* read.

While any popular text is "fair game" for being re-written in some way, it is not the case that anything goes in the re-writing. Authors are expected to stay close to the narrative "design" they have chosen: for example, "in canon," where the author remains true to the nature, characters, foci and settings of a media text while adding new storylines or exploring relationships between characters; an alternative universe design; or cross-over fic, where characters and plots from two different original sources appear in the same story (e.g., *Star Wars* mixed with *Lord of the Rings*; cf. Thomas in 2007); and so on. Authors are expected to signal how their work builds formally on the work of others (typically with an opening disclaimer acknowledging who "owns" which characters and settings). Reviewers almost uniformly know to position themselves as supportive and collegial in their feedback, balancing expressions of pleasure in the story with gentle, constructive suggestions for further improving the narrative (Black, 2007).

There is a discursive expectation that authors will aim to write well-crafted stories that attend to standard grammar, spelling and punctuation conventions, and authors have a plethora of "writing advice" sites that spell out "socially recognized ways" of producing good quality fanfic available to support them. These advise how to avoid creating overblown, non-credible "Mary Sue" characters (Black, 2005), and how to provide a good balance of dialogue and description, develop a plotline that isn't too hackneyed, ensure that characters or problems are introduced with sufficient explanation or foreshadowing, that character names are managed in ways that avoid reader confusion, and so on.

Identity Analysis

Black emphasizes the importance of identity and "presentation of self" within fanfiction writing (Black, 2005, 2007) and is especially interested in studying ways in which "adolescents with limited English proficiency construct identities in online English and text-dominated spaces" (Black, 2006: 170). Identity is "the ability to be recognized as a 'kind of person,' such as an anime fan, within a given context" (Black, 2007: 118). From among diverse available options Black uses discourse analysis techniques drawn from Gee (2001) to investigate how fanfic writers engage in (multiple) identity work in their narratives and profile pages within Fanfiction.net.

Black analyzes the "author notes"—notes to readers that come before the start of each story or chapter within a story—and reviewer comments in the body of fanfic produced by a young ESL migrant to Canada writing under the pen name of Tanaka Nanako. The analysis represents Nanako's growing proficiency with English and narrative writing and her developing sense of self as an accomplished writer. In her earliest fics (at age 14) these notes comprise apologies for English spelling and grammatical errors. Later, they begin eliciting reviewer feedback on English grammar and plot development. Black's analysis of Nanako's author notes portray her as having developed a culturally hybrid writing identity spanning the anime fanfic she writes based on the popular television series *Card Captor Sakura*, her pre-migration insider knowledge of Chinese culture, her Canadian immigrant identity, and the development of a carefully contrived linguistic hybridity within her narratives. She blends Japanese terms that have high social cache within anime fanfiction, along with Chinese Mandarin dialogue for her Chinese characters into her English medium fics. Her ability to draw on resources from three languages is highly regarded by her readers (Black, 2005: 123).

New Ethos Stuff, "Post-genre" Writing and Classrooms

Fan fiction writing offers young authors a space in which to develop dimensions of writing that are valued in school. Black (2007: 133) describes Fanfiction.net as an "affinity space" within which "members are using digital literacy skills to discover, discuss, and solve writing and reading-related problems, while at the same time pursuing the goals of developing social networks and affiliating with other fans." As described earlier in this paper, affinity spaces are places or sets of places where people can affiliate with others based primarily on shared activities, interests, and goals (Gee 2004). Participants in affinity spaces can access archived resources, dispersed and shared knowledge, collaborative help and expert advice in forms ranging from FAQs, "walkthroughs" and "guides," to one-on-one conversations and feedback. Educational researchers and theorists working in games studies, cyberculture studies, as well as in the study of fanfiction and other forms of collaborative writing, are increasingly pondering the extent to which principles and procedures organic to affinity spaces might be appropriable within formal learning settings with a view to enhancing teaching and learning there.

Narrative writing grounded in collaborative reworkings of television series or movie plotlines is often dismissed by teachers as "poor writing" and lacking in imagination and creativity. It is rarely considered in relation to larger social practices of intertextuality and "media mixing," which afford growing kudos in work and leisure contexts beyond the school and, in the case of intertextual sophistication, within formal education itself (Jenkins, 2004; Lankshear and Knobel, 2002). Fanfiction research may usefully inform the work of educators in multiple ways. These include helping teachers to better understand and respond to students' classroom narratives, providing insights into mass participation in forms of popular culture that increasingly engage the energies of people across the social spectrum within their out-of-school and post-school lives, and drawing attention to the extent to which conventional genre boundaries (e.g., narrative) and norms for expertise are under challenge from "new literacies" (Thomas, 2007; see also, Jenkins, 2006).

The collaborative nature of fanfic writing and reviewing, the importance of identity, affinity spaces and intertextuality in most fanfic, the post-genre narrative forms of fanfic, and so on make fanfic practices a rich field of study for researchers interested in applying insights from fanfic practices to classrooms. Indeed, fan practices in general are fruitful foci for further research. Examples of worthwhile areas for research include:

• Examining a range of fan practices, such as game walkthroughs, Lego models of an online game event, fan wikis, and critiquing the limitations of the genres currently taught and valued in schools as media of expression and meaning-making.

• Asking how might traditional narrative analysis be extended or reworked to better accommodate an analysis of writing practices, networks and affinities.

• Asking how might one go about "researching" a particular affinity space.

• Examining the "goodness of fit" between Goldhaber's theory of attention economics, with its constitutive social class system comprising stars and fans, and social practices within fanfic affinity spaces (see Goldhaber, 1997).

• Investigating what successful instances of positive collaboration between companies and fanfic writers might have to tell us about effective ways to navigate copyright issues within fan practices.

• Asking in what ways might classrooms better accommodate collaborative writing and linguistic hybridity.

Conclusion

New literacies are a substantial and far-reaching historical phenomenon whose challenging presence to conventional literacies has set in train a dialectic we believe will play out during the decades ahead. The outcome we envisage will be some kind of "resolution" that transcends currently contending categories of practice and not a simple displacement of one by the other. Meanwhile, it is human beings as the *enactors* of literacies who carry this dialectic. It is through them at the level of individuals, members of groups (of interests and affinities), and bearers of institutional roles engaging in literacies as social practices that this literacy dialectic plays out.

This phenomenon begs deep and rich understanding. It is worthy of understanding in its own right as a social-historical process of major significance—indeed, of significance on an epochal scale. As a Time-Warner executive once remarked to us: "This is as good as Gutenberg." It is also worthy of understanding as a means for enabling people and institutions to work toward humanizing this dialectic as far as possible, to push it in directions of progressive resolution. Such understandings call for sustained research and theoretical work, notably within education and at interfaces between education and social practices within settings and institutions beyond schools and universities. A key component of this research and theoretical development will focus on new literacies as more or less discrete practices and in relation to established literacies. This chapter has advanced a basis from which to envisage a research agenda for sociocultural studies of new literacies and offered some examples of research along what we see as fruitful lines.

References

Alvermann, D. ed. (2002). *Adolescents and Literacies in a Digital World.* New York: Peter Lang.

Alvermann, D. E., M. Hagwood, A. Heron-Hruby, P. Hughes, K. Williams, & J. Yoon (2007). Telling themselves who they are: What one out-of-school time study revealed about underachieving readers . *Reading Psychology*, (28)1: 31–50

Anderson, C. (2006). *The Long Tail: Why the Future of Business Is Selling Less of More.* New York: Hyperion.

Andrews, G. (2006). Land of a couple of dances: Global and local influences on freestyle play in Dance Dance Revolution. *Fibreculture*. 8. Retrieved July 18, 2006, from: http://eight.fibreculturejournal.org/fcj-048–land-of-a-couple-of-dances-global-and-local-influences-on-freestyle-play-in-dance-dance-revolution/

Appadurai, A. (1996). *Modernity at Large: Cultural Dimensions of Globalization*. Minneapolis: University of Minnesota Press.

Bigum, C. (2003). The knowledge producing school: Moving away from the work of finding educational problems for which computers are solutions. Retrieved February 16, 2006, from: deakin.edu.au/education/lit/kps/pubs/comp_in_nz.rtf (no longer available; see instead: kps.wikispaces.com).

Black, R. W. (2005). Access and affiliation: The literacy and composition practices of English language learners in an online fanfiction community. *Journal of Adolescent & Adult Literacy*. 49 (2): 118–128.

Black, R.W. (2006). Language, culture, and identity in online fanfiction. *e-Learning*. 3(2): 170–184.

Black, R. W. (2007). Digital design: English language learners and reader reviews in online fiction. In M. Knobel and C. Lankshear (eds), *A New Literacies Sampler*. New York: Peter Lang. 115–136.

Black, R. W. (2008). Just don't call them cartoons: The new literacy spaces of anime, manga, and fanfiction. In J. Coiro, M. Knobel, C. Lankshear and D. Leu, (eds.), *Handbook of Research on New Literacies*. Mahwah, NJ: Erlbaum. 583–610.

Burn, A. (2006). Playing roles. In D. Carr, D. Buckingham, A. Burn, and G. Schott (eds), *Computer Games: Text, Narrative and Play*. London: Polity. 72–87.

Burn, A. (2008). The case of rebellion: Researching multimodal texts. In J. Coiro, M. Knobel, C. Lankshear and D. Leu, (eds.), *Handbook of Research on New Literacies*. Mahwah, NJ: Erlbaum. 151–178.

Burn, A. (2009). *Making New Media: Creative Production and Digital Literacies*. New York: Peter Lang.

Castells, M. (2000). *The Rise of the Network Society* (2nd edn). Oxford: Blackwell.

Chandler-Olcott, K. and Mahar, D. (2003). "Tech-savviness" meets multiliteracies: Exploring adolescent girls' technology-mediated literacy practices. *Reading Research Quarterly*. 38(3): 356–385.

Cope, B., Kalantzis, M. and Lankshear, C. (2005). A contemporary project: An interview. *E-Learning* 2(2): 192–207. Retrieved April 3, 2006, from: www.wwwords.co.uk/elea/content/pdfs/2/issue2_2.asp#7.

Csikszentmihalyi, M. (1990). *Flow: The Psychology of Optimal Experience*. New York: Harper and Row.

Csikszentmihalyi, M. (1996). *Creativity: Flow and the Psychology of Discovery and Invention*. New York: Harper Perennial.

Engeström, Y., Miettinen, R., and Punamäki, R. (1999). *Perspectives on Activity Theory*. Cambridge: Cambridge University Press.

Gee, J. (1997). Foreword: A discourse approach to language and literacy. In C. Lankshear, *Changing Literacies*. Buckingham: Open University Press. xiii–xix.

Gee, J. (2001). Identity as an analytic lens for research in education. *Review of Research in Education,* 25: 99–125.

Gee, J. (2004). *Situated Language and Learning: A Critique of Traditional Schooling*. London: Routledge.

Gee, J. P. (2007). Pleasure, learning, video games and life: The projective stance. In M. Knobel and C. Lankshear (eds), *A New Literacies Sampler*. New York: Peter Lang. 95–114.

Goldhaber, M. (1997). The attention economy and the net. *First Monday*, 2(4). Retrieved July 2, 2000, from: firstmonday.org/htbin/cgiwrap/bin/ojs/index.php/fm/article/view/519/440.

Gonzalo, F. (2003). Simulation versus narrative. In M. Wolf and B. Perron (eds), *The Video Game Theory Reader*. London: Routledge. 221–236.

Hammer, J. (2007). Agency and authority in role-playing "texts." In M. Knobel and C. Lankshear (eds), *A New Literacies Sampler*. New York: Peter Lang. 67–94.

Hodas, S. (1996). Technology refusal and the organizational culture of schools. In R. Kling (ed.), *Computerization and Controversy: Value Conflicts and Social Choices*. 2nd ed. San Diego: Academic Press. 197–218.

Ito, M. (2005). Otaku media literacy. Retrieved June 22, 2006, from: http://www.itofisher.com/mito/publications/otaku_media_lit.html.

Ito, K. (2007). Possibilities of non-commercial games: The case of amateur role-playing games designers in Japan. In S. de Castell and J. Jenson (eds), *Worlds in Play: International Perspectives on Digital Games Research*. New York: Peter Lang. 129–142.

Jenkins, H. (1992). *Textual Poachers: Television, Fans, and Participatory Culture*. New York: Routledge.

Jenkins, H. (2004). Why Heather can write. *Technology Review*. Feb. 6. Retrieved December 27, 2005, from: technologyreview.com/articles/04/02/wo_jenkins020604.asp.

Jenkins, H. (2006). *Fans, Bloggers, and Gamers: Exploring Participatory Culture*. New York: NYU Press.

Kaptelinin, V. and Nardi, B. (2006). *Acting with Technology: Activity Theory and Interaction Design*. Boston, MA: MIT Press.

Lam, W.S.E. (2000). L2 literacy and the design of the self: A case study of a teenager writing on the Internet. *TESOL Quarterly*. 34(3): 457–482.

Lankshear, C. and Bigum, C. (1999). Literacies and new technologies in school settings. *Pedagogy, Culture and Society*, 7(3): 445–465.

Lankshear, C. and Knobel, M. (2002). DOOM or Mortal Kombat? Bilingual literacy in the "mainstream" classroom. In L. Diaz Soto (Ed.), *Making a Difference in the Lives of Bilingual/Bicultural Children*. New York: Peter Lang. 31–52.

Lankshear, C. and Knobel, M. (2006). *New Literacies: Everyday Practices and Classroom Learning* (2nd ed). Maidenhead & New York: Open University Press.

Latour, B. (2005). *Reassembling the Social: An Introduction to Actor-Network-Theory*. New York: Oxford University Press.

Leander, K. (2005). Imagining and Practicing Internet Space-times with/in School. Keynote paper presented to the National Council of Teachers of English Assembly for Research Mid-winter Conference, Columbus, OH. February 18–20.

Leander, K. (2007). "You won't be needing your laptops today": Wired bodies in the wire-less classroom. In M. Knobel and C. Lankshear (eds), *A New Literacies Sampler*. New York: Peter Lang. 25–48.

Leander, K. (2008). Toward a connective ethnography of online/offline literacy networks. In J. Coiro, M. Knobel, C. Lankshear and D. Leu, (eds), *Handbook of Research on New Literacies*. Mahwah, NJ: Erlbaum. 33–66.

Leander, K. and Lovvorn, J. (2006). Literacy networks: Following the circulation of texts, bodies, and objects in the schooling and online gaming of one youth. *Cognition & Instruction*. 24(3): 291–340.

Leander, K. and McKim, K. (2003). Tracing the everyday "sitings" of adolescents on the Internet: A strategic adaptation of ethnography across online and offline spaces. *Education, Communication, & Information*. 3(1): 211–40.

Leander, K. and Mills, S. (2007). The transnational development of an online role player game by youth: Tracing the flows of literacy, an online game imaginary, and digital resources. In Blackburn, M. and Clark, C. T. (eds), *Literacy Research for Political Action*. New York: Peter Lang. 177–198.

Leander, K. and Sheehy, M. (eds) (2004). *Spatializing Literacy Research and Practice*. New York: Peter Lang.

Livingstone, S., Van Couvering, E., and Thumim, M. (2008). Literacies: Disciplinary, critical, and methodological issues. In J. Coiro, M. Knobel, C. Lankshear and D. Leu, (eds.), *Handbook of Research on New Literacies*. Mahwah, NJ: Erlbaum. 103–132.

Merchant, G. (2007). Mind the gap(s): Discourses and discontinuity in digital literacies. *E-Learning*, 4(3): 241–255.
O'Reilly, T. (2005). What is web 2.0?: Design patterns and business models for the next generation of software. Retrieved April 4, 2006, from: oreilly.com/web2/archive/what-is-web-20.html.
Perkel, D. (2006). Copy and Paste Literacy: Literacy practices in the production of a MySpace profile. Unpublished paper. Retrieved September 12, 2007, from: ischool.berkeley.edu/~dperkel/media/dperkel_literacymyspace.pdf.
Sawyer, K. (2006). *Explaining Creativity: The Science of Human Innovation*. New York: Oxford University Press.
Scribner, S. and Cole, M. (1981). *The Psychology of Literacy*. Cambridge, MA: Harvard University Press.
Shaffer, D. (2005). Epistemic Games. *Innovate,* 1(6). Retrieved November 1, 2005, from: innovateonline.info/pdf/vol1_issue6/Epistemic_Games.pdf.
Shaffer, D. (2006). *How Video Games Help Children Learn*. New York: Palgrave Macmillan.
Shaffer, D., Squire, K., Halverson, R. and Gee, J. (2005). Video games and the future of learning. *Phi Delta Kappan*, 87(2): 105–111.
Squire, K. (2006). From content to context: Videogames as designed experiences. *Educational Researcher* 35(8): 19–29.
Squire, K. (2008). Video game literacy: A literacy of expertise. In J. Coiro, M. Knobel, C. Lankshear and D. Leu, (eds), *Handbook of Research on New Literacies*. Mahwah, NJ: Erlbaum. 635–670.
Steinkuehler, C. (2006). Massively multiplayer online videogames as participation in a Discourse. *Mind, Culture, and Activity* 13(1): 38–52.
Steinkuehler (2007). Cognition and literacy in massively multiplayer online games. In J. Coiro, M. Knobel, C. Lankshear and D. Leu, (eds), *Handbook of Research on New Literacies*. Mahwah, NJ: Erlbaum. 611–634.
Thomas, A. (2007). Blurring and breaking through the boundaries of narrative, literacy and identity in adolescent fan fiction. In M. Knobel and C. Lankshear (Eds), *A New Literacies Sampler*. New York: Peter Lang. 137–166.
Trainor, J. (2003). Critical cyberliteracy: Reading and writing *The X-Files*. In J. Mahiri (ed.) *What They Don't Learn in School: Literacy in the Lives of Urban Youth*. New York: Peter Lang. 123–138.
Wellman, B. (2001). Physical place and cyber-place: Changing portals and the rise of networked individualism. *International Journal for Urban and Regional Research* 25 (2): 227–52.

FIFTEEN

Digital Remix: The New Global Writing as Endless Hybridization (2009)

Colin Lankshear and Michele Knobel

Biography of the Text

We have been interested in remixing practices of various kinds for many years. But it was not until we heard Lawrence Lessig talking at a 2005 conference in Oslo that we became convinced of the importance of remix as a key concept for thinking about new literacies. Lessig asserted, with much conviction, that while those of us over the age of 40 think of writing as writing with text, for many young people this is only one way to write and not even the most interesting way to write. Rather, the more interesting ways for legions of young people involve expressing ideas by using images, sound and video. This, however, can involve significant risks with respect to issues of copyright and intellectual property: themes that Lessig addressed with great energy in his civic activism as much as in his writing. We made "remix" the theme for a substantial chapter in the second edition of *New Literacies* and further developed our ideas for a pre-conference institute at an International Reading Association meeting in Toronto in 2007. This talk addressed remix as the art and craft of endless hybridization, and a version was subsequently published in the *Journal of Adolescent and Adult Literacy* in 2008. We subsequently reworked the text for a book chapter published in Spain (Lankshear and Knobel 2010). An amended version of that chapter is presented here.

Introduction

"Remix" is the practice of taking cultural artifacts and combining and manipulating them into a new kind of creative blend. Until recently this concept was associated almost entirely with recorded music. It referred to using audio editing techniques to produce "an alternative mix of a recorded song that differed from the original and involved taking apart the various instruments and components that make up a recording and remixing them into something that sounds completely different" (Seggern, n.d.). This practice of remixing became very popular during the 1990s across a range of musical genres—notably, in hip hop, house and jungle music, but also in mainstream pop, and rhythm and blues, and even in heavy metal music (ibid.). Remixes sometimes simply provided a speedier version of a song, or a leaner, more stripped back sound, or an elongated song to keep people dancing longer. Once digital sound became the norm, however, all manner of mixing and "sampling" techniques were applied using different kinds of hardware devices or software on a computer (Hawkins, 2004).

This remains the dominant conception of remix. Recently, however, the concept has been expanded in important and interesting ways associated with activism contesting copyright and intellectual property legislation. Beginning with music remix, digital remixing has been the object of high profile and punitive legal action based on copyright law. The legal backlash against popular practices of remix has helped fuel an organized oppositional response to what is seen as unacceptable levels of constraint against the public use of cultural material—including a fascinating moment on 2 May 2007, centering on the dissemination of code integral to overriding digital rights management restrictions on copying certain kinds of DVDs (see, for example, http://everydayliteracies.blogspot.com/2007/05/red-vs-blue-today-surely-goes-down-as.html).

The concept of remix and remixing has become a rallying point for organized response to existing copyright arrangements: namely, within arguments developed by Lawrence Lessig (2004, 2005) for the need to establish a Creative Commons (creativecommons.org). Lessig argues that digital remix constitutes a contemporary form of writing that is reaching the stature of a mass everyday cultural practice.

Lawrence Lessig on Digital Remix as Writing

Lessig (2005) claims that at a very general level all of culture can be understood in terms of remix, where someone creates a cultural product by mixing meaningful elements together (e.g., ideas from different people with ideas of one's own), and then someone else comes along and remixes this cultural artifact with others to create yet another artifact. Whenever we comment on a film or a book and discuss it with others we are taking the original author's creativity and remixing it in our own life, using it to extend our own ideas or to produce a criticism. Lessig (2005) says that every single act of reading and choosing and criticizing and praising culture is in this sense remix, and it is through this general practice that cultures get made. History shows us, for example, that remix isn't specific to digital times but has always been a part of any society's cultural development (see, for example, Pettitt's analysis of remix in Shakespeare's work, 2007). More specifically Lessig refers to a practice of creative writing within the school curriculum in parts of North America whereby students read texts by multiple authors, take bits from each of them, and put them together in a single text. This is a process of taking and remixing "as a way of creating something new" (Lessig, 2005: n.p.).

At the broadest level, then, remix is the general condition of cultures: if there is no remix there is no culture. We remix language every time we draw on it, and we remix meanings every time we take an idea or an artifact or a word and integrate it into what we are saying and doing at the time. At a more specific level we now have digital remix enabled by computers. This includes, but goes far beyond, simply mixing music. It involves mixing digital images, texts, sounds, animation, and many other kinds of found artifacts. Young people in countries around the globe are taking up digital remix on a massive scale and it is becoming increasingly central to their practices of making meaning and expressing ideas. Lessig argues that for many young people these practices are ways of *writing*:

> When you say the word writing, for those of us over the age of 15, our conception of writing is writing with text . . . But if you think about the ways kids under 15 using digital technology think about writing—you know, writing with text is just one way to write, and not even the most interesting way to write. The more interesting ways are increasingly to use images and sound and video to express ideas. (Lessig in Koman, 2005: n.p.)

Lessig (2005) provides a range of examples of the kinds of digital remix practices that are "the more interesting ways [to write]" for young people. These include remixing clips from movies to create "faux" trailers for hypothetical movies; setting remixed movie trailers to remixed music of choice that is synchronized to the visual action; recording a series of anime cartoons and then video-editing them in synchrony with a popular music track; mixing "found" images with original images in order to express a theme or idea (with or without text added); and mixing images, animations and texts to create cartoons or satirical posters (including political cartoons and animations), to name just a few types.

We accept this conceptual extension of "writing" to include practices of producing, exchanging and negotiating digitally remixed texts, which may employ a single medium or may be multimedia remixes. (We also recognize as forms of remix various practices that do not necessarily involve *digitally* remixing sound, image and animation, such as paper-based forms of fanfiction writing and fan-producing manga art and comics, which continue to go on alongside their hugely subscribed digital variants.)

Typical Examples of Remix Practices

As indicated above, types of currently popular remix include:

- Photoshopping remixes (e.g., Lostfrog.org)
- Music and music video remixes (e.g., Danger Mouse's "Grey Album" and the Grey video)
- Machinima remixes (e.g., Koinup.com/on-videos)
- Moving image remixes (e.g., Animemusicvideos.org)
- Original manga and anime fan art (e.g., DeviantArt.com)
- Television, movie, book remixes (e.g., Fanfiction.net)
- Serviceware mashups (e.g., Twittervision.com)

This list isn't exhaustive. It simply illustrates the range of remix practices possible using digital files, software and online networks and archives.

"Photoshopping" as image remix

Adobe's famous digital image editing software, Photoshop, has been appropriated as a verb for diverse practices of image editing, many of which constitute forms of remix. Affordable image editing software and enhanced online storage capacities, along with image-friendly website hosting services mean that photoshopping has quickly become a popular online practice, engaging a wide range of contributors with diverse levels of artistic and technical proficiency. Image remixing can take various forms. These include adding text to images, creating photo montages that mix elements from two or more images together (including prankster-type remixes that place the head of a famous person on, for example, the body of someone caught in a compromising situation), changing the image content itself in some way (e.g., removing someone's hair or body parts, adding a fifth leg to a dog), and changing image properties (e.g., changing the colors or image focus, fiddling with brightness levels or shading, etc.). Some of the most common uses of image remixing include

- for fun (including hoaxes) (e.g., Worth1000.com, SomethingAwful.com),
- for expressing solidarity or affinity (e.g., Lostfrog.org), and
- for making political points (e.g., Antiwarposters.com).

Music and music video remixes

At its most basic level, music remixing involves taking bits and pieces of existing songs and splicing them together. Originally, this mixing work required two or more vinyl record turntables and a "mixer" (a machine that allowed the artist to alter the tempo, dynamics, pitch and sequencing of songs), or access to music studios to physically splice two-track tapes to create a single multi-track recording (Hawkins, 2004). With software like GarageBand and Cakewalk "the tracks from any song, regardless of original tempo, can be digitally altered to work over a huge range of tempos and keys" (ibid.: viii), and can be mixed and remixed in countless ways.

Early music remixes included "scratching" ("manually moving the vinyl record beneath the turntable needle," Wikipedia, 2007: n.p.); sampling, taking snippets from one or more songs and weaving them together to create a changed or entirely new song; and sequencing, the ordering and repetition of samples or added sounds or tracks. A popular example of remixing, as judged by mainstream media, is Danger Mouse's "Grey Album," which was released in 2004. In this work, Danger Mouse mixed together samples of instrumental tracks from the Beatles' *White Album* with rapper Jay-Z's a-cappella *Black Album* to create 12 distinctively different songs (see Ayers, 2006). In August 2004, Laurent Fauchere and Antoine Tinguely—Swiss film directors—mixed archival and commercial footage of the Beatles (e.g., television studio live performances, concert footage, clips from the movie, *A Hard Day's Night*) with footage from a Jay-Z performance and computer-generated imagery and live-actor footage made expressly for the music video to create what has come to be known as the "Grey Video." This video was made to promote one of the singles from the *Grey Album* but subsequently ran into legal troubles due to its use of unauthorized clips.

Professional dance music DJs and remix musicians tend to have access to vocal and instrumental tracks for songs that greatly facilitate mixing. Computer and software developments in the 1990s saw the rise of Do-It-Yourself underground music remixers. These amateurs did not have studio-approved access to individual tracks within songs and worked instead with entire, intact songs, often splicing together "wildly different" songs in their work (Wikipedia, 2007). This in turn shaped a distinctive approach to remixing music that has been taken up by professional remixers, too (see, for example, amateur and professional remixes based on the theme song to *Doctor Who* at whomix.trilete.net). A number of singers and bands encourage underground remixing and make mixable versions of their work available online for downloading and tinkering with. It can be argued that developments in video and sound editing software around the same time also saw music fans beginning to make their own music videos as tributes to favorite songs or bands. A search for "fan music videos" on Youtube.com suggests these kinds of music videos are now a popular and well-established practice. These fan videos include sampling clips from movies or television shows, creating movies within video games, using flash animation (or stop motion animation, claymation, etc.), roping friends and family into participating in a live-action video, and so on, all set to a favorite song or used to create a tribute (to a band, a movie, a television series, etc.).

AMV

Anime Music Videos (AMV) are a distinct branch of fan music clips. These music clips always use anime—animated Japanese cartoons—as their visual resources. The anime can come from anime movies, television series, or be original creations of the AMV maker. AMV are often tribute videos; that is, the clips used within an AMV collectively summarize key aspects of a favourite anime series. AMV can also be conceptual and focus on a particular dimension of a favourite anime series, such as a relationship between two different characters, values such as strength, courage, determination, loyalty, or archetypal themes such as the struggle between good and evil. AMV can also simply be a celebration of anime itself, set to a favorite song. For example, the Newgrounds portal for AMV contains material like Chuck Gaffney's remix of clips from several anime shows like *Inuyusha, Dragon Ball-Z,* and *Sailor Moon,* among others, set to the chorus of Alphaville's song "Big in Japan" (newgrounds.com/portal/view/136982).

One of our research informants, DynamiteBreakdown, is a 17–year-old student who spends much of his spare time working on AMV, with some of his projects requiring months of time and hundreds of anime clips to complete. Dynamite's work is well regarded and his "Konoha Memory Book" AMV won the "popular vote" at the 2006 Anime Expo in Los Angeles (topping the list in several judging categories, but competition rules permitted only one official award per video). Dynamite mainly focuses on the *Naruto* anime series and creates AMV that are largely conceptual in nature. He also uses many of his anime to summarize a complete series, and sometimes includes text or other devices to help viewers interpret his video clips. Referring to one of his favorite personal AMV creations he says:

> For "Before We Were Men", I tried my hardest to make this video stand out above all the other Naruto V. Sasuke AMVs that are out there. I wanted to show all the things that the two had gone through up to the fight that they have near the end of the series. Also I tried to throw in a bit of fan service with the text [i.e., words like "passion", "angst" appearing at specific points in the video] and the ending along with keeping the theme of the Video feed effect at the beginning and end.

Machinima

"Machinima" is the term used to describe the process by which fans use video game animation "engines" to create movies. A game engine is the core software that makes a game run. It provides the various "functionalities" needed in a game, like rendering graphics, scripting, animation, sound, collision detecting, networking, a scene graph, and so on (Kelland, Morris, and Lloyd, 2005). This includes making the most of the computer-generated imagery (CGI) in the game, as well as the physics of the game world (e.g., who and what can fly, what can and can't be blown up, who can become invisible and how) to create new animated texts. The movie action is generally captured by software that can record onscreen action (e.g., Fraps, CamStudio, Machinimation), although some games and 3D worlds have recording functions built into them (e.g., "The Movies," Second Life). In the recent past this kind of animation work demanded extremely expensive, high-end 3D graphics and animation engines and was found mostly within the realm of professional animators.

Creating machinima involves telling a story using tools found within the game engine such as camera angle options, script editors, level editors, and the like, along with resources, such as characters, backgrounds, themes, characters and character movement ranges, settings, lighting, game world physics, etc. available within the game. The resulting clips or "takes" are spliced together using movie editing software (e.g., iMovie, Sony Vegas). Titles and credits can also be added.

According to Machinima.com, a popular how-to website and archive of machinima animations,

> you don't need any special equipment to make Machinima movies. In fact, if you've got a computer capable of playing Half-Life 2, Unreal Tournament 2004 or even Quake [all three are popular video games], you've already got virtually everything you need to set up your own movie studio inside your PC. You can produce films on your own, or you can hook up with a bunch of friends to act out your scripts live over a network. And once you're done, you can upload the films to this site and a potential audience of millions. (Machinima.com, 2006: n.p.)

The term "machinima" is also used to describe the genre of animation generated by this process. These animations may be fanfics and extend a game narrative in some way, or the game may simply provide tools and resources for producing an entirely unrelated text. Machinima need not be amateurish in quality, either. Animations like *Hardly Workin'* and *Red vs. Blue* have won film festival awards around the world (Kelland, Morris, and Lloyd, 2005). That being said, increasingly sophisticated video games and the development of user-friendly video-editing software have seen machinima move more towards the everyday. Those new to the machinima creation process can now access online tutorials and interviews with renowned machinima makers for insider tips on how to create one's own high-quality animations (see also Hancock and Ingram, 2007). The popularity of this kind of animation remixing has seen the launch of games that directly and openly encourage remixing, like Lionhead Studios' "The Movies" (lionhead.com/Games/TheMovies) or Epic Games' "Unreal Tournament." Machinima as a genre has also directly influenced music video production, with MTV making a regular spot available for machinima music videos in 2005. In August 2006, the Coca-Cola company launched a ground-breaking machinima advertisement—called, "Coke Side of Life"—that used the Unreal Tournament video game to create a "Grand

Theft Auto"-like setting (with "Grand Theft Auto" being a highly controversial video game in its own right) to tell a story about the importance of being kind to others. Similarly, game companies themselves, like Blizzard Entertainment, run annual competitions for machinima made using their games.

Fan fiction

Fan fiction—or "fanfic" to its aficionados—is the name given to the practice where devotees of some media or literary phenomenon like a television show, movie, video game, or book write stories based on its characters. Most fanfic is written as narrative, although songfic and poetryfic are also popular forms and some fan fictions are carried as manga drawings and animations. "Costume play" or cosplay—dressing up as favourite manga and anime characters—and live action roleplays based on a favorite popular culture text are also gaining in popularity (see, for example, Cosplay.com).

Fanfic writing can be classified into a number of different types. The most common of these include in-canon writing, alternative universe stories, cross-overs, relationshipper (or shipper) narratives, and self-insert fanfic:

• *In-canon writing* maintains the settings, characters and types of plotlines found in the original media text as far as is possible, and simply adds new "episodes" or events to the original text (e.g., a new "episode" of the television show, *Xena: Warrior Princess*, that maintains the characters and setting as faithfully as possible and that builds directly on the narratives and character histories and adventures already developed within the series itself). Pre-sequels and sequels are popular versions of in-canon writing.

• In *alternative universe stories* characters from an original media text are transposed into an entirely new or different "world" (e.g., placing key characters from the Star Wars movies into a Lord of the Rings universe, or an entirely new, invented universe).

• *Cross-overs* bring characters from two different original media texts together in a new story (e.g., Spiderman brought together with the characters from the sci-fi television series, "Stargate SG–1").

• *Relationshipper (or "shipper") narratives* focus on establishing an intimate relationship between two (often minor) characters where none existed or was downplayed in the original text. These texts can focus on heterosexual relations (e.g., between *Star Trek's* Admiral Kathryn Janeway and Chakotay characters), or homoerotic/homosexual relations between characters (e.g., between *Star Trek's* Captain Kirk and Mr. Spock). The latter kind of fanfics are also referred to as "slash fiction."

• In *self-insert fanfic* writers insert themselves as recognizable characters directly into a narrative (e.g., many young female fanfic writers write themselves into the Harry Potter series in place of Hermione, one of Harry's closest friends; many writers invent a character that is a mix of themselves and attributes from popular culture characters and insert this hybrid character into their text).

Fan art

Fan art can take any form and respond to any kind of generative text. Perhaps the most easily found fan art online, however, focuses on manga and anime. Manga (or "Japanese comics") is a stylized graphic genre that itself can be described as a remix of traditional religious scroll-illustration style meets 18th century Dutch art meets Disney animation (cf., Amano and Wiedemann, 2004). Anime is the animated version of manga. Popular manga include *Fruits Basket* and *Sorcerers and Secretaries*; popular anime television series include *InuYasha, Yu-Gi-Oh!* and *Dragonball-Z*. Award-winning anime movies include *Spirited Away* and *Barefoot Gen*. Manga anime in all its forms can provide material for remixing as fan art. Fan art can take the form of single images, faithfully copied from the original manga text, to entire manga stories that mix existing characters and storylines from a range of manga and anime worlds along with original work (see, for example, work archived on DeviantArt.com and TheOtaku.com). Discussion and email lists abound that are devoted to presenting and reviewing fan art work.

Amateur manga drawing was a popular practice in Japan long before manga crossed into the English-language market, and *otaku* attended manga markets or comics conventions and distributed and discussed their own manga images with other fans. This practice was echoed in English-speaking countries as manga and manga events become more readily available to fans. Many of these face-to-face production and review networks subsequently have moved to online spaces, and have become an important source of feedback on drawing techniques (e.g., fine-tuning perspective, facial expressions, hair etc.) and plot developments for *otaku* manga writers and artists. Manga fans are particularly serious about their artwork and regularly form "circles" or distributed groups (especially online) devoted to constructively critiquing each other's manga drawings. Most highly prized within these circles are *original* drawings, rather than copies of existing manga artwork. Kelly Chandler-Olcott and Donna Mahar provide excellent examples of the kind of art-focussed critique that takes place between manga fans in their case study of Eileen, a 13–year-old aspiring manga artist. Eileen scans and posts an original drawing she had done to an email discussion list, and receives the following feedback:

> The background is kinda simple, which is actually a pretty good idea. You might want to add something towards the bottom of the picture to balance all the items you have floating around at the top.... Also, his chest is either really small, or really smushed. Either way, it's not a good look with large biceps (those are the ones on the top of the arms, right? I get confused sometimes). Not to be crude,

> but he needs more shading in the crotch area. It seems there's nothing there from knee to knee. Otherwise I love the expression, specially the grin. It totally sets the mood to scare some people. Or freak them out, whatever. And like usual, nice shiney hair, Very pretty. (Mailing list posting, December 7, 2001; in Chandler-Olcott and Mahar, 2003: 377)

How-to manga drawing and animation courses and tutorials available online (e.g., howtodrawmanga.com), and commercial companies are releasing affordable manga drawing and anime software packages developed especially for fans and hobbyists (e.g., eFrontier's "Manga Studio" and "Anime Studio").

Serviceware Mashups

The term "mashup" (or "mash up"), originally used in the context of music remixing, is now widely applied to the process of merging at the program level two or more application interfaces (APIs) with each other and/or with available databases. This creates new software or online-interface serviceware applications out of services and data that already exist, leveraging them to perform (often highly) specific tasks, or to meet particular purposes that cannot otherwise be met via existing applications and services. This is a form of customizing and tailoring existing resources to meet niched purposes, perhaps most commonly understood at present by reference to the emergence of "apps" for mobile phones and tablets (as well as on the internet).

Some typical examples of established serviceware mashups include Panoramio.com, Twittervision.com and Wikipediavision. Panoramio.com combines Flickr-style photo hosting with Google Maps, so that users can find photos taken in particular places, or discover where a particular photo was taken. Twittervision mashes together the Twitter micro blogging API with Google Maps to show where in the world 'tweets' are being made in close to real time. Wikipediavision (lkozma.net/wpv) is similar. It shows, in close to real time, from where in the world changes are being made on Wikipedia.org. Other serviceware mashup examples include Twittervision.com, which mashes together Twitter—an on-the-go personal information service—with Google Maps to show who is "Twittering" and where at any given moment; and Panoramio.com, which takes Flickr-style photo hosting and mashes it with Google Maps so that posted photos are accompanied by a map showing where they were taken.

Digital Remix as Endless Hybridization

Since the possibilities for remixing digitized cultural material are practically infinite, the new global writing that is digital remix shares in common with conventional writing the potential for endless hybridization.

Hybridization

In biology there are two main kinds of hybridization:

• The result of interbreeding between two animals or plants of different "orders": either different species within the same genus (e.g., mules, hinnies, ligers, zedonks, in the case of animals), or between different sub-species within a species (e.g., Bengal tiger and Siberian tiger).

• Crosses within a single species to obtain characteristics that are not found at all or are not found consistently in the "parent" populations. The aim is to get desirable characteristics consistently by deliberate re-arrangement of genetic material to create new "breeds."

Some hybrids are infertile, but many are not.

These features can be applied analogously to cases of digital remix. If we claim in this case that "family" within the conventional biological taxonomy encompasses particular types of expressive media and services, then the concepts of "genus" and "species" help us to trace fertile interbreeding at both levels (see Table 15.1).

Genus	Species + Species	Hybrids	Examples*
Movies	Video games + movie editing for advertising purposes	Machinima advertising hybrids	"Game On" (Volvo commercial) Coca-Cola's GTA-style "Coke Side of Life" commercial
	Video games + movie editing for entertainment purposes	Machinima fan hybrids	"Red vs. Blue" "Illegal Danish Super Snacks" "A Few Good G-Men"
	Video games + movie editing for commercial entertainment purposes	Machinima media hybrids	"Make Love not Warcraft" (a *South Park* episode)
	Movie editing + commercial movies	Movie trailer hybrids	*Mary Poppins* as a horror movie trailer *The Shining* as a feel-good movie trailer

Table 15.1: Genus and species analysis of some remix hybrids
(Note: examples can be found using a Google.com search)

Continued next page...

Genus	Species + Species	Hybrids	Examples*
Music Video	Popular song + movie editing	Anime Music Videos	"Konoha Memory Book" "Narutrix"
	Popular song + movie editing	Machinima music videos	"Still Seeing Breen"
Storytelling	Popular text + fan spin	Fan fiction	*Pirates of the Caribbean* in canon fanfics *InuYasha* alt universe fics *Game universe* fanfics
Still image	Existing photo + photo editing	Photoshop hybrids	LOLtrek Lostfrog.org All Your Base See also: KnowYour-Meme.com
Music	Pop rock + rap	Remix hybrid	*The Grey Album* (DJ Danger Mouse)
	Folk + rap	Remix hybrid	*The Score* (The Fugees)
	Television theme song + other song genre	Remix hybrid	*Doctor Who* remixes (whomix.trilete.net)
Serviceware	Existing service-ware (e.g., Twitter.com) + existing serviceware (e.g., Maps.google.com)	Customizable service hybrid	Twittervision.com Panoramio.com

Table 15.1 continued: Genus and species analysis of some remix hybrids
(Note: examples can be found using a Google.com search)

Endless(ness)

In the sense that each new mix becomes a meaning-making resource (affordance) for subsequent remixes, there is no "end" to remixing. Each remix in principle expands the possibilities for further remix.

In reality, however, many remixes prove to be "infertile." They are not re-mixed, may not even be viewed, read or listened to more than a few times. This may be completely immaterial to the producers, for whom the full significance of the work might consist merely in bringing a creation to fruition, as an expression of fan appreciation, as self-expression, as another "self-identity constitutive move."

Indeed, there are multiple potential indices of "fertility" or, at least, of non-sterility. Mere total number of "views" per uploaded video comprises a measure of fertility, at least in one sense of "reproduction"—since viewing entails making a copy. And the number of views is certainly one measure recognized by practitioners of remix as evidence of objective (beyond subjective) attainment. AMV videos uploaded to YouTube and viewed more than half a million times can be considered fertile, for example. Longevity is another measure of fertility in the sense that current remixes that reference previous remixes in a layering of significance signal the fertility of an earlier remix. The "All Your Base" set of photoshopped remixes is a good example of long-lived, fertile remix (see knowyourmeme.com/memes/all-your-base-are-belong-to-us). This particular set of images grew out of a clip of the opening sequence to the Japanese video game, *Zero Wing*, which had been uploaded to the internet sometime in 2000 or 2001. The syntactic and semantic hiccups within the English subtitles of this clip tapped deeply into what a *Times Magazine* article identified as "geek kitsch" humor (Taylor 2001). One phrase, uttered by the leader of the invading force—"All your base are belong to us"—especially caught on and resulted in an hilarious set of photoshopped images that reproduced this phrase within a variety of settings (e.g., rewording of the iconic "Hollywood" sign, on billboards and road signs, on food products, as part of television game shows). The phrase and images from this original photoshopped set subsequently have appeared in more recent remixes. For example, providing the syntax for newspaper headlines (e.g., "All your x are belong to us"), as images in other remixed photo sets (e.g., the plane pulling a banner in the Lost Frog remix, the phrase and accompanying image appearing in a copyright resistance movement online), and in the form of remixed photo sets located within new settings (e.g., the Danish production, "All Your Iraq Are Belong to Us") (see Knobel and Lankshear, 2007).

As the operating principle of culture, remix is endless. Interesting questions include those about what gets remixed, how items get remixed, when remix begets innovation, and the directions this takes.

Aspects of the "Art" of Remix

When we talk about the "art" of remix we have generally in mind the aesthetics, appreciation, form and composition dimensions of remix practices. These are centrally concerned with the questions of what makes a remix "good" or of "high quality" and of the kinds of elements or components (including their modes) that go into effective and fertile remixes.

At this level, "art" ranges over the conceptualization of a production (where the concept comes from, how it comes together, what makes it a strong concept, etc.); the design for realizing the concept as elegantly and pleasingly as possible; norms and criteria and other aspects of "a tradition" whose observance is seen as integral to "good" work, and so on. Certainly, trial and error seems to play a role in developing criteria for judging quality. As DynamiteBreakdown explains:

> I started using Naruto Episodes and I would produce like 1 a night, but they weren't amazing. After "We Will Fight for Her," one of my first major AMV projects, I spent a LOT more time on AMVs.

And

> I use them [opening and closing sequences from original television anime] a lot since the animation in them is superb and sometimes it's perfect for scenes; but the credits are looked down upon just as badly as being able to see subtitles. That [i.e., not using opening and closing sequences and subtitles] would be the only way I could be taken seriously on AMV.org.

Phade's 2002 Guide to making AMV remixes suggests:

> Before you get started on your own video, you must first figure out what is a good anime music video and what makes it good. Figuring this out isn't some 10 minute revelation. You have to watch many videos over and over again that are considered good. Watch them closely. Watch them several times in a row. Why is the video good? Be sure to view a minimum of Phade's Required Viewing [hyperlink]. Figure it out and then try to do what they did.
>
> Then watch some mediocre videos. Why are they mediocre? What makes them mediocre? What could the creator have done to make the video better? (Now don't get too cocky here about what they could have done better. There are be plenty of legitimate reasons why they didn't make it as good as they could: insufficient equipment, insufficient time, insufficient footage, not enough effort, or they just plain got tired of making the video and just wrapped it up.) At any rate, notice what could have been done better and try to avoid what they did: learn what not to do. I'm not going to give you a list of mediocre videos; I'm sure you've found some on your own. =) (Phade, 2002: n.p.)

A second guide at the same site (Kalium, 2004) addresses itself to AMV theory. In it, Kalium addresses concepts like "synch"—or, the connection between music and video, without which you have anime and music but no real connection between them—and the musical, lyrical and mood dimensions of synch (or sync); like "concept"—or one's vision for the video; what one wants viewers to think and understand, or how one wants them to feel—and the storytelling, exploration and examination dimensions of the concept; and like "effects," in their meaning, composition, appearance, and. dimensions. It provides examples for the various concepts addressed and, as such, essays some explicit guidelines for expert performance—guidelines that could be built into formal educational considerations of aesthetic creativity.

So far as popular cultural practices of remix are concerned the questions of how aficionados get to know what is good, how to emulate that, and how far they take steps to emulate it are interesting and to date have not been subjected to much inquiry. Various lines are open. For example:

1. Numbers of views and ratings as guides to what are good (i.e., a market appreciation model). What are the views and ratings saying?

2. Hard core aesthetic theory and the like (e.g., film theory, art theory, design theory). This might or might not include trying to apply theories like social semiotics. Kalium's AMV theory guide mentioned above might be seen as a stripped back semiformal version of more traditional kinds of aesthetic theory.

3. Folk/practitioner theory; that is, what do fans and fan practices tell us? For example, what do amateur remixers who create popular AMVs say about what makes a good AMV? What do people's machinima "favorites" lists on YouTube tell us?

Aspects of the "Craft" of Remix

The "craft" of remix requires knowing the "technical stuff" of remixing. Photoshopped remixes, for example, have been greatly facilitated by the development of more user-friendly software, but still require a basic set of technical skills that can be honed through trial and practice. These skills include "preparation" know-how, like being able to scan hardcopy images and convert them to digital formats, being able to transfer images from a digital camera or memory card to a computer, knowing how to access free image archives online, knowing how to download images from the internet, knowing, at times, how to convert image files into different or compatible file formats (e.g., *.jpg, *.tff, *.gif). Photoshoppers also need to know how to use digital image software (e.g., Adobe's Photoshop, Jasc's Paintshop Pro). Within digital image software environments, photoshopping craft includes knowing how to use marquee tools and crop functions to select specific sections of an image and transfer them to another; knowing how to use a repertoire of image adjustment tools and functions, like blur, fill, clone stamp, shape selection, and color matching functions, as well as, magic wands, palettes, and so on. Technical know-how can be gained via tutorials built into the software itself, and beginners have access to a range of online and book-based guides to help with mastering a wide range of tools and functions (see, for example, Corel, 2007; Perkins, 2006; Worth1000, 2007). That being said, technical competence is not a guarantee of effective image remixing. For example, producing a less-than-proficient yet conceptually clever image remix often wins out over slickly produced but unimaginative image remixes in many online forums (Knobel and Lankshear, 2007).

In the case of fanfiction, craft includes a range of technical know-how, along with being able to participate effectively within a fanfic forum or "affinity space" (cf. Black, 2008). Technical know-how needed for posting fanfic online includes, as mentioned earlier, working out how to register as a member of the group, how to post stories written either by means of word processing software located on one's computer hard drive or entirely online at, for example, Google Docs (docs.google.com). It includes knowing how to rate one's fanfic (e.g., general or mature audience), how to find one's fanfic archive space and the archives of others. It may even include knowing some basic HTML coding language for including hyperlinks within one's profile page or notes to readers. Participating effectively in a fanfic affinity space includes knowing how to review other people's fanfics in constructive and supportive ways; in some cases, it includes knowing how to write collaboratively across distances; posting stories regularly and taking reviewer feedback into account; acknowledging when a character has been borrowed—whether from a commercial source or from a friend's own fanfic writing; among others (see Black, 2007, 2008; Lankshear and Knobel, 2006).

A range of online and offline resources exist for enhancing the craft of other remix practices. These include the FAQs (answers to frequently asked questions) posted on remix community forums or websites, fan art how-tos (e.g., howtodrawmanga.com), detailed walk-throughs for creating short machinima movies (e.g., Hancock and Ingram, 2007; Hawkins 2005; Marino 2004), guides to music remixing (e.g., Hawkins 2004), digital video editing handbooks (e.g., Kenworthy, 2005; Videomaker, 2004), guides to creating AMV (e.g., AMV.org, 2007), among others.

Some Conceptual and Theoretical Links to Literacy Education

Defining literacies

We define literacies as "socially recognized ways in which people generate, communicate and negotiate meanings, as members of Discourses, through the medium of encoded texts."

By "socially recognized ways" we mean something close to the concept of "practice" as it was developed by Scribner and Cole (1981) in relation to literacy. They defined practices as "socially developed and patterned ways of using technology and knowledge to accomplish tasks" (Scribner and Cole 1981: 236). That is, when people participate in tasks that direct them "to socially recognized goals and make use of a shared technology and knowledge system, they are engaged in a social practice" (ibid.). Practices comprise technology, knowledge and skills organized in *ways* that participants recognize, follow, and modify as changes emerge in tasks and purposes as well as in technology and knowledge.

This is what we see going on everywhere, and graphically, in today's literacy scene. New socially recognized ways of pursuing familiar and novel tasks are emerging and evolving apace—and with a good deal of consciousness on the part of people who are building and evolving them as this is going on. Interestingly, much of this conscious building and refining is being done by "tech savvy" people—who are often young. This is why we have appealed to Scribner and Cole's account of practice, rather than some of the more recent accounts within literacy studies. Scribner and Cole put technology right in the foreground of their account of "practice." This visibility often slipped subsequently into the background as conceptions of literacy practices increasingly centered on *texts* and their linguistic-semiotic dimensions. We want to put the technology squarely back in the frame. Our focus here is on diverse *social practices* of remix. (Material in this and the two subsequent paragraphs repeats sections from pages 286-287 above.)

Encoding involves much more than "letteracy." Encoding means rendering texts in forms that allow them to be retrieved, worked with, and made available independently of the physical presence of an enunciator. The particular kinds of codes employed in literacy practices are varied and contingent. In our view, someone who "freezes" language as a digitally encoded passage of speech and uploads it to the internet as a podcast is engaging in literacy. So, equally, is someone who photoshops an image—whether or not it includes a written text component.

Social practices of literacy are *discursive*. Discourse can be seen as the underlying principle of meaning and meaningfulness. We "do life" as individuals and as members of social and cultural groups—always as what Gee calls "situated selves"—in and through Discourses, which can be understood as meaningful co-ordinations of human and non-human elements. Meaning-making draws on knowledge of Discourses; that is, on insider perspectives—these often go beyond the literal—beyond what is "literally" in the sign. Part of the importance of defining literacies explicitly in relation to Discourses, then, is that it speaks to the meanings that insiders and outsiders to particular practices can and cannot make respectively. It reminds us that texts evoke interpretation on all kinds of levels that it may be possible only partially to "tap" or "access" *linguistically*.

Technical-Discourse-Evaluative View of Literacy Education

Advocates of social practice approaches to literacy and literacy education as against psycholinguistic and skills-based approaches often insist on the multi-dimensional nature of literacy. The various dimensions of mature, rounded literacy practices are identified and named in varying ways by sociocultural theorists. Our preference is for recognizing what might be called "technical," "discourse" and "evaluative" dimensions. Being literate in the sense we think is well exemplified by serious aficionados of popular remix practices entails being proficient within each dimension.

The *technical* dimension involves knowing one's way around the processes and tools for encoding the meaning one seeks to articulate. In fan fiction this might mean literal print encoding, although within online spaces it will also include aspects like setting up an account, logging on, editing online, saving files and so on. In the case of making machinima, the technical dimension can be highly complex and demanding, including knowing how to play a game to high levels in order to have access to sufficiently developed characters, props and settings, but also how to perform myriad operations to change texture, modify a character's features, synchronize gestures and speech, edit and splice video clips, and so on. This overlaps with much of the craft aspect of remix.

The *discourse* dimension involves bringing cultural knowledge to bear on the tasks or purposes of the practice in which one is engaged—how to mobilize and co-ordinate the meaning elements. This overlaps significantly with much of the art aspect of remix, although a good deal of the craft aspect is in here as well. It is about knowing what kind of situated practice we are in, what the rules, norms and criteria are that apply to that practice, what kinds of shared meanings circulate within members of the practice and the kinds of signs, symbols, sounds (meaning tokens) that bespeak these, and how different co-ordinations out of these can convey different meanings. For example, what is mobilized and coordinated to make the "Hopkin image on a toasted sandwich for sale on eBay" image work as a remix, or DynamiteBreakdown's "Konoha Memory Book"?

The *evaluative* dimension has to do with knowing how to enhance or improve the practice in order for it to better fulfill the interests of those who engage in it and who are impacted by it. There are internal and external aspects to this evaluative dimension. From an *internal* standpoint the purposes of a literacy practice are taken as given and the evaluative dimension is concerned with realizing these purposes more fully, efficiently, richly, or whatever. This might consist in refining techniques to get a better finish, or simplifying processes to make the practice more inclusive, or developing better guides, and so on. From an *external* standpoint the evaluative dimension may involve revising purposes to make them more responsive to people's needs, more altruistic, more representative and so on. From an internal standpoint remixing elements of Google maps and Wikipedia's editing data feed enhances understanding of the encyclopedia as a dispersed, collaborative project. From an external standpoint, turning remix away from purposes of the kind that turned the ill-fated Star Wars Kid into a psychically injured young person (for more, see en.wikipedia.org/wiki/Star_Wars_Kid) can only enhance remix as a family of social practices.

In the various examples of remix practices and the remix artifacts we have described it is easy to see these three dimensions integrated into what the practitioners have done. They provide good concrete cases of what theoretical concepts and distinctions look like within what are increasingly familiar practices for school-aged learners. Moreover, it would be odd to think of others *teaching* the practitioners to integrate the three dimensions into their work—precisely because the various remix "literacies" have been acquired and refined as situated practices that "come at" novice remixers "whole." Rather, the cases work better in hindsight, as examples of what is already evident to remixers and that, as learners in formal settings, they might be able to relate to other learning contexts when connections are made by a teacher or by peers.

Powerful Tools: Manipulation and Distribution

Gee (2007: 33) observes that humans feel "expanded and empowered when they can manipulate powerful tools in intricate ways that extend their area of effectiveness." He further notes that many of the tools young people increasingly have access to today are "smart tools" that have knowledge built into them in ways that enable them to "collaborate" with the tool users to do complex things that the tool user either could not do alone or could not do as effectively. The tool user and the tool "each have knowledge that must be integrated together" if a purpose is to be achieved (Gee, 2007: 34). The smart tool permits the user to have experiences of extending themselves into a world and to manipulate aspects of that world in a fine-grained way in pursuing understanding, mastery, or creative production as a purpose (to have success in one's purpose, in other words).

Gee extends this concept of smart tools beyond material artifacts like smart computer programs to include also intellectual tools (he mentions geometry). When we understand "things" like "concepts" and "bits of theory," and "theorems," "distinctions," "categories" and the like as tools that we can use, and when we get some dexterity in using them, we likewise feel expanded and empowered when going about our worlds of practice, including our educational worlds of practice.

Kalium's AMV theory guide referred to earlier is an example of an abstract "intellectual" smart tool that AMV aficionados can use in conjunction with their digital tools to realize their goals. It tunes the creator into the need to attend simultaneously to "synch," "concept," and "effects" in going about the task. As one becomes increasingly adept at translating these concepts and norms into concrete "moves" in the context of whatever genre one is working in—a process aided by receiving feedback from more experienced peers after posting a creation online—one experiences greater power in taking on more difficult assignments, all the while becoming more knowledgeable about and proficient in the art and craft of remix.

Such smart tools are, however, most readily and effectively acquired and mastered in contexts of situated practice of producing for authentic audiences under conditions where support, expertise and feedback are available just in time and just in place, where this is constructive rather than punitive, and where it is recognized that advancement is by "levels," as in a game, and not all or nothing, as in a pass or fail high stakes test. Classroom pedagogy stands to learn much from remix affinities and how they enable learning and achievement.

Learning To Be

Finally, it is instructive to consider the extent to which and ways in which the processes involved in learning to be a proficient remixer enact the requirements for what learning scientists call "deep learning." This is because learning in the context of becoming a remix practitioner is precisely a matter of "learning to be" (a remixer) and not simply "learning about" (remix). Gee (2007: 172) describes deep learning as "learning that can lead to real understanding, the ability to apply one's knowledge, and even to transform that knowledge for innovation." He argues that pursuing deep learning requires moving beyond learning *about*—"what the facts are, where they came from, and who believes them"—to learning *to be*—which involves "design" in the sense of understanding how and when and why knowledge of various kinds is useful for and sufficient for achieving particular purposes and goals. According to Gee (2007: 172)

> Deep learning requires the learner being willing and able to take on a new identity in the world, to see the world and act on it in new ways. Learning a new domain, whether physics or furniture making, requires learners to see and value work and the world in new ways, in the ways in which physicists or furniture makers do. One deep reason this is so is because, in any domain, if knowledge is to be used, the learner must probe the world (act on it with a goal) and then evaluate the result. Is it "good" or "bad," "adequate" or "inadequate," "useful" or "not," "improvable" or "not"?
>
> Learners can only do this if they have developed a value system—what Donald Schön calls an "appreciative system"—in terms of which such judgments can be made. Such value systems are embedded in the identities, tools, technologies, and worldviews of distinctive groups of people—who share, sustain, and transform them—groups like doctors, carpenters, physicists, graphic artists, teachers, and so forth through a nearly endless list.

This is precisely what learning to be a remixer, as described above, involves. It is writ large in the social relations of participation in remix affinity spaces, in project collaborations, publication of guides and walkthroughs, the operation of feedback and rating systems, and the attitudes of practitioners like DynamiteBreakdown who quickly understand the need to develop an appreciative system that honors the art and craft of remix.

Bibliography

Amano, M. and Wiedemann, J. (2004). *Manga Design*. New York: Taschen.

AMV.org (2007). How-To Guides. Retrieved May 5, 2007, from: animemusicvideos.org/guides.

Ayers, M. (2006). The cyberactivism of a dangermouse. In M. Ayers (ed.), *Cybersounds: Essays on Virtual Music Culture*. New York: Peter Lang. 127–137.

Black, R. (2007). Digital design: English language learners and reader reviews in online fiction. In M. Knobel and C. Lankshear (Eds), *A New Literacies Sampler*. New York: Peter Lang. 115–136.

Black, R. (2008). *Adolescents and Online Fan Fiction*. New York: Peter Lang.

Chandler-Olcott, K. and Mahar, D. (2003). "Tech-savviness" meets multiliteracies: Exploring adolescent girls' technology-mediated literacy practices. *Reading Research Quarterly*. 38(3): 356–385.

Corel (2007). Paint Shop Pro® Photo XI Tutorials. Retrieved May 5, 2007, from: corel.com/servlet/Satellite/us/en/Content/1192639386269.

Gee, J. (2007). *Good Video Games and Good Learning: Collected Essays on Video Games, Learning and Literacy*. New York: Peter Lang.

Girish (2009). Grazr.com—Create free widgets online. *Sites To Use*. Retrieved July 22, 2009, from: sitestouse.com/grazr-com-%E2%80%93–create-free-widgets-online.

Hancock, H. and Ingram, J. (2007). *Machinima for Dummies*. New York: Wiley.

Hawkins, E. (2004). *The Complete Guide to Remixing*. Boston, MA: Berklee Press.

Hawkins, B. (2005). *Real-Time Cinematography for Games*. Hingham, MA: Charles River Media.

Kalium (2004). Kalium's AMV Theory Primer. Retrieved 18 July, 2008, from: animemusicvideos.org/guides/kalium/index.html.

Kelland, M., Morris, D. and Lloyd, D. (2005). *Machinima: Making Movies in 3D Virtual Environments*. Boston, MA: Thomson.

Kenworthy, C. (2005). *Digital Video Production Cookbook: 100 Professional Techniques for Independent and Amateur Filmmakers*. Palo Alto, CA: O'Reilly Media.

Koman, R. (2005). Remixing culture: An interview with Lawrence Lessig. Retrieved April 22, 2006, from: oreillynet.com/pub/a/policy/2005/02/24/lessig.html.

Knobel, M. and Lankshear, C. (2007). Online memes, affinities and cultural production. In M. Knobel and C. Lankshear (eds), *A New Literacies Sampler*. New York: Peter Lang. 199–227.

Lankshear, C. and Knobel, M. (2006). *New Literacies: Everyday Practices and Classroom Learning*. 2nd edn. Maidenhead & New York: Open University Press.

Lankshear, C. and Knobel, M. (2010). Remix digital: La nueva escritura global como hibridación sin limites. In E. Lucio-Villegas and A. Guardas (eds), *El Valor de la Palabra: Alfabetizaciones, Liberaciones y Ciudadanías Planetarias*. Valencia, Spain: Ediciones de Centre de Recursos I Educació Continuá.

Lessig, L. (2004). *Free Culture: How Big Media Uses Technology and the Law to Lock Down Culture and Control Creativity*. New York: Penguin.

Lessig, L. (2005). *Re:MixMe*. Plenary address to the annual Network for IT-Research and Competence in Education (ITU) conference, Oslo, Norway. October.

Machinima.com (2006). What is Machinima? *Machinima.com*. Retrieved February 14, 2006, from: machinima.com/article.php?article=186 (no longer available; see instead: machinima.org).

Marino, P. (2004). *3D Game-Based Filmmaking: The Art of Machinima*. Scottsdale, AZ: Paraglyph Press.

Pandoralicious (2009). Retrieved July 22, 2009, from: sites.google.com/site/pandoralicious.

Perkins, M. (2006). *Beginner's Guide to Adobe Photoshop*. Amherst, MA: Amherst Media.

Pettitt, T. (2007). Opening the Gutenberg parenthesis: Media in transition in Shakespeare's England. Paper presented at the "media in transition 5: creativity, ownership and collaboration in the digital age" Conference, MIT, Boston. Retrieved July 1, 2007, from web.mit.edu/comm-forum/mit5/papers/Pettitt.Gutenberg%20Parenthesis.Paper.pdf.

Phade (2002). Phade's Guide to Good Anime Music Videos. Retrieved July 18, 2008, from: anime-musicvideos.org/guides/PhadeGuide.

Scribner, S. and Cole, M. (1981). *The Psychology of Literacy*. Cambridge, MA: Harvard University Press.

Seggern, J. (n.d.). *Postdigital remix culture and online performance* (Exhibition at University of California at Riverside). Retrieved July 1, 2007, from: ethnomus.ucr.edu/remix_culture/remix_history.htm (No longer available).

Taylor, C. (2001). All your base are belong to us. *Time*. 157(9): 4.

Videomaker (2004). *Videomaker Guide to Digital Video and DVD Production*. Third Edition. Focal Press.

Wikipedia (2007). Music Remix. Retrieved November 24, 2007, from: en.wikipedia.org/wiki/Remix.

Worth1000 (2007). Photoshop Tutorials. Retrieved September 24, 2009, from: worth1000.com/tutorials.

SIXTEEN

Becoming Research Literate Via DIY Media Production (2010)

Colin Lankshear and Michele Knobel

Biography of the Text

This chapter describes an approach to teaching and learning that we have been developing since 2004. It grew out of an invitation to offer an intensive mode summer course in Newfoundland to a cohort of 35 learners enrolled in a Masters degree program with a literacy specialization. The venue for the course was a ski lodge, which had two generous floors of interior space, as well as attractive out door spaces with picnic tables and seating. We arrived to find participants had self-organized in groups more or less based on where they came from on the island and/or on the basis of friendships formed in previous courses. With no prior knowledge of the cohort and the cultural ways of the program at large we simply adopted the intact groups and ran with a team-based approach to learning.

At the time, wireless access in hotels and motels was still rare, and we always traveled with an Apple wireless base station so that we could both be online when there was just one internet access point in our accommodation. The ski lodge had internet access and part of the course dealt with literacy and new technologies. Being open to what our Australian colleague Michael Doneman had long called "pedagogy on the fly," we set up the base station at the lodge (which at that time had a dial-up connection) and invited participants to bring their laptops. We bought a few wireless internet adapters that participants whose machines lacked inbuilt wireless capability could use. In no time each group had at least one or two machines online, and we quickly adapted the original syllabus to include a substantial do-it-yourself, "hands on" new literacies component, getting participants involved in blogging as well as in situ online searching as the course proceeded.

Each semester and each summer, in our respective universities and courses, we have developed variations on the approach that "grew itself" that summer. The following chapter describes the most recent iteration as of summer 2010.

Introduction

This chapter describes a context in which participants acquire multiple "literacies" simultaneously and conjointly through a learning approach that is in some ways unconventional within formal academic programs in higher education. We discuss some of the ways in which this approach reflects a range of wider contemporary trends and conclude by briefly considering its potential for bridging differences across contemporary learning cultures.

Nature, Purpose, and Context of the Learning

The version of the course described here is an intensive mode block course that runs over 4 weeks. Weeks 2 and 4 are face to face (6 hours a day, five days a week), and Weeks 1 and 3 are "off site" or non face to face. The course has two broad purposes:

(a) To address the theme of "new" literacies/digital literacies/new media in theory and in practice.

(b) To provide an introduction to literacy research and to researching literacy—how to *locate* and *use* literacy research effectively as teachers, and how to *do* research. The goal is to enable teachers taking the course to become informed consumers of research and to have some experience of being producers of research.

Typically, participants are all teachers or education administrators.

We combine the process of learning to become "research literate" with the process of learning some "new" literacies and approaches to understanding concepts and theories associated with new literacies. Participants learn how to read and to write research proficiently in combination with learning to theorize and to practice new literacies. These elements are integrated as tightly as possible. We think of this as a kind of productive interactivity between becoming competent theorists and practitioners in the area of "new" literacies, and becoming proficient readers/consumers and writers/producers of research in the area of new literacies.

The Theoretical and Conceptual Foundations of the Course

The conception and practice of the course builds on a mix of elements from sociocultural theory and elements of theories and speculations about contemporary social, technological, and economic change.

From the standpoint of sociocultural theory, learning does not focus on children, or minds, or schools but, rather, "on human lives seen as trajectories through multiple social practices in various social institutions" (Gee, Hull and Lankshear, 1996: 4). For learning to be efficacious, what a learner learns at any point in time "must be connected in meaningful and motivating ways with 'mature' (insider) versions of related social practices" (ibid.). Learning is ultimately accountable to proficient performance in social practices and what Gee (1996) calls "Discourses" and Wittgenstein (1953) called "forms of life." These are meaningful ways of doing and being that integrate and "co-ordinate" purposes, tools, ways of speaking and writing, actions and skills, knowledge and understanding, ways of dressing and interacting, and so on within material contexts and situations. From a sociocultural perspective learning is about becoming "Discourse competent." In the case of our courses, participants are being challenged to get started on becoming Discourse competent as teacher researchers (of their own learning) simultaneously with starting to become Discourse competent as digital media creators.

Gee draws on this perspective when he distinguishes between "deep learning" and the kind of surface learning that often results from an emphasis on decontextualized subject content within formal education. By "deep learning," Gee means learning that can generate "real understanding, the ability to apply one's knowledge and even to transform that knowledge for innovation" (Gee, 2007: 172). He argues that if we want to encourage deep learning it is necessary to move beyond "learning about" and, instead, to focus more on "learning *to be*" (ibid.; our italics). He claims that deep learning requires that learners be "willing and able to take on a new identity in the world, to see the world and act on it in new ways" (ibid.). In part, this points to the *materiality* and *situatedness* of deep learning, where ideas and "content" are grounded in specific tasks, interactions, purposes, actions, outcomes, and the like. In addition, however, if one is learning to be a historian, or a music video creator, it is necessary to see and value things about the world and one's work or activity in the ways that historians and music video creators do. Among other things, this is because

> in any domain, if knowledge is to be used, the learner must probe the world (act on it with a goal) and then evaluate the result. Is it "good" or "bad," "adequate" or "inadequate," "useful" or "not," "improvable" or "not"? (Gee, 2007: 172)

Gee argues that this involves learners developing the kind of value system that Donald Schön (1983) calls an "appreciative system" as a basis for making such judgments. Appreciative systems

> are embedded in the identities, tools, technologies, and worldviews of distinctive groups of people—who share, sustain, and transform them—groups like

> doctors, carpenters, physicists, graphic artists, teachers, and so forth through a nearly endless list. (Gee, 2007: 172)

These ideas resonate with John Seely Brown and Richard Adler's (2008) account of "social learning." By "social learning," Brown and Adler mean learning based on the assumption that our understanding of concepts and processes is constructed socially in conversations about the matters in question and "through grounded [and situated] interactions, especially with others, around problems or actions" (2008: 18). From a social learning perspective, the focus is more on *how* we learn than simply on *what* we learn. It shifts the emphasis from "the content of a subject to the learning activities and human interactions around which that content is situated" (ibid.). This is just the kind of engagement and process a DIY (do-it-yourself) media creator experiences when, for example, she interacts with peers to resolve (what turns out to be) a file-compatibility or file-conversion problem in the course of creating an anime music video or a machinima movie, or when he responds to feedback about how to enhance the quality of a music remix.

Social learning puts the emphasis squarely on "learning to be" (Brown and Adler, 2008: 18; Gee, 2007: 172). According to Brown and Adler (2008: 19),

> mastering a field of knowledge involves not only "learning about" the subject matter but also "learning to be" a full participant in the field. This involves acquiring the practices and the norms of established practitioners in that field or acculturating into a community of practice.

Current theories and speculations about technological and economic change also contribute important ideas to our understanding of learning under contemporary conditions. John Hagel and John Seely Brown's (2005) account of an emerging paradigm shift in our everyday thinking about how to mobilize resources for getting things done has important implications for thinking about education and learning.

In their discussion of emerging models for mobilizing resources, Hagel and Brown (2005: 1) observe that in the course of their daily lives people perceive and act on the basis of " 'common sense' assumptions about the world around us and the requirements to meet our goals" (ibid.). Such assumptions collectively make up "common sense models" for judgment, decision-making and action within everyday routines. Hagel and Brown claim that each major technology shift generates a new common sense model and that in the context of contemporary technology innovations—notably, the microprocessor and packet-switched electronic networks dating from the 1970s—we are now "on the cusp of a shift to a new common sense model" that will reshape many facets of our lives (ibid.).

They describe this emerging new common sense model in terms of a shift away from "push" approaches toward "pull" approaches. This shift can in turn be understood in terms of a convergence between the twin needs to confront uncertainty (itself partly a consequence of recent technological innovations) and to promote sustainability, on the one hand, and the opportunities technological innovations offer for meeting these same needs, on the other. Hagel and Brown's argument has particular relevance to educators, because education/learning is a major sphere of resource mobilization, and to the extent that the projected shift from "push" to "pull" plays out, education/schooling will be impacted in far-reaching ways.

Throughout the twentieth century the dominant common sense model for mobilizing resources was based on the logic of "push." Resource needs were anticipated or forecast, budgets drawn up, and resources pushed in advance to sites of anticipated use so they would be in place when wanted. This "push" approach involved intensive and often large-scale planning and program development. Indeed, Hagel and Brown see programs as being integral to the "push" model. They note, for example, that in education the process of mobilizing resources involves designing standard curricula that "expose students to codified information in a predetermined sequence of experiences" (2005: 3). Conventional education, in fact, is a paradigm case of the push model at work.

According to Hagel and Brown we are now seeing early signs of an emerging "pull" approach within education, business, technology, media, and elsewhere, that creates *platforms* rather than programs: platforms "that help people to mobilize resources when the need arises" (2005: 3). More than this, the kinds of platforms we see emerging are designed to enable individuals and groups to do more with fewer resources, to innovate in ways that actually create new resources where previously there were none, and to otherwise add value to the resources to which we currently have access. Pull approaches respond to uncertainty and the need for sustainability by seeking to expand opportunities for creativity on the part of "local participants dealing with immediate needs" (2005: 4). From this standpoint, uncertainty is seen as creating opportunities to be exploited. According to Hagel and Brown (2005: 4), pull models

> help people to come together and innovate in response to unanticipated events, drawing upon a growing array of highly specialized and distributed resources. Rather than seeking to constrain the resources available to people, pull models strive to continually expand the choices available while at the same time helping people to find the resources that are most relevant to them. Rather than seeking to dictate the actions that people must take, pull models seek to provide people on the periphery with the tools and resources (including connections to other people) required to take initiative and creatively address opportunities as they arise. . . . Pull models treat people as networked creators (even when they are customers purchasing goods and services) who are uniquely positioned to

> transform uncertainty from a problem into an opportunity. Pull models are ultimately designed to accelerate capability building by participants, helping them to learn as well as innovate, by pursuing trajectories of learning that are tailored to their specific needs.

Jay Cross (2006: 38) summarizes what he sees as the key differences between the characteristics of "push" and "pull" as competing logics by means of a series of contrasts. Where "push" assumes you can predict demand and anticipate needs accordingly, "pull" assumes that the world is unpredictable and emphasizes responding rather than anticipating. Where "push" systems are rigid and static, "pull" approaches are flexible and dynamic. Whereas "push" glues components together, producing monoliths, "pull" operates on the basis of small pieces that are loosely joined. Most significantly for education, "push" provides a program; "pull" offers what Cross calls a "learnscape," in the sense of a platform for learning.

In parallel, with respect to learning specifically, "push" gives us training, where "pull" privileges learning; "push" is built on curriculum and "pull" on discovery; "push" provides courses and "pull" offers support for performance. In place of the "push" approach through training programs, "pull" is based on establishing "collaboration platforms." "Push" provision tends to be mandated, whereas "pull" emphasizes self-service. "Push" makes provision "just in case," but "pull" works on the principle of resources being available "just in time"—to be looked for and found as needed and when needed, rather than being handed resources in case there's a need for them.

Cross's idea of a collaboration platform has particular significance for our classes. The key to collaboration platforms is the enabling capacity of what are widely referred to as Web 2.0 services and applications. These enabling services and applications are not packages, or artifacts, or consumables, but *resources* that have to be *performed*: "things" like search engines, or wikis, or blog services, or user-content management services like YouTube, or photosharing facilities like Flickr, to name just a few. You aren't given any product *per se* when you register with and access such services. Rather, you have the opportunity to "drive" or use them, and what you get from them depends on how you "perform" them. For example, machinima creators using wikis to publish their videos can embed YouTube videos in wiki pages, hyperlink to relevant how-to tutorials on Machinima.com, use tagging functions to sort and classify their videos as they want, and marry the wiki with a discussion forum that includes troubleshooting and getting-started support advice—all at little or no monetary cost or little need for pricey online digital media storage.

The efficiency, efficacy, or value-producing capacity of many Web 2.0 enabling services or resources is a function of large scale *leverage*. What makes Google's search engine so powerful and effective is not just that it is user friendly or convenient but, rather, that so many people use it and in doing so they have built the data base upon which Google draws when returning search results. Such services facilitate and mobilize *participation* and *collaboration*, often involving literally millions of people whose contributions add up to something massive—as in the case of Amazon.com's bibliographic data base and product reviews, or Wikipedia's information base as a searchable source.

Since the time that "Web 2.0" was first mooted as a business model (cf. O'Reilly, 2005), the concept has widened to accommodate widespread social interaction and collaborative relatedness among people who meet and interact on the "webtop" (as distinct from those who work offline on their computer desktops). The logic is peer-to-peer, interactive, collaborative, and participatory. Web 2.0 embraces all manner of affinity spaces (Gee 2004), social networking spaces, interactive and collaborative production spaces, peer-to-peer sharing spaces, collaborative online working spaces, collaborative production spaces and so on. This sense of Web 2.0 emphasizes the importance of Web services that facilitate working collaboratively with friends and strangers across time and distance. Specific examples of Web 2.0 applications that encourage such collaboration include sites like Google Docs that support truly collaborative writing, Fanfiction.net and the recursive role reviews can play there with respect to authors improving their narrative writing through input from friends and strangers, and voice over internet services that enable free conference calls across widely dispersed physical locations, to name just a few.

Scholars like Donna Alvermann (2010), Rebecca Black (2008), David Buckingham (2003), Andrew Burn (2009), Julia Davies and Guy Merchant (2009), James Gee (2003, 2004, 2007), Henry Jenkins (2006; Jenkins et al., 2006), Marc Prensky (2006), Will Richardson (2006), Katie Salen (2008), John Seely Brown and Richard Adler (2008), and Constance Steinkuehler (2008), among many others, have discussed at length how online resources and popular cultural affinities have converged in ways that enable and sustain modes of learning very different from the predominantly "push" approach of conventional schooling. In a nutshell, then, in the kind of course we are describing here we aim to bring together (i) learning about new literacy theory and (ii) learning about research methods and research literature within contexts where participants—working in teams, as collaborating novice media creators and researchers—have opportunities to begin learning to be(come) new media literacy producers *and* qualitative teacher researchers in "hands on" ways, and where they have (just in time and just in place) access to expertise and resources when they need them in order to be able to "go on."

The Teaching and Learning Approach within the Courses

In the following pages we will be trying to convey a sense of working with cohorts of students who predominantly have low to at best very modest levels of "tech savviness"—it is not unusual for individuals to tell us they have never sent an email attachment, for example—and who have absolutely no prior experience of the forms of media creation expected of them in this course. The most prior experience that any cohort members to date have had by way of working with media applications used in the course has been making short sound files in Audacity and importing short video clips into Windows MovieMaker or images into Photostory (approximately 5 percent of the student body total). None has had any prior experience of conducting academic research according to the kinds of criteria employed by competitive research funding organizations or involved in doing a research thesis.

Moreover, we will be trying to convey a sense that the emphasis here is on learning, not on being taught. We remain as "hands off" as possible. Rather, we try to follow the kinds of principles operating when, for example, exponents of fan practices like anime music video remixing learn their art and craft. We provide some templates and visual guides that are rough equivalents to the kinds of walk-throughs and "main steps" guides available on the internet. The students have texts that address key concepts, criteria, procedures, and so on, pertinent to do-it-yourself media creation and to academic research. We do not speak to these, however, in the way that traditional lectures and tutorials focus on "the reading" for that day or week. Rather, the emphasis is on participants getting a sense of the content by reading about it within the context of hands on activity. The emphasis everywhere and always is on peers interacting with peers, having a go, committing something to text or to hard drive, and for us to feed back in ways that raise questions or offer suggestions rather than "showing" or "telling." On the rare occasions where "explicit instruction" occurs, it is much more in the vein of trying to explain how various elements relate to one another than of providing detailed information on specifics. Typically, if there is a need for explicit treatment of some point or concept we will oversee participants conducting an online library search or Google Scholar search (scholar.google.com) and then draw on our experience to help point them toward sources most likely to be reliable or of "good academic quality"—making explicit what to look for next time they have to do it by themselves (e.g., academic repute, citations, academic peer review, in the case of scholarly sources; evidence of expertise; year of publication; etc.).

The image readers should have in mind for the learning going on here is of small groups of learners working in teams on situated "hands on" tasks. The groups have a media artifact to create—such as a stop motion animation, an anime music video, a machinima video, a music remix, and so on—that should show as much fidelity as possible to the standards, "ways" and values of affinity groups associated with the practice. They have a formal research report to write that shows as much fidelity as possible to the standards, "ways" and values of academic researchers. This report proceeds from the research purpose of identifying, describing, and reflecting upon in light of some relevant theory, key patterns evident in the data they have collected during the process of creating their media artifact; these patterns having emerged from their hands on attempts to understand and use appropriate forms of data analysis, such as use of open coding, development of emergent categories, pattern matching (of one variety or another), and so on. The image should be of constant interaction, movement, tool use, conversation, negotiation, trying to "nut things out" in pairs or threes or as a group, as well as periods of intense individual activity to complete a task or a segment of writing, or whatever, and bringing it to "the team table" for feedback, suggestions, and so on. The image should be of us as instructors moving from group to group on a reasonably regular basis, when we are not in Google docs (docs.google.com) reading what they are writing while they are writing it, or running our own online searches or trawls through texts to see what kinds of resources or troubleshooting advice could be useful to address issues that have arisen within various teams. Much of the point and purpose of the approach is to give participants space and structure within which to experience, *from the inside*, their activities of learning to be media creators and researchers as a matter of mastering *systems*. They have to come to see these engagements as a matter of participating in Discourses (e.g., of a certain kind of photosharing or music remixing or video animation; of a certain kind of qualitative research), and to see these Discourses as *systems*—coordinations of interlocking elements of the kinds Gee (1996, 1997, 2008) has described so well.

(i) Before the Course Begins

As early as possible, and ideally two to four weeks ahead of time, we create an email address list for all participants, and we create a basic website for the course using Google sites (sites.google.com). Google sites is an extremely "light" and functional resource. Anyone with a Google account can put a basic website together with a few clicks and a bit of keying. We make no attempt to "jazz" the website up, beyond using colored font for sub-headings. Our sites comprise minimal text in the body of the page and maximum use of "attachments" at the foot of the page. Apart from course texts, any resources we recommend or make available are provided either as documents available for downloading (readily uploadable from our computers) or as hyperlinks to sites and resources already available on the web. It is easy to have a working website that is quickly updatable (as easy as adding text to a word processing document) in place within 30 minutes or so.

The first attachment on the site is always the course syllabus, which contains a description of the modus operandi of the course, as well as the usual syllabus fare: tasks, timetable, reading, assessment procedures, and so on. This emphasizes the team-based collaborative nature of the course and invites participants to do any team-building work they want to ahead of time. It also stresses the importance of wireless internet access during the face to face sessions, and encourages everyone to bring a wireless-capable mobile computer (netbook, laptop, tablet, etc.) with them. It also informs participants that during the first week they will be learning to create a digital media artifact and that they will be collecting data on their activity as they go along. So they are encouraged to bring any data collection devices—voice recorders or mp3 players with recording functions, digital cameras, video recorders, thumb drives, external disk drives for extra storage space, mobile phones with cameras, note pads or books, colored pens, etc.—with them from the first day. A list of options for popular digital media artifact production is offered, including such items as stop motion animation, machinima video creation, photoshopping and photosharing, music remixing, music video remixing, podcasting, and so on. While teams are encouraged to select an option from the list, they are welcome to come up with alternatives if they prefer. The bottom line is that whatever option is chosen it must be something that they have no prior experience of creating and minimum prior experience pertinent to creating it.

Once the website is in place and as far ahead of schedule as possible an email message is sent to all participants, pointing them to the website. This means they can start getting a feel for the course and start on any preliminary reading or preliminary tasks. (For example, participants are usually asked to produce a statement of what they understand by certain key concepts and ideas and bring it on the first face to face day. This constitutes a kind of "baseline" for prior knowledge, against which they can assess their "head" learning by the end of the course and trace the evolution in their thinking, understanding, and recourse to literature). They can also begin thinking about the options for digital media production and, in cases where potential teams are in place, work towards firming up an option so that they are ready to start on the first face to face day.

Establishing contact early means that any issues, queries, or concerns that occur to participants can be aired and we get a chance to sort things out. For example, participants occasionally express concerns that they are "computer illiterate" and are fearful they will not be able to do the work. In such cases we inform them that the course website contains links to examples of work produced from scratch by people just like them, who expressed similar concerns at the outset. We assure them that they will *not* fail to complete an acceptable artifact. At the same time, we openly tell them that in some cases we have not produced many of these artifacts ourselves, so beyond asking generic questions like, "Have you searched online for information?" or "Have you tried X, or Y?" we will not be able to *show* them how. In cases where we *have* produced such artifacts we will be acting *as if* we have not. So they will be learning in much the same way that many of their own students learn as members of popular cultural affinities—from peer interactions, online searching, participation in affinities, hit and miss, and so on. We always advise the groups that if they really get stuck they can ask their children or the children of friends. We explain to students that everyone experiences much frustration, considerable uncertainty, and numerous short term "failures." However, the bottom line is that teams are always able to learn enough in the first face to face week to produce a digital media artifact of which they invariably feel proud. In any event, those who have expressed misgivings over their ability to complete the coursework overwhelmingly "hang in" and arrive on the first face to face day with some prior reading done, their preliminary tasks in hand, and with their laptops and other assorted gear ready to roll.

(ii) The "Gist" of the Syllabus Statement

As already mentioned, the syllabus statement makes it very clear that during the first face to face week participants will be *learning to be* participants in media creation, while at the same time *learning to be* participants in academic research in their roles as data collectors—collecting data relevant to the research task of investigating how they learn. Clearly, they cannot become *full* participants during such a short initiation. This requires a lengthy apprenticeship and a genuinely keen interest. Rather, the goal is to get some distance on the way toward proficiency and, especially, to get a grounded sense of as many as possible of the kinds of qualities and capabilities full participants in the practice command: the kinds of things proficient performers of the identities in question are good at doing and caring about, and about how people get to be good at them.

When they arrive "on site" cohort members know that during the first face to face week they must progressively get into role in a situated way as media producers and data collectors. They will have to begin speaking and thinking and acting and caring and proceeding like anime music video or stop motion animation creators speak, think, act, care, and proceed within a hands on context of actively creating media. As emphasized in the syllabus:

> Basically, what will happen is that participants will arrive at the first face-to-face week having completed the reading and preliminary tasks described immediately below. On the first morning of the first face-to-face day each participant will join a small team (up to 6 people) on the basis of the kind of media artifact they most want to learn to produce. As soon as these teams are established they will start the process of learning how to "do their own media" and how to "be a kind of media creator." (Syllabus, July 2010)

Likewise, participants know from the outset that throughout the entire course they must progressively get into role in a hands-on and situated way as qualitative teacher researchers:

> As soon as each team starts learning how to do their own media and how to be a kind of media creator they will start gathering data that will constitute a

> "record" of how the learning occurred. Participants will, in effect, be investigating their own team's learning approaches and processes, making records of one kind or another of how things were got right, how things went wrong and were eventually got right, and so on. Participants will be doing this simultaneously with learning how to create the product. They will be studying themselves doing their learning "in situ." So, for example, when "a light comes on" during the process—something wasn't working and someone discovers how to do it—you might get them to briefly recall how they arrived at their discovery and make an audio recording of what they say. Or, if someone is good at doing something that another person wants to master, you might audio and video record a "talk through" or a "think aloud" of how it is done. Or if someone googles how to do something or how to solve a problem and finds a resource online that explains this, you can record the URL of the resource so that you can come back to it later, and so on. Alternatively, team members might take turns at doing observations and recording fieldnotes and the learning process goes ahead. In other words, at the same time as participants are learning to be a certain kind of media creator they will be starting off learning to be a certain kind of qualitative teacher researcher. (Syllabus, July 2010)

And

> … Whereas the previous [face to face] component involved hands-on learning of how to produce a media artifact and hands-on learning of how to collect data, this part of the course will involve hands-on learning of how to interrogate data within the process of producing an evidence-based account of a learning experience … It is important to emphasize the "hands-on" and "in role" nature of this process. Participants will be "handling" data they have previously collected within the act of being novice qualitative researchers; coding it in the ways that researchers code data; looking for patterns in the data that indicate some of the main features and characteristics of their collective learning experience. Participants will also be "handling" various concepts to get a sense of where these concepts come from, why they have been developed, what kinds of experiences and processes they refer to, and so on. Finally, participants will be getting a "hands-on" sense of how research teams actually work in real life settings: how they interact with each other to generate written reports and to make formal presentations to peers (including producing presentation artifacts like slide shows with embedded "evidence" that support the claims they are making). (Syllabus, July 2010)

The syllabus provides a set of learning objectives, specifies set texts and recommended reading, outlines preliminary tasks, includes a provisional schedule and timelines (that will operate until teams define their own schedules and rhythms), and gives suggestions for what participants might usefully bring to the face to face sessions. There is also a statement of criteria to be met in their written work and team presentations to the cohort at the end of the four week block.

Something of the tenor of the information provided in the syllabus statement may be indicated by the following excerpts:

> The course is based on the premise that a lot of young people (as well as a lot of not so young people) spend a lot of time and energy learning very successfully to do and be things they are interested in, but where this mode of learning is very different from what they encounter in school. This course aims to create conditions for participants to engage intensively in this kind of learning—peer-based, relatively non-formal, relatively "uncurricular," non-individual and "in role."

and

> The course is partly about mastering the art of "overcoming anxiety" and learning to trust oneself and one's fellow team members to "come through." It is also partly about learning to have faith in a cohort's collective capacity to do things we might not previously have thought possible—and that at many points in the course we will think ARE impossible. But they are not impossible—although you will have to take that on faith!!

and

> It is important for participants to recognize that the instructors know perfectly well that they are asking a lot of the participants. It is equally important for participants to know that the instructors will not let them down. By the end of the first face-to-face week each team *will* have managed to create a digital media artifact, even though many participants may regard themselves as "computer illiterate" or as "digital klutzes" at the start of the course. Anyone who is tempted to think of themselves in this way and who are inclined to worry about it just needs to know that over the past several years large numbers of people have arrived at the first face-to-face day with their first ever laptop that they bought the day before and have had to get someone to show them how to turn it on. Equally, the instructors learned to troubleshoot Windows Vista (which involved a LOT of troubleshooting) without ever having had that operating system themselves (and they haven't had it since).

(iii) The learning space

There are two bottom lines for the learning space. The first is that there must be ample space that approximates in kind to the types of spaces employed in mature versions of the social practices in question: popular cultural media creation and qualitative research. As Lawrence Lessig (2005) recognizes, an adolescent's bedroom is a typical site of popular cultural media creation. This has less to do with the presence of a bed than with having space to run a computer, having some uninterrupted time in which to work, and so on. It also has to do with enjoying a sense of autonomy, so that one has "free scope and rein" to be creative—one can spread out a bit, make some mess, leave stuff in place until taking it up the next time, and the like. So far as the research dimension of the work is concerned, anything that approximates to the kind of space and facilities associated with research centers, and research and development spaces, will be suitable. Ideally, this includes "peripherals" like tea and coffee making facilities, space to lounge around in and shoot the breeze, and so on. Numerous authors have spoken about instances in which talk occurring at the coffee maker or the water cooler has generated creative solutions to problems and ideas for development. We prefer to have this kind of space and these kinds of peripherals available for our cohorts.

The second bottom line, of course, is high speed, unfettered internet access. Ideally we want to ensure that 30 or more people can simultaneously be using the internet in spaces as different as Google docs, Blogger.com, YouTube.com, the university's online library, NewLits.org, Wikipedia.org, and using Google's search engine at large.

The most "bounteous" space we have worked in to date has been the ski lodge, with its multiple levels of inside/outside open space, equipped with tables and chairs and bar stools, BBQ tables and so on—although a generously endowed teacher resource center and a hotel with conference facilities are not far behind. Interestingly, a high school outside of school hours, where we have been able to use the library, foyers, and multiple classrooms has also proved bounteous, with the sole exception of internet filtering. Lifting this restriction, and providing free access to the staff common room, would make this an ideal space as well. The same holds for on-campus university space of similar character and facilities.

(iv) The collaboration platform

As previously indicated, the idea of a collaboration platform displaces that of a training program in the familiar sense. It involves identifying and "pulling on" applications and services that enable forms of collaboration through which participants can discover and use what they need in order to conjointly realize their learning purposes as social learners (Brown and Adler, 2008; Cross, 2006; Gee, 2007).

Our collaboration platform has two main components. First, there is a face to face component comprising self-selected teams using physical space, tools, and material resources (including human resources) conducive to the tasks of creating media and doing hands on research. That is, the kind of physical space and material resources involved are structured and arranged in such a way as to positively enable collaboration within and between groups. Second, and crucially, there is a virtual component that is quintessentially "Web 2.0" in character, and overwhelmingly "Google" in application. This virtual component extends to the whole of the internet, but is "channeled" to a large extent through a suite of free and relatively robust Google applications. It is important to emphasize that this is *not* any kind of learning management system, whether corporate—along the lines of Blackboard, WebCT, etc.—or open source—along the lines of Moodle. Rather, it is a loosely joined, ad hoc collection of services and applications that can be added to, taken up in whatever proportions, and driven by participants. There are no content filters, no webmasters, no intellectual proprietors, no controls. If any institutional content filters or other kinds of controls arise we try to override them. For example, some schools actually block resources like Wikipedia and applications like Google docs! Our personal preference is for physical sites like conference facility hotels, lodges, and resource centers where there are no internet filters. On occasions where filters have become an issue we have been impressed by initiatives and logics developed by teams to overcome or bypass them (such as by identifying what can best be done at home, or by finding ways to employ a proxy browser to bypass local controls).

With respect to the virtual dimension of the collaboration platform, we use combinations of the following resources and applications according to need, work rhythms, and participant preferences.

• *Google docs* (docs.google.com). Anyone with a Google account can set up a Google document, which functions very like an online Word document, and invite other people to write collaboratively within the same document regardless of where they are. Multiple people can be writing and editing the document simultaneously (although too many working at once will cause glitches to occur when it comes to saving the most recent version of the document), and previous versions of each document can readily be accessed. In cases where it may be important to be able to identify individual contributions, groups in the past have used simple devices like assigning different font colors to different contributors. Contributors to the document can be sent email notification of updates from within Google docs, and short messages inviting feedback, explaining changes, and so on can be included with the notification.

• *Google sites* (sites.google.com) offer a "light" website that is very quickly and easily established and updated, and can readily be made into a collaborative resource by sharing a collective username and password. A Google site provides abundant archiving space, and a site established for a cohort can serve as a holding and/or publishing space for the cohort and its entire corpus of work.

• *Google scholar* (scholar.google.com) is a specialized search tool for academic sources. It provides bibliographic information, a citation count generated within its database and can be customized to automatically locate and link to resources inside a specified electronic library archive (see below).

• *Online library access to electronic journals and data bases, with Google scholar preferences activated* (scholar.google.com/scholar_preferences). Activating Google scholar preferences means that if an electronic resource (e.g., a journal article) located by a search is available through the university library to which participants have online access, they will be able to download the resource simply by clicking on a hyperlink, rather than having to record the details of the resource and then run a separate online library database search.

• *Google books* (books.google.com) walks a fine line with copyright enforcers who argue Google is infringing on authors' intellectual property with this service. It does, however, provide useful and considerable online access to book content, albeit on a "hit and miss" basis. Sometimes the pages you want are not available; sometimes entire chapters that are exactly what you want *are* available. An easy search using book title or author name quickly leads to the book being sought. Thereafter, it is a matter of whether the available content provides what you want or not. The odds are well worth taking a chance on: the nearest library holding a particular book might be a long way distant.

• *Gmail* (mail.google.com) provides an easy way to establish a Google account that enables access to Google sites and Google docs, and is a wonderfully abundant and searchable email service. Like all the other Google services and applications it is free.

• *A collaborative course blog using Blogger* (blogger.com), which can also be accessed via one's Google account. Blogger can be set up to allow posting via email or mobile phone, too. By sharing the blog user name and password it is possible for anyone in the cohort to post and comment on the blog, or the blog can be designed as a collaborative forum, with each user posting under their own name or alias. A cohort blog can be used as an all purpose site for posting information, requests for feedback, raising issues, and doing any other useful work that invites and enables collaboration. Wikis and social network sites like Ning.com and Socialgo.com are further options, and can be integrated with wikis and other online resources (e.g., JingProject.com, for creating their own how-to tutorials).

• *The rest of the internet*—that is, anywhere a conventional Google search or a more specialized Google scholar search may lead to for the purposes of furthering the team's collaborative work.

(v) The Learning Approach

Invariably, courses begin with a whole group session of up to 90 minutes in which people introduce themselves, teams are formalized, and we reiterate the emphasis on social learning and the fact that while participants will be maximally "hands on" we will be maximally hands off. Perhaps the key principle is: "Google is your friend." This refers less to the Google suite of applications employed than to the idea that "when you want to know something, SEARCH." Teams quickly learn that asking us is a last resort and that when they *do* ask they will be inviting suggestions for further searching rather than direct answers. A key driver here is our own commitment to participants becoming self-sufficient users of a range of resources who are confident of their own ability to nut out how to do something new, solve problems and get things done. When issues or problems seem pervasive across groups we will call for short, full group sessions to tap collective thinking and get solutions as quickly as possible (sometimes a group has found a solution, so we get them to explain; the hope is always to *elicit* rather than provide a response). Examples here include additional explication of what it means to write really detailed fieldnotes that describe what's going on, rather than summarize it, or having a group demonstrate to everyone else how to use a file conversion service like Zamzar.com to convert flash video files to .wav files.

In addition, and as near as possible to points of application, we lead short "walk through" sessions on things like how and why to activate Google scholar preferences, how to conduct natural language and phrase-based searches on Google (along with some Boolean terms), setting up collaborative Google documents, and so on. We also point the whole group to key guiding templates as near as possible to their points of application, and quickly walk through these to explain how we see them operating, and to emphasize trying to use them first and then seek feedback or clarification. Interaction between groups is encouraged, and the course website provides generous exemplars of high quality work done by previous cohorts. We find that teams are quick to make good use of these exemplars. For example, they often use them to see how a textbook description of a research method or technique has been implemented in practice and reported by others. Likewise, they use them as guides for laying out their fieldnotes or transcriptions of recorded conversations about how, say, group members solved a problem.

We encourage writing as often and as much as possible from the outset, and committing material to Google docs as soon as can be. Early on, we move around the groups frequently, looking for points where quick interventions might be most effective. We assume that what the groups will most need from us are productive guides to working as well as possible with theory and with collecting and analyzing data and discussing analytic results in conversation with prior research and theory. Hence, we focus most on looking for points at which feedback and reference to expert resources and exemplars are most needed and will be most useful. Not surprisingly, teams quickly find they can make good progress on their media creation and generating data on the processes involved. As we get a sense of their data and their media progress we can point to ways for organizing the data, make suggestions about when and how to begin some preliminary analysis of the data (during the "at home" week between the two face to face components), and the like. We try to stay as close as possible to Gee's (2007: 27, 40) principle of "performance before competence." In every cohort we find that performance begets *confidence* and teams quickly produce. From that point our role becomes one largely of trying to help them maximize *quality*. Obviously, it is always easier to do this once something is in place—just in time (at the point of need or application) and just in place.

Typical Outcomes

We have described a four week intensive course comprising around 60 face to face hours and two full time weeks "at home." Perhaps the best way to provide a general sense of typical outcomes is by pointing to some of the digital media creations produced by the teams and to indicate some typical excerpts from written work. On average (taken from courses conducted in 2009 and 2010) teams produce the equivalent of a 2 minute stop motion animation, a 3 minute music remix, a 2–3 minute machinima movie, or a 3–4 minute music video, and a written research report of around 6,500–7,000 words. We judge that with a small amount of mentoring from published academics the best reports of their research would be publishable in any number of peer reviewed journals. We have yet to see a report that is not at least the equivalent of an assignment for which we would typically give a B+ or A grade elsewhere. We think it would be fair to say that in the vast majority of cases any individual participant who wanted to do a qualitative research thesis would be "good to go" by the completion of the course. Across the various courses we have operated on the basis of this kind of learning approach we recall less than five occasions where team members have complained of an individual not "doing their share"—from a total of 750 to 1000 participants across the years. It is common for participants to report that they have worked harder in these courses than in any other academic course, and we have often found participants logged into Google docs between 1am and 2am—including during semester courses ahead of a day's teaching.

Some typical examples of team-based media creation include the following:

(a) Stop motion animation
- Lego Alliance, available at: http://www.youtube.com/watch?v=9hpJCJb4p0w
- The Escape, available at: http://www.youtube.com/watch?v=JfqWVqZumiw

(b) Machinima
- Bar Room Brawl, available at: http://www.youtube.com/watch?v=Ny_Ek5DpXtI

(c) AMV
- Sakura: The Climb, available at: http://www.youtube.com/user/mountainclimbers1#p/a/u/0/XmVtw8QoLbg

(d) Photoshopped celebrity reporting spoof
- Caught in Cape Breton, available at: http://www.youtube.com/watch?v=4R0kF5iE8Ao

With respect to the conventional writing component of the course, the following brief excerpts taken from some of the teams' reports might usefully provide a sense of their written work.

(i) An example of working with theory and concepts:

> Collaboration can be defined in a variety of ways depending on the context in which the concept is being used. There has been much debate in the academic literature over the meaning of the term "collaboration" and a proper definition which clearly describes all aspects of the concept.
>
> The problem with defining a complex concept like "collaboration" is that definitions can be limiting, or can be too broad. According to Clark and colleagues (1996), dialogue is at the center of collaboration. However, this definition is limiting because it does center on the word "dialogue" as the primary means of collaboration. It does not recognize the collective work members together can produce. Alternatively, collaboration can be characterized as understanding the work of one another, not working on the same research material (John-Steiner, Webber & Minnis, 1998, p. 775).
>
> John-Steiner, Webber and Minnis (1998) define collaboration differently. Their definition focuses on the ability of experts to share not only their knowledge and talents, but also their resources in a collaborative manner, and not as individuals per se. The final product is a reflection of all participants' contributions (John-Steiner et al., 1998, p. 776). This is in contrast to Clark and colleagues, whose definition of collaboration focuses on individuals understanding each other's work, but not necessarily contributing anything to it. (Clavaglia, Landry, and Stone, 2010: 9)

(ii) An overview of the data collection approach employed:

> Data were collected by a group of three participants over a period of five days. Each day, team members worked for approximately six hours per day resulting in a total of 30 hours of observations, field notes, head notes, post facto notes taken by each team member individually at differing times. Field notes proved to be of the utmost importance because they documented specific written details and accounts of what was occurring, what was being created, and the emotions that group members had felt. Field notes included writing legibly, marking the time down regularly, using codes, and developing a shorthand language. Head notes were also used as an important tool for the data collection. Head notes are "mental notes researchers make while systematically watching an event within a context where writing notes in the heat of the moment is impossible" (Lankshear & Knobel, 2004, p. 228). Group members would observe an event and make mental notes in their minds. At times, it was difficult to write an observation down as events were taking place, so team members made head notes and would write details as post facto notes which were written as soon as possible after the

> observed event. Some notes were typed and other notes were written down. Group members copy and pasted the URLs of websites that were useful sources of information and took screen shots of those websites. (Donovan, Hawley, and Whitty, 2010: 12)

(iii) An overview of the data analysis approach employed:

> The data analyzed for this project comprised our collective field notes with reference made at certain points to our supplementary photographs and field-based videos whenever something in these notes required additional elaboration. The primary method used was pattern finding, drawn from ethnographic approaches to research. Pattern finding is a process of identifying patterns discernible across pieces of information. According to David Fetterman, an internationally recognized ethnographer, in the pattern finding process the researcher begins with a "mass of undifferentiated ideas and behaviour, and then collects pieces of information, comparing, contrasting, and sorting gross categories and minutiae until a discernible pattern of behavior becomes visible' (Fetterman 1989, p. 92). We used four questions based on Fetterman's approach to guide our analysis of data:
> • What's going on here?
> • Who is doing what?
> • Have I seen this particular event or action before? Is it significant? Why or why not?
> • What things are happening or being done more than once? What does this mean or suggest?
> Applying these questions to our data meant that we began to notice similarities in what we had collected. We colour coded and identified patterns as they emerged from the data. Our identification of patterns was guided by our review of the literature; that is, we expected to find certain patterns in our data that corresponded to key characteristics of what constitute *new* literacies (e.g., participation, collaboration, distributed knowledge). At the same time, we remained open to finding additional insights into what constitutes a *new literacy* practice in our data as well. (Beck et al., 2009: 11–12).

(iv) Use of a table within the process of developing categories during data analysis. See Table 16.1 below.

Some Concluding Remarks

A foremost aim in our work with students is to encourage writing. Participants often acknowledge being reluctant academic writers, even though they are prolific writers in other areas of their lives. The course described here tries to link writing to the material performance of an authentic academic role, and to encourage unselfconscious academic writing by making it as natural an extension as possible of being in role as a researcher. We enlist participants from the outset in writing the kinds of research critiques and reports that we do in our own work as researchers,

Table 16.1: Sample preliminary fieldnote analysis (Clavaglia, Landry and Stone, 2010, field notes).

Problem Solving through Messing About	Problem Solving through an Affinity Space	Problem Solving through Group Collaboration	Problem Solving through a Search Engine	Problem solving through Other Means
I go back into www.youtube.com to search for Haiti, started with search "President Obama on Haiti." Parts of the speech I think we can use so I copy and paste the url into www.zamzar.com. I got an error message "File has no extension." I asked Holly and Janice if they got this error. They didn't. so I went back to youtube.com. I repeated the process, opening up a new Explorer window and it worked. While waiting for the file to convert I go back to www.youtube.com to search for more speeches (see Ann, 7:31,32,33&34)	Interesting—could not find videos on www.youtube.com so I went to google, put search "Garth Brooks" "the change" which brought me to video.google.com (help from the prof on the refined search with quotation marks) & found both songs there, but had to try to convert 2 different versions b/c that error "File has no extension" but the next 2 worked. I converted them in www.zamzar.com then saved them on thumb drive (see Ann, 9:45)	Holly attempted to video Janice showing how to implement the song into Garageband, but she failed to press the record button so they did it twice (see Ann, 1:3)	Holly did the same thing and found a user friendly site (in her opinion), which included a step-by-step video. Holly was pleased and went ahead w the download but the website was unavailable (see Ann, 3:13)	Holly clicked on www.mediaconverter.org and searched "Interlake for Haiti—Haiti News Clips Collection" in the search, then tried to convert and download. This didn't work, so Holly went through chapter 2 in our DIY Media text. This took a long time and she wasn't finding anything so she tried going through the sign up on the site—found out it costs $4 so went back to the book to look for free software converter, checked index in our book, couldn't find it (see Ann 3:11)

along with contributory forms of writing integral to research—writing field notes in the heat of the moment, creating tables to organize, summarize or help reduce data, transcribing verbatim recorded speech for coding, and so on. This kind of writing is much easier to do in situ, with models and on demand feedback in place.

This ease is augmented by the focus on collaborative writing. Participants tell us, unprompted, that this enables them to produce text that they do not think they would be able to produce individually and certainly not within such a short time frame. They report experiences of being "emboldened" to write; they take more risks; and they can produce more fluently, confidently and rapidly—which means there is something "there" to respond to. Often we get to "what is right" by progressively responding to something that is "not right"—but that something has to be there in the first place. There is a saying in software programming—known as Linus' Law (see en.wikipedia.org/wiki/Linus'_Law)—that "with enough eyeballs, all bugs are shallow." With enough beta testers and co-developers, glitches in programming code will be identified quickly and someone will soon come up with the solution. We think we see something like this happening in the collaborative writing approach in the courses. In addition, even though individuals may take primary responsibility for particular parts or areas of the overall report (divvied up on the basis of template structure), each "part" has to co-ordinate with every other "part," so that by the end all participants have experience of working in each "part" and get an understanding of each "part" and how all the parts relate. The result is that any participant would be able to generate "text"—whether written report text or digital remix text or video text—in any area of the overall products with sufficient competence to know how to go on. Moreover, while we often require university students to produce *individual* work, all the way through to completing a Ph.D. thesis, the truth is that in academic professional life as well as in the professional lives of media producers and creators, production is often, *if not typically*, collaborative.

In several places, Gee (2004, 2007, 2008) discusses the relationship recognized by sociocultural learning theorists between effective learning and particular forms of talk and patterns of participation and interaction. He emphasizes several important aspects of this relationship, including the following.

• Distinctive forms of content are associated with distinctive forms of language, and the connections are made public in talk.

• Learners need to learn to interpret, analyze, debug, and explain their experiences (and the connections between goals and reasoning within experience)—all of which can be rendered public through talk.

• Learners need to learn from the interpreted experiences of others, their peers and more expert people in the domain.

• Communities of practice are, in part, formed through the sorts of talk that allow for public sharing and joint modeling, building and problem solving. (see Gee, 2007: Ch. 10)

Referring to the example of *Yu-Gi-Oh!*, a manga, anime series, computer game and card game popular among children as young as 5 or 6 years of age, Gee identifies the very specific form of talk—a "specific register or social language"—required by participants. This language is grounded in the meanings, functions and purposes of *Yu-Gi-Oh!* as a social activity. Unlike the situation so often observed in classrooms, where some participants have the "right" social language for school learning and others don't, young *Yu-Gi-Oh!* players at large can "discuss and debate with others in this language on common ground" (Gee, 2007: 158). This is not simply because they have *talked*, however. To a large and important extent it is because "they have played the game and that play has given them embodied, situated meanings for the language" (ibid.).

Participant involvement in the two discursive domains at the core of our course—researching and engaging in new media literacies—illustrates similar tendencies to those identified by Gee in relation to subjects in studies he has researched, such as the *Yu-Gi-Oh!* card game players. Indeed, we see this in spoken and written language alike. For example, in one study, a team of people with no previous experience learned to assemble and program a Lego kitset robot to perform a simple task. When they analyzed the data they had collected (using fieldnotes, and audio and video recordings of their activity and conversations) during the one week in which they assembled and programmed the robot, they saw, graphically, shifts in their technical language. Within two days they were using technical terms and phrases they had never previously encountered when going about their work (e.g., shifting from "that sun dial thing" to the more technically accurate "sensor"). Arriving at the basics of a shared language made their work go more rapidly and effectively, and everybody understood each other because they were acquiring the concepts and phrases and terms within concrete situations linked to material processes mediated by expert language modeling (in the form of manuals, online guides, posts in forums, YouTube talk-throughs and demonstrations, etc.). The same is apparent in the research dimension where theoretical and methodological concepts and procedures are rendered public and shared through talk within the teams and between team members and the instructors.

Likewise, as foreshadowed above, the courses provide abundant evidence of situated written language, or situated writing. Participants begin writing from the outset but always within material contexts and with tangible artifacts and objects of the kinds that bona fide research teams encounter. Researchers have to *write* like researchers, and this takes many different forms. In data collection for example, they have to know how to generate usable field notes; they have to know how and when to capture verbatim speech; they have to know how to transcribe speech reliably; how to recognize an appropriate opportunity for a screen capture on a computer and to make and store and retrieve it. In data organization they have to know how to label and classify and summarize in order to aid data retrieval at the point of analysis. In data analysis one has to be able to render patterns and categories through use of codes and charts and tables and the like. Writing about triangulation of data, for example, becomes lucidly functional in the presence of data pertaining to a specific event or instance when team members collectively bring spoken and observational data to bear upon it, and when they bring complementary perspectives to bear upon it grounded in what they have seen and heard.

Of course, participants are in no ways experts or well-prepared independent digital media producers or researchers by the end of the course. They have simply had an intensive introduction to academic research Discourse and discourse (Gee, 1996). By the same token, as we have tried to show here, it seems fair to say that they make noteworthy entrees into forms of reading and writing that are integral to the postgraduate professional formation of teachers under contemporary conditions, and do so within a very short time frame.

Finally, many authors have commented on the difference many young people experience in classrooms between the learning culture they experience there and the kind of peer-based learning culture they experience within popular cultural affinity spaces (e.g., Alvermann, 2010; Burn, 2009; Jenkins et al., 2006; Thomas, 2007). The point is not that education should emulate familiar and preferred learning approaches and styles from non-academic/non-scholastic spaces. Rather, the point is that if it is possible to appropriate principles and procedures associated with efficacious learning from outside formal institutions in ways that enhance the efficacy of formal educational learning, other things being equal we would be silly not to. In a nutshell, the approach we have described tries to integrate as richly and fully as possible into a formal learning context as many of the "rhythms" and "ways" we have experienced while learning and otherwise participating within online popular cultural affinity spaces. We want to encourage teachers and administrators who may have had less experience of this kind of phenomenon to get "a feel for it" within a formal learning context. The logic here is simple: if it works for learning to become researchers there are no grounds for thinking it cannot work for academic learning more generally. Our thinking is that if teachers and administrators can have strong lived experience of academic learning as social learning in the kind of way we are trying to foster in these courses, it may go part of the way toward ameliorating some of the tensions and gaps young people face between their "life world" learning and their institutionalized learning.

References

Alvermann, D. (ed.) (2010). *Adolescents' Online Literacies: Connecting Classrooms, Digital Media, and Popular Culture*. New York: Peter Lang.

Beck, S., Coley, C., Conway, K., Hoven, D. and Maynard, P. (2009). From collaboration to affinity space: Learning a new literacy. Steady Brook, Newfoundland: Research Report for Summer Program. Mimeo.

Black, R. (2008). *Adolescents and Online Fanfiction*. New York: Peter Lang.

Brown, J. Seely, and Adler, R. (2008). Minds on fire: Open education, the long tail, and learning 2.0. *Educause* (January/February): 17–32.

Buckingham, D. (2003) *Media Education: Literacy, Learning and Contemporary Culture*. Cambridge: Polity Press.

Burn, A. (2009). *Making New Media: Creative Production and Digital Literacies*. New York: Peter Lang.

Clark, C., Moss, P., Goering, S., Herter, R., Lamar, B., Leonard, D., Robbbins, S., Russell, M., Templin, M., and Wascha, K. (1996). Collaboration as dialogue: Teachers and researchers engaged in conversation and professional development. *American Educational Research Journal*, 33(1): 193–231.

Clavaglia, J., Landry, A., and Stone, H. (2010). Breaking down the barriers: Participating in new literacies. North Sydney, Nova Scotia: Research report for Summer Program. Mimeo.

Cross, J. (2006). *Informal Learning: Rediscovering the Natural Pathways that Inspire Innovation and Performance*. San Francisco: Pfeiffer.

Davies, J. and Merchant, G. (2009). *Web 2.0 for Schools: Learning and Social Participation*. New York: Peter Lang.

Donovan, S., Hawley, J. and Whitty, S. (2010). Social learning resulting from the production of a media artifact. North Sydney, Nova Scotia: Research report for Summer Program. Mimeo.

Fetterman, D. (1989). *Ethnography: Step by Step*: Thousand Oaks, CA: Sage.

Gee, J. (1996). *Social Linguistics and Literacies: Ideologies in Discourses*. New York: Routledge.

Gee, J. (2003). *What Video Games Have to Teach Us about Learning and Literacy*. New York: Palgrave.

Gee, J. (2004). *Situated Language and Learning*. New York: Routledge.

Gee, J. (2007). *Good Video Games and Good Learning: Collected Essays on Video Games, Learning and Literacy*. New York: Peter Lang.

Gee, J. P. (2008). Lucidly Functional Language. New Literacies: A Professional Development Wiki for Educators. Developed under the aegis of the Improving Teacher Quality Project (ITQP), a federally funded partnership between Montclair State University and East Orange School District, New Jersey. Retrieved July 25, 2010, from: newlits.wikispaces.com/Lucidly+Functional+Language

Gee, J., Hull, G., and Lankshear, C. (1996). *The New Work Order: Behind the Language of the New Capitalism*. Boulder, CO: Westview Press.

Hagel, J. and Brown, J. Seely (2005). From push to pull: Emerging models for mobilizing resources. Unpublished working paper, October. Retrieved January 21, 2010, from: edgeperspectives.com.

Jenkins, H. (2006). *Fans, Bloggers, and Gamers: Exploring Participatory Culture*. New York: New York University Press.

Jenkins, H., with Purushotma, R., Clinton, K., Weigel, M., and Robison, A. (2006). *Confronting the Challenges of Participatory Culture: Media Education for the 21st Century*. Chicago: MacArthur Foundation.

John-Steiner, V., Weber, R., and Minnis, M. (1998). The challenge of studying. *American Educational Research Journal*, 35(4): 773–783.

Lankshear, C. and Knobel, M. (2004). *A Handbook for Teacher Research*. New York: Open University Press.

Lessig, L. (2005). Re:Mix:Me. Paper presented at the 2005 Annual ITU Conference, 'Creative Dialogues.' Oslo, Network for IT-Research and Competence in Education (ITU), University of Oslo.
O'Reilly, T. (2000). What is Web 2.0? Design patterns and business models for the next generation of software. Retrieved August 3, 2010, from: oreilly.com/web2/archive/what-is-web-20.html.
Prensky, M. (2006). *Don't Bother Me Mom—I'm Learning!* New York: Paragon.
Ranciere, J. (1991). *The Ignorant Schoolmaster: Five Lessons in Intellectual Emancipation.* Trans. K. Ross. Palo Alto, CA: Stanford University Press.
Richardson, W. (2006). *Blogs, Wikis, Podcasts, and Other Powerful Web Tools for Classrooms.* Thousand Oaks, CA: Corwin Press.
Salen, K. (2008). Toward an ecology of gaming. In K. Salen (ed.), *The Ecology of Games: Connecting Youth, Games, and Learning.* Cambridge, MA: The MIT Press. 1–20.
Schön, D. (1983). *The Reflective Practitioner.* New York: Basic Books.
Steinkeuhler, C. (2008). Cognition and literacy in massively multiplayer online games. In D. Leu, C. Lankshear, M. Knobel, and J. Coiro (eds), *Handbook of Research on New Literacies.* Mahwah, NJ: Erlbaum. 611–634.
Thomas, A. (2007). *Youth Online: Identity and Literacy in the Digital Age.* New York: Peter Lang.
Wittgenstein, L. (1953). *Philosophical Investigations.* Oxford: Basil Blackwell.

Name Index

Subject Index